D1282234

THE SEA URCHIN: DEVELOPMENTAL BIOLOGY

Papers by
Paul Gross, Tom Humphreys, Everett Anderson, et al

MSS Information Corporation
19 East 48th Street New York, N.Y. 10017

TABLE OF CONTENTS

CREDITS AND ACKNOWLEDGMENTS

Anderson, Everett, "Oocyte Differentiation in the Sea Urchin, *Arbacia Punctulata*, with Particular Reference to the Origin of Cortical Granules and Their Participation in the Cortical Reaction," *The Journal of Cell Biology,* 1968, 37:514-539.

Aronson, Arthur I.; and Fred H. Wilt, "Properties of Nuclear RNA in Sea Urchin Embryos," *Proceedings of the National Academy of Sciences*, January, 1969, 62:186-193.

Brookbank, John W., "DNA Synthesis and Development in Reciprocal Interordinal Hybrids of a Sea Urchin and a Sand Dollar," *Developmental Biology*, 1970, 21:29-47.

Denny, Paul C.; and Patricia Reback, "Active Polysomes in Sea Urchin Eggs and Zygotes: Evidence for an Increase in Translatable Messenger RNA after Fertilization," *The Journal of Experimental Zoology*, October, 1970, 175:133-140.

Emerson, Charles P., Jr.; and Tom Humphreys, "Regulation of DNA-Like RNA and the Apparent Activation of Ribosomal RNA Synthesis in Sea Urchin Embryos: Quantitative Measurements of Newly Synthesized RNA," *Developmental Biology*, 1970, 23, 86-112.

Epel, David, "Protein Synthesis in Sea Urchin Eggs: A 'Late' Response To Fertilization," *Proceedings of the National Academy of Sciences*, April, 1967, 57:899-906.

Fansler, B.; and L. A. Loeb, "Sea Urchin Nuclear DNA Polymerase: II. Changing Localization during Early Development," *Experimental Cell Research*, 1969, 57:305-310.

Hinegardner, Ralph T., "Growth and Development of the Laboratory Cultured Sea Urchin," *The Biological Bulletin*, December, 1969, 137:465-475.

Humphreys, Tom, "Efficiency of Translation of Messenger-RNA before and after Fertilization in Sea Urchins," *Developmental Biology*, 1969, 20:435-458.

Jenkins, Kenneth D.; and Paul C. Denny, "Effects of Fertilization on the Ribosomal Subunit Pool of Sea Urchin Eggs," *Biochimica et Biophysica Acta*, 1970, 217:206-208.

Kedes, Laurence H.; Paul R. Gross; Goffredo Cognetti; and Anne L. Hunter, "Synthesis of Nuclear and Chromosomal Proteins on Light Polyribosomes during Cleavage in the Sea Urchin Embryo," *Journal of Molecular Biology*, 1969, 45:337-351.

Kijima, S.; and Fred H. Wilt, "Rate of Nuclear Ribonucleic Acid Turnover in Sea Urchin Embryos," *Journal of Molecular Biology*, 1969, 40:235-246.

Loeb, L. A.; and B. Fansler, "Intracellular Migration of DNA Polymerase in Early Developing Sea Urchin Embryos," *Biochimica et Biophysica Acta*, 1970, 217:50-55.

Loeb, L. A.; B. Fansler; R. Williams; and D. Mazia, "Sea Urchin Nuclear DNA Polymerase: I. Localization in Nuclei during Rapid DNA Synthesis," *Experimental Cell Research,* 1969, 57:298-304.

Moav, Boaz; and Martin Nemer, "Histone Synthesis. Assignment to a Special Class of Polyribosomes in Sea Urchin Embryos," *Biochemistry*, 1971, 10:881-888.

Sachs, Martin I.; and Everett Anderson, "A Cytological Study of Artificial Parthenogenesis in the Sea Urchin *Arbacia Punctulata,*" *The Journal of Cell Biology*, 1970, 47:140-158.

Selvig, Susan E.; Paul R. Gross; and Anne L. Hunter, "Cytoplasmic Synthesis of RNA in the Sea Urchin Embryo," *Developmental Biology*, June, 1970, 22:343-365.

Stavy, Lary; and Paul R. Gross, "Protein Synthesis *In Vitro* with Fractions of Sea Urchin Eggs and Embryos," *Biochimica et Biophysica Acta*, 1969, 182:193-202.

Timourian, Hector; and George Watchmaker, "Protein Synthesis in Sea Urchin Eggs: II. Changes in Amino Acid Uptake and Incorporation at Fertilization," *Developmental Biology*, November, 1970, 23:478-491.

Wilt, F. H.; H. Sakai; and D. Mazia, "Old and New Protein in the Formation of the Mitotic Apparatus in Cleaving Sea Urchin Eggs," *Journal of Molecular Biology*, 1967, 27:1-7.

Regulation of Protein Synthesis in Fertilization And Early Development

*PROTEIN SYNTHESIS IN SEA URCHIN EGGS: A "LATE" RESPONSE TO FERTILIZATION**

By David Epel

Fertilization of echinoderm eggs results in a complex series of metabolic activations, resulting in greater than 30 changes within the first ten minutes after insemination. The most prominent of these are modifications in structure,[1] increases in respiration rate[2] and coenzyme content,[3] increases in substrate uptake,[4–6] transient proteolytic activity,[7] and a decrease in external pH.[8] These changes are possibly related to enzymic activations leading to the synthesis of lipids, proteins, and nucleic acids required for cell division and differentiation.

Studies on the temporal sequence of these diverse physiological reactions could provide insights into the mechanisms and interrelationships of these changes. Thus far, such temporal data are available on the light-scattering and external pH changes[9] (resulting from structural changes in the cell cortex), activation of NAD kinase[3] (resulting in NADP and NADPH synthesis), and activation of respiration.[2]

The present report concerns the temporal relationships of the above events to the postfertilization increase in protein synthesis, and also attempts to resolve contradictory results regarding amino acid incorporation in unfertilized eggs. Previous studies have shown that the rate of protein synthesis is low or negligible in unfertilized eggs, and increases markedly after fertilization.[6, 10–12] This increased rate is apparently dependent on mRNA already present in unfertilized egg cytoplasm, since it is unaffected by either actinomycin D[13] or enucleation.[14] Current hypotheses regarding activation mechanisms of this increased protein synthesis, which need not be mutually exclusive, implicate (1) structural changes in ribosomes resulting from protease activation at fertilization,[15] (2) synthesis of a factor(s) controlling mRNA translation rate,[16] or (3) energy-dependent processes involved in mRNA attachment.[17]

In the present study, effects of changes in cellular amino acid permeability following fertilization[5, 6] were minimized by "preloading" unfertilized eggs with radioactive amino acid, and incorporation kinetics measured by sampling at close intervals following fertilization. The results indicate that the activation of protein synthesis is actually a "late" response to fertilization, since increased synthesis does not begin until six to ten minutes after insemination. The results also show that unfertilized eggs transport and concentrate leucine and valine, and incorporate these amino acids into protein; that a sizeable amount of added leucine is converted to compounds not involved with protein synthesis; and that the rate of this conversion is accelerated by fertilization.

Materials and Methods.—Handling of gametes: Shedding of gametes of *Lytechinus pictus* (Pacific Bio-Marine Co.) was induced by intracoelomic injection of 0.5 *M* KCl, and cell counts were made by the dilution–capillary tube method.[18] The eggs, maintained at 16°C, were washed 4–5 times by decantation with millipore-filtered sea water, and then "preloaded" with C^{14}-amino acid before fertilization by incubation in μM solutions of the isotope for 4–10 min. Exogenous isotope was then removed by four washes with sea water and gentle centrifugation in a hand centrifuge, and insemination effected by addition of 10 μl undiluted sperm per 10 ml of egg suspension.

Isotope incorporation: One-half-ml samples of a suspension of preloaded eggs were added to 5.0 ml 5.5% TCA containing 0.05 M unlabeled amino acid. The eggs were centrifuged, and 1.0 ml supernatant liquid was removed for isotopic analysis of the TCA-soluble fraction as described by Berg.[19] For preparation of TCA-insoluble protein 1 mg bovine serum albumin was added to the TCA-egg suspension, which was then heated at 90°C for 20 min, filtered through Gelman A or Whatman GF/B glass fiber filters, and the residue washed four times with 5% TCA–0.05 M C^{12}-amino acid, twice with ethanol:ether:chloroform (2:2:1), and twice with ether. The filters were then glued to planchets and counted in a Nuclear-Chicago gas flow counter with mica window.

Autoradiography: Preloaded eggs were fixed in alcohol:acetic acid (3:1), washed three times with 95% alcohol, taken through toluene, and embedded in Tissuemat. Sections of 5 μ thickness were spread on gelatin-subbed slides, rehydrated through toluene and alcohol to water, immersed for 10 min in boiling 5% TCA, washed thoroughly, dipped in Ilford K5 nuclear emulsion, and left for 3 weeks at 5°C before development.

Chromatography: The ether-extracted TCA-soluble fraction was chromatographed on Whatman no. 1 paper by ascending chromatography for 16 hr with 2,4 lutidine:collidine:water, 1:1:1, plus 1% diethylamine,[20] and radioactivity measured with a Nuclear-Chicago actigraph radiochromatogram scanner. Acid hydrolysates of the TCA-soluble and insoluble fractions were prepared by boiling in 6 N HCl under reflux conditions, the solution was dried in a flash evaporator, and the redissolved residue chromatographed as above.

Isotopes: Specific activities of the uniformly labeled C^{14}-amino acids (International Chemical and Nuclear Corp.) were: C^{14}-leucine, 210 mc/mM; C^{14}-valine, 190 mc/mM; C^{14}-phenylalanine, 300 mc/mM.

Results.—Amino acid uptake and incorporation in unfertilized eggs: Unfertilized eggs, incubated for five minutes in 0.5–7.3 $\times$ 10^{-6} M C^{14}-labeled leucine or valine, respectively contain 8.1–1.3 times more amino acid, on a volume basis, than was originally present in the external medium. Data not herein presented show that the transport system approaches saturation at micromolar levels, that leucine is transported at a greater rate than is valine, and that the uptake rate is initially linear.

In all experiments, unfertilized eggs were found to incorporate appreciable amounts of C^{14}-labeled leucine, valine, or phenylalanine into TCA-insoluble protein (Figs. 1 and 5). This incorporation does not result from a small population of exceptionally active cells, since autoradiographs of unfertilized eggs showed that *all* cells incorporated C^{14}-leucine into hot TCA-insoluble material.

Protein synthesis following fertilization: Figure 1 depicts the results of an experiment in which cumulative incorporation of C^{14}-leucine into protein was measured at close intervals following fertilization. The data show a ten-minute lag period between sperm addition and the first increase in rate of protein synthesis, the rate then increasing for seven to ten minutes to a constant rate five times that of unfertilized eggs. Similar kinetics were found in nine separate experiments, using eggs from seven different females and either C^{14}-labeled leucine, valine, or phenylalanine. In these experiments the relative rate of protein synthesis had increased 5 to 15-fold 20 minutes after fertilization. The lag period between insemination and increased protein synthesis varied between six and ten minutes, with an average of 7.8 $\pm$ 1.2 minutes.

The same lag period is observed when cumulative protein synthesis is measured with different methods of protein precipitation. In one experiment, the kinetics of C^{14}-valine incorporation were compared in samples which were either heated or not heated in 5 per cent TCA. The former method hydrolyzes sRNA, releasing amino acyl-sRNA and polypeptidyl-sRNA from the TCA-insoluble residue.[21]

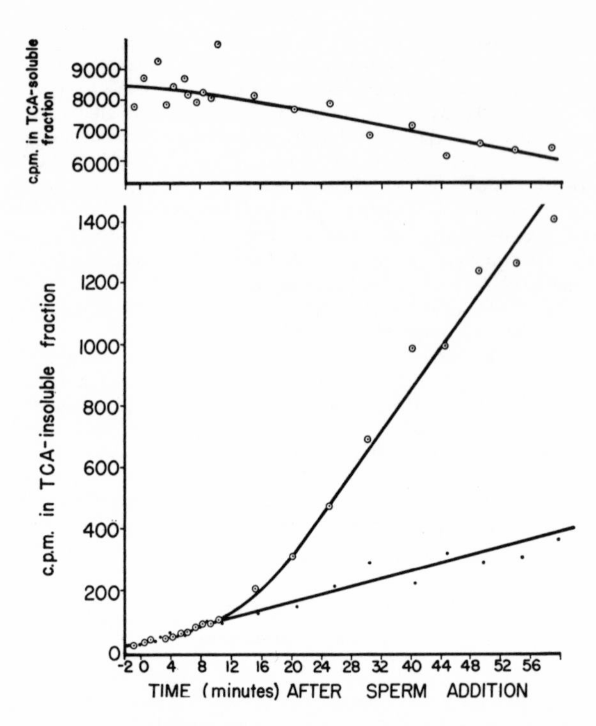

FIG. 1.—Cumulative incorporation of C^{14}-leucine by unfertilized (*dots*) and fertilized (*circles*) eggs (3880 cells/sample). Upper part of figure shows cpm/sample in the TCA-soluble fraction of fertilized eggs.

Conversely, elimination of the heating step should result in retention in the TCA-insoluble residue of these sRNA-linked compounds. The results of this experiment, shown in Figure 2, indicate similar kinetics in both cases. The same is also true if TCA-soluble basic polypeptides are precipitated with the TCA-tungstic acid procedure of Gardner *et al.*[22]

Auxiliary metabolism of leucine following fertilization: Data in the upper part of Figure 1 show that although the TCA-soluble fraction decreases in parallel with the increase in TCA-soluble material, this decrease is not stoichiometric (2200-cpm decrease in TCA-soluble vs. 1400-cpm increase in TCA-insoluble material). This lack of stoichiometry, also found with valine, suggests a significant conversion of these amino acids to substances other than protein. Chromatographic analyses show this to be true, and indicate that this conversion is of sufficient magnitude to affect the measurement of rate of protein synthesis.

Figure 3 depicts a radiochromatographic tracing of the TCA-soluble fraction of fertilized eggs which were preloaded before fertilization with C^{14}-leucine. The major peak coincides chromatographically with leucine, whereas the substances responsible for the two minor peaks, *X1* and *X2*, have not yet been identified. These compounds do not migrate with α-ketoisocaproic acid, nor are they peptides, as their chromatographic behavior is unaltered by 16 hours' boiling in 6 *N* HCl.

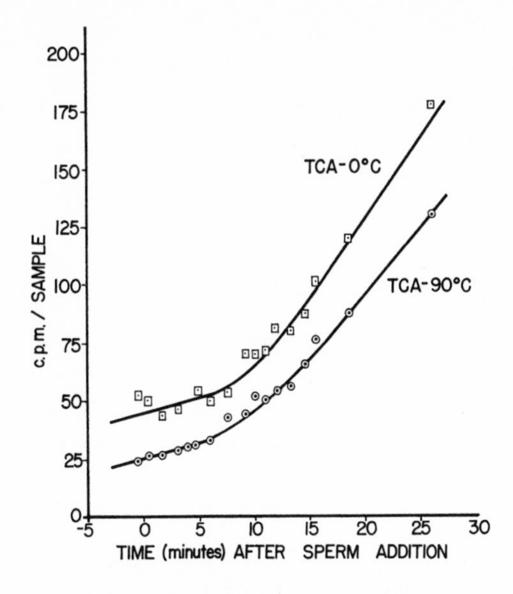

Fig. 2.—Cumulative incorporation of C^{14}-valine by fertilized eggs into hot (*circles*) and cold (*squares*) TCA-insoluble material (2720 cells/sample).

They are not protein constituents, as only C^{14}-leucine is released upon acid hydrolysis of labeled protein.

Figure 4 shows that these compounds gradually appear in unfertilized eggs, and that their rate of formation is accelerated by fertilization (2.5–3.4 times in two experiments). Unfertilized eggs, preloaded and then incubated for two hours at 16°C, convert 30 per cent of the TCA-soluble C^{14}-leucine into *X1* and *X2*. If these eggs are now fertilized, the duration of the period between sperm addition and increased protein synthesis is the same as in eggs fertilized immediately after loading with C^{14}-leucine, and free of *X1* and *X2* at the time of fertilization. Hence it appears that these two unidentified substances are not intermediate products in the synthesis of protein from leucine, and that their accelerated rate of formation after fertilization is unrelated to the postfertilization increase in rate of protein synthesis.

Analysis of rate of protein synthesis: Because of the extensive leucine conversion, analysis of the rate of leucine incorporation must be corrected for the actual amount of C^{14}-leucine in the TCA-soluble fraction. In two experiments, measurements were made of radioactivity of the TCA-soluble and TCA-insoluble fractions, and the amount of leucine actually present in the TCA-soluble fraction was determined by radiochromatography. The results of one of these experiments are shown in Figure 5. The bottom curve represents leucine incorporation into protein, the upper one the amount of C^{14}-leucine determined by radiochromatography to be present in the TCA-soluble fraction. Figure 6 shows an analysis of the rate of incorporation calculated from Figure 5 by (1) slope analysis (curve *A*) and (2) by calculation of the percentage of the C^{14}-leucine incorporated into protein per two-minute interval (curve *B*). For comparison with other fertilization reactions, the figure also depicts the previously determined temporal sequence of four other fertilization-induced changes.[9]

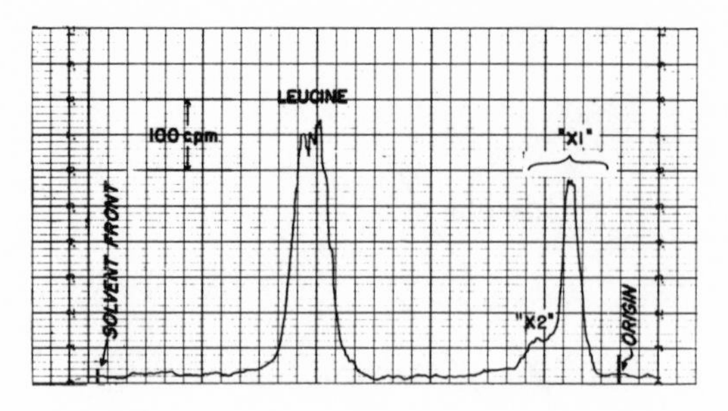

Fig. 3.—Recorder tracing of radiochromatogram of TCA-soluble fraction of eggs preloaded with C^{14}-leucine and sampled at 30 min after fertilization.

Discussion.—The reported experiments demonstrate that *unfertilized* eggs of *L. pictus* incorporate leucine and valine into protein. Furthermore, autoradiographic data show that this protein synthesis occurs in all cells of the population, so that the observed incorporation cannot be attributed to the presence of immature oöcytes.

These findings confirm and extend recent data of Tyler *et al.*[10] obtained with the same species. They differ, however, from results with other species of sea urchins, whose unfertilized eggs reportedly synthesize little or no protein, and whose rate of protein synthesis increases more than 100-fold upon fertilization.[11, 12] These conflicting results might reflect species differences, or might result from not taking into account the lower permeability and incorporation rates of unfertilized eggs. Since fertilization results in four to eightfold increases in amino acid uptake,[5, 23] and 5 to 15-fold increases in protein synthesis rate, measurements of synthesis which do not consider these permeability changes could indicate apparent increases of 20 to 120-fold.

Protein synthesis has also been reported in unfertilized echiuroid,[24] polychaete,[25] and amphibian[26] eggs. The metabolic significance of this synthesis is presently unclear (cf. ref. 26). Kavanau, over 15 years ago, observed a depletion of free amino acids, and a concomitant increase in protein, in unfertilized sea urchin eggs stored for 24 hours after ovulation.[27] He suggested that this might represent a continuation of the ripening process, dependent in the female animal on utilization of exogenous amino acids of the body fluids, but proceeding after ovulation through utilization of the endogenous amino acid pool.

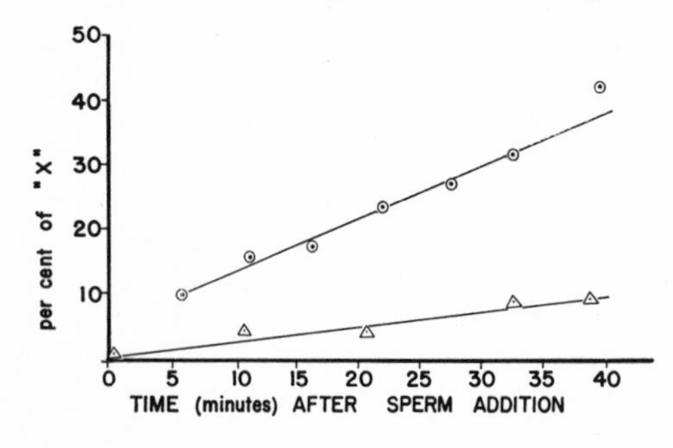

Fig. 4.—Percentage of total radioactivity of the TCA-soluble fraction found in X1 and X2 in unfertilized (*triangles*) and fertilized (*circles*) eggs.

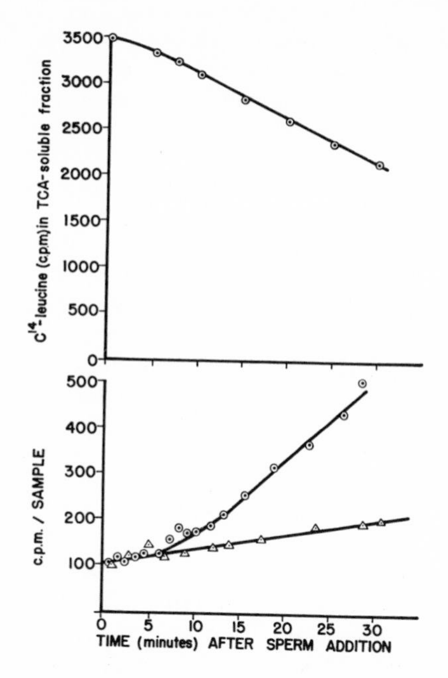

Fig. 5.—*Lower figure:* Cumulative incorporation of C^{14}-leucine in unfertilized (*triangles*) and fertilized (*circles*) eggs (14,000 cells/sample). *Upper figure:* Actual cpm of C^{14}-leucine in TCA-soluble fraction of fertilized eggs.

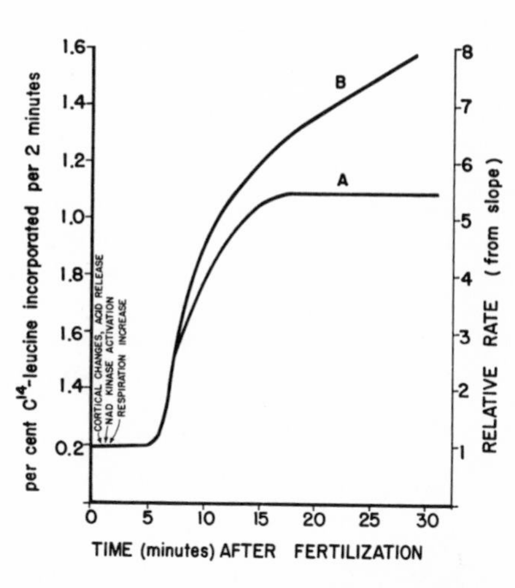

Fig. 6.—Relative rates of protein synthesis. Procedure as described in text. Temporal sequence of other changes determined in separate experiments (ref. 9).

Estimates of relative rates of protein synthesis are complicated by the observed conversion of leucine to other products. The increased rate of conversion after fertilization, of considerable interest in itself, suggests that fertilization increases amino acid catabolism (cf. ref. 27).

Two alternative modes of analyzing relative rate of protein synthesis were presented in Figure 6. The first (*A*), derived from slope analysis of the cumulative formation of radioactive protein, would be valid if amino acid compartmentation exists, and if the pool involved in protein synthesis is small and essentially saturated by added C^{14}-leucine. The second alternative (*B*), calculated from the percentage of C^{14}-leucine incorporated per two-minute interval, would be valid if C^{14}-leucine were always in equilibrium with leucine utilized for protein synthesis (whether compartmentation existed or not). Since the later phases of these kinetics correspond almost exactly to the kinetics observed in *in vitro* systems prepared at various times after fertilization,[11] this latter approach probably reflects more accurately the *in vivo* rate.

The most interesting aspect of these experiments is the finding of a temporal lag between the cortical events and increased protein synthesis. This lag indicates that the structural and metabolic changes following fertilization are not simultaneously activated. Rather, the changes associated with cortical granule breakdown, acid excretion, NAD kinase activation, and increased respiration appear to be "early" responses to fertilization, whereas increased protein synthesis appears to be a "late" response. The results presently suggest no obvious correlations of these early events with the increased rate of protein synthesis. Furthermore, the absence of correlation with increased respiratory activity implies that the postfertilization burst in O_2 consumption[2] does not result from the energy demands of protein synthesis. This absence of correlation is further strengthened by the lack of immediate effect of puromycin on respiration[28] and ATP levels.[23]

The results of this kinetic study are pertinent to proposed mechanisms of increased protein synthesis after fertilization. The five to nine-minute lag between cortical changes and increased protein synthesis does not appear to support the hypothesis that protease activation at the time of cortical granule breakdown[7] is the sole factor increasing protein synthesis.[15] Rather, the observed lag suggests a chain-type reaction system, where a number of reactions must occur (or new products accumulate) before increased mRNA readout can be initiated. Clues to the kinetic behavior of these other reactions, inferred from the kinetics of protein synthesis (Fig. 6*B*) indicate that in addition to the lag phase there is also an acceleration phase during which protein synthesis increases towards its full postfertilization rate.

The present results, while not eliminating involvement of proteases, suggest that multiple factors,[29] such as the synthesis of a rate-controlling substance in fertilized eggs[16] or the involvement of an energy-linked process in mRNA-ribosome attachment,[17] are controlling translation rate. Whatever the factors, one proof of their operation *in vivo* should be their temporal conformation to both the lag and acceleration phases.

I wish to thank Miss Sigrid Elsaesser for expert technical assistance, and Dr. Meredith Gould for preparing the autoradiographs. I also appreciate the many pertinent comments of Drs. John P. Phillips, C. B. Van Niel, and H. Hilgard. The valuable criticisms of Dr. Van Niel and Dr. Norman Wessels aided greatly in the preparation of this manuscript.

* Supported by a grant from the National Science Foundation (NSF GB-4206).

[1] Runnstrom, J., *Protoplasma*, **4**, 388 (1928).

[2] Ohnishi, T., and M. Sugiyama, *J. Biochem.*, **53**, 238 (1963); Epel, D., *Biochem. Biophys. Res. Commun.*, **17**, 69 (1964).

[3] Epel, D., *Biochem. Biophys. Res. Commun.*, **17**, 62 (1964).

[4] Piatigorsky, J., and A. H. Whiteley, *Biochem. Biophys. Acta*, **108**, 404 (1966).

[5] Mitchison, J. M., and J. E. Cummins, *J. Cell Sci.*, **1**, 35 (1966).

[6] Gross, P. R., and B. J. Fry, *Science*, **153**, 749 (1966).

[7] Lundblad, G., *Arkiv. Kemi*, **7**, 127 (1954).

[8] Runnstrom, J., *Biochem. Z.*, **258**, 257 (1933).

[9] Epel, D., in *Molecular Aspects of Development*, ed. R. Deering and M. Trask (Washington: Government Printing Office, in press); Epel, D., and B. C. Pressman, manuscript in preparation.

[10] Tyler, A., J. Piatigorsky, and H. Ozaki, *Biol. Bull.*, **131**, 204 (1966).

[11] Hultin, T., *Exptl. Cell Res.*, **25**, 405 (1961).

[12] Nakano, E., and A. Monroy, *Exptl. Cell Res.*, **14**, 236 (1958); Monroy, A., and M. Vittorelli, *J. Cell Comp. Physiol.*, **60**, 285 (1962); Sofer, W. H., J. F. George, and R. M. Iverson, *Science*, **153**, 1644 (1966).

[13] Gross, P. R., L. I. Malkin, and W. A. Moyer, these PROCEEDINGS, **51**, 407 (1964).

[14] Brachet, J., A. Ficq, and R. Tencer, *Exptl. Cell Res.*, **32**, 168 (1963); Denny, P. C., and A. Tyler, *Biochem. Biophys. Res. Commun.*, **14**, 245 (1964).

[15] Monroy, A., R. Maggio, and A. Rinaldi, these PROCEEDINGS, **54**, 107 (1965); Mano. Y., *Biochem. Biophys. Res. Commun.*, **25**, 216 (1966).

[16] Candelas, G. C., and R. M. Iverson, *Biochem. Biophys. Res. Commun.*, **24,** 867 (1966).
[17] Hultin, T., *Devel. Biol.*, **10,** 305 (1964); Marcus, A., and J. Feeley, these PROCEEDINGS, **56,** 1770 (1966).
[18] Shapiro, H., *Biol. Bull.*, **68,** 363 (1935).
[19] Berg, W., *Exptl. Cell Res.*, **40,** 469 (1965).
[20] Dent, C. E., *Biochem. J.*, **43,** 169 (1948).
[21] Zubay, G., in *Procedures in Nucleic Acid Research,* ed. G. L. Cantoni and D. R. Davies (New York: Harper and Row, 1966), p. 455.
[22] Gardner, R. S., A. J. Wahba, C. Basilio, R. S. Miller, P. Lengyel, and J. F. Speyer, these PROCEEDINGS, **48,** 2087 (1962).
[23] Epel, D., unpublished results.
[24] Gould, M. C., *Am. Zool.*, **5,** 635 (1965).
[25] Winesdorfer, J. E., *Am. Zool.*, **5,** 635 (1965).
[26] Smith, L. D., R. E. Ecker, and S. Subtelny, these PROCEEDINGS, **56,** 1724 (1966).
[27] Kavanau, J. L., in *Embryonic Nutrition,* ed. D. Rudnick (Chicago: Univ. of Chicago Press, 1958), p. 11.
[28] Giudice, G., *Devel. Biol.*, **12,** 233 (1965).
[29] Wright, B. E., *Science,* **153,** 830 (1966).

Protein Synthesis in Sea Urchin Eggs

II. Changes in Amino Acid Uptake and Incorporation at Fertilization

HECTOR TIMOURIAN AND GEORGE WATCHMAKER

INTRODUCTION

A number of investigators have shown that radioactively labeled amino acids are incorporated more rapidly into the protein of the sea urchin egg after fertilization than before. The uptake of some amino acids is increased after fertilization (Timourian and Denny, 1964; Mitchison and Cummins, 1966; Tyler *et al.*, 1966), but the increased rate of incorporation into protein is not totally accounted for by increased permeability to the amino acids. When sea urchin eggs were preloaded with the radioactive amino acid, so that both unfertilized and fertilized eggs had the same initial amount, an increased amount of label was found in the protein fraction after fertilization (Nakano and Monroy, 1958; Epel, 1967). It has also been shown that radioactively labeled amino acids are incorporated more rapidly into protein in *in vitro* preparations of cell-free homogenates and subcellular fractions from fertilized eggs than those from unfertilized eggs (Hultin and Bergstrand, 1960; Hultin, 1961; Timourian and Denny, 1964; Timourian, 1967; Stavy and Gross, 1967, 1969). Furthermore, it has been shown that the number of ribosomes engaged in protein synthesis (the fraction found in polysomes) increases after fertilization (Monroy and Tyler, 1963; Stafford *et al.*, 1964; Rinaldi and Monroy, 1969; MacKintosh and Bell, 1969b; Denny and Reback, 1970).

Since the unfertilized egg synthesizes protein (MacKintosh and Bell, 1969a, b; Epel, 1967; Tyler *et al.*, 1968), it is now fairly well agreed that there is an increase in the rate of protein synthesis after fertilization and not an activation as was previously assumed. However, neither the degree of increase nor the time after fertilization at which it occurs have been well characterized.

The question of the time of occurrence of the increase in rate of protein synthesis has been considered by Epel (1967) in experiments in which the unfertilized eggs were preloaded with leucine-^{14}C to circumvent the changes in permeability at the time of fertilization. His results showed that an increase in the rate of protein synthesis occurred at 6–10 minutes after fertilization. On the other hand, Rinaldi and Monroy (1969) found a detectable increase in the ribosomes engaged in protein synthesis (polysomes) within 2 minutes after fertilization. It is surprising that the increase in polysomes can be detected much earlier because the quantitative measurement of polysomes is less sensitive than the measurement of incorporation of radioactively labeled amino acids. In this paper, we present evidence that the rate of protein synthesis increases as early as 1–2 minutes after fertilization. Furthermore, we show that the lag observed by Epel (1967) can be repeated in our experiments, but that it is due not to a lag in the incorporation of protein synthesis; but to a release of valine-^{14}C (both free and incorporated into protein). This release of egg materials at the time of fertilization coincides with the release of other materials from the egg, for example, β-1,3-glucanase (Epel *et al.*, 1969b).

MATERIALS AND METHODS

Egg materials. Eggs of the sea urchin *Strongylocentrotus purpuratus* were obtained from excised gonads. The jelly coat was removed by washing the eggs with artificial seawater at pH 5. They were then strained through bolting silk and washed several more times in artificial seawater. Fertilization, development, and incubation of eggs were also carried out in artificial seawater. The artificial seawater was made according to the formula of Tyler and Tyler (1966). To fertilize the eggs, 10 μl of concentrated semen was added to 50 ml of the egg suspension.

Incubation of eggs. To determine uptake and incorporation, 0.1 ml of the radioactively labeled amino acid was added to 100 ml of a suspension containing approximately 10,000 eggs per ml. A total of 20 μCi of L-valine-^{14}C (uniformly labeled, specific activity greater than 200 mCi/mmole), or a total of 0.25 mCi L-valine-^{3}H (specific activity 2 Ci/mmole), was used. The incubations were done in 250-ml Erlenmeyer flasks with 2-ml calibrated dispenser heads (overflow type from CaLab, Oakland, California). The eggs were kept in suspension during the incubation by gentle manual swirling.

When the unfertilized eggs were preloaded, they were washed twice by centrifugation through 40 ml of a precooled isotonic sucrose cushion (30% 1.1 *M* sucrose and 70% 0.55 *M* KCl).

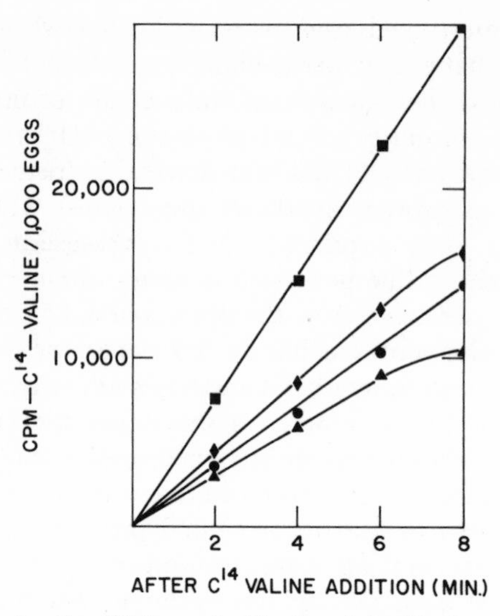

FIG. 1. Uptake of valine-^{14}C by unfertilized and fertilized eggs. Eggs were incubated with valine isotope before fertilization ●——●, at the time of sperm insemination ▲——▲, and 5 ◆——◆ and 10 ■——■ minutes after fertilization.

Stopping the reaction. At the specified times for sampling, 2 ml of the egg suspension was placed on top of a 40-ml precooled sucrose cushion as described above and immediately pelleted by centrifugation at 1500 rpm in a refrigerated centrifuge. This procedure washes the amino acid in the surrounding seawater, removes the seawater, and stops the uptake and protein synthesis. Preliminary experiments, in which uptake and incorporation were quenched by the addition of 0.001 *M* valine-^{12}C to the sucrose cushion and to the homogenizing media showed that our procedure is subject to an error of no more than 0.2 of a minute.

The preparation of homogenates. After the eggs were pelleted, the sucrose and the band of seawater containing the radioactive amino acid were aspirated from the top of the tube. The egg pellet was then homogenized in 2 ml of 0.1 *M* Tris-HCl buffer (pH 7.8) containing 0.005 *M* magnesium acetate, 0.18 *M* KCl, and 0.12% Triton X-100. The homogenization was done in a Potter-Elvehjem type homogenizer with a Teflon pestle. To obtain the soluble protein the homogenates were centrifuged for 90 minutes at 40,000 rpm in a 40.3 rotor with plastic adaptors.

Determination of radioactivity. The incorporation of radioactively labeled amino acid into the protein fraction was measured by a modification of the technique of Mans and Novelli (1961), which consists of precipitating macromolecules on filter paper with trichloroacetic acid (TCA). In each case, 0.2 ml of either the homogenate or the supernatant after centrifugation was applied to 3-in × 1-in paper strips. The filter papers were washed once with 5% cold TCA, once with 95°C TCA, three times with 95% ethyl alcohol, once with a 100% ethyl alcohol, and twice with acetone. For determination of total uptake the papers were not washed in TCA. The paper strips were air-dried and placed with 20 ml of Bray's solution in glass vials for scintillation counting in a Nuclear Chicago spectrophotometer. Efficiency for ^{14}C was near 50%; for tritium it was about 5%.

The eggs were counted according to the method of Tyler and Tyler (1966). All determinations were calibrated to cpm per 1000 eggs. All determinations were made in duplicate. Papers without homogenate but washed together with the ones containing homogenate were used to determine the background for each experiment. Background counts (35–45) were then subtracted from the measured values.

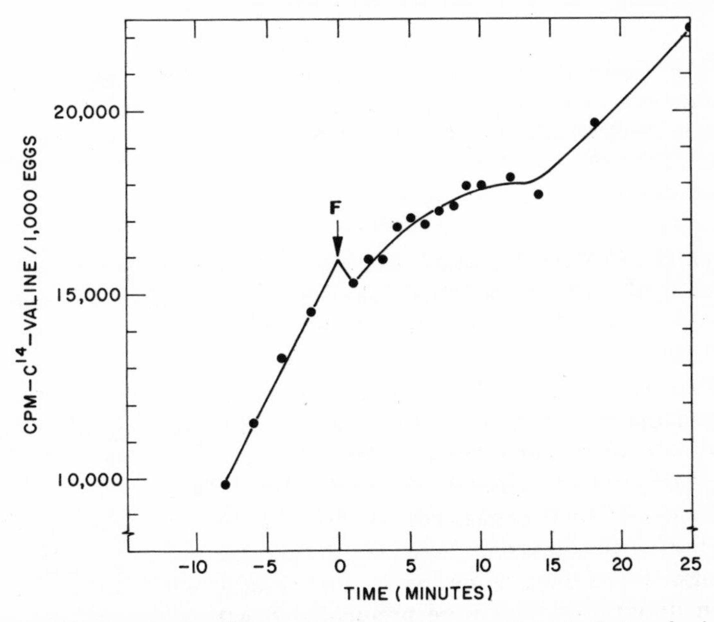

Fig. 2. Uptake of valine-^{14}C by eggs continually exposed to the isotope beginning at 18 minutes before fertilization. *F*, fertilization time.

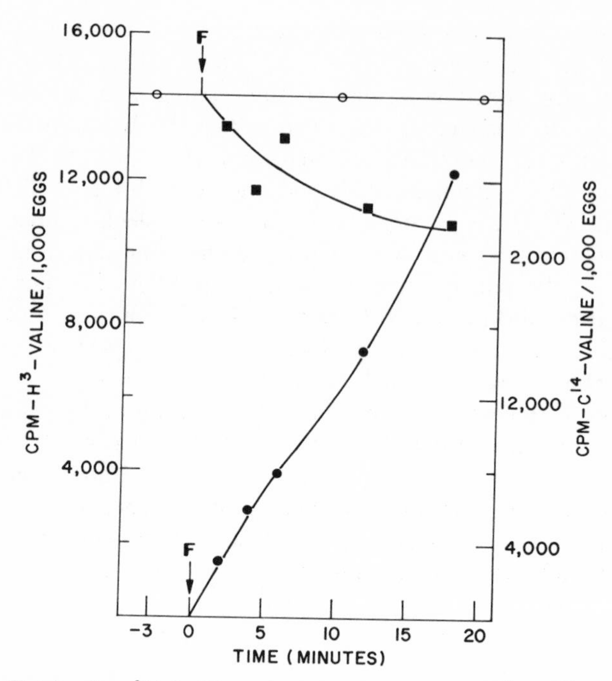

FIG. 3. Total valine-^{3}H inside unfertilized ○ —— ○ and ■ —— ■ fertilized eggs preloaded with the isotope for 60 minutes before fertilization. Uptake of valine-^{14}C by a group of the same eggs exposed to the isotope at the time of fertilization ● —— ●. *F*, fertilization time.

RESULTS

Uptake of valine-^{14}C. The initial rates of valine uptake were determined by exposing unfertilized eggs, eggs at the time of fertilization, and eggs 5 and 10 minutes after fertilization, to valine-^{14}C (Fig. 1). The rate of uptake decreased at fertilization and remained lower than the rate of the unfertilized egg for at least 5 minutes. The increased uptake reported in the literature (Timourian and Denny, 1964; Mitchison and Cummins, 1966; Tyler *et al.*, 1966) cannot be discerned until 10 minutes after fertilization. When the rate of uptake was measured in eggs continuously exposed to valine-^{14}C before and during fertilization, there was a decrease not only in the rate but also in the total valine inside the egg after fertilization (Fig. 2). When unfertilized eggs were preincubated with valine-^{3}H and then exposed to valine-^{14}C at the time of fertilization, some of the valine-^{3}H already in the egg was lost at the time of fertilization, while simultaneously valine-^{14}C was taken up (Fig. 3). It is concluded from the above experiments that some valine is lost from the egg at the time of fertilization and that its rate of uptake does not increase until 5 to 10 minutes after the time of fertilization.

The incorporation of labeled valine into protein. Eggs continuously exposed to valine-^{14}C during the time of fertilization showed increased rate of valine incorporation 5 to 6 minutes after fertilization (Fig. 4). This applied to both the total incorporation and the ratio of incorporated radioactivity to uptake (radioactivity inside the egg). However, this delay between time of fertilization and increase in rate of incorporation is not real, as can be shown experimentally as follows. When unfertilized eggs were preincubated with valine-^{3}H and the total amount of ^{3}H into the protein fraction was measured, an increased incorporation is detectable 4–6 minutes after fertilization (Fig. 5). However, when we plot incorporation into protein in terms of the amount of valine inside the egg, the increased rate is detectable within the first 2 minutes after fertilization (Fig. 6).

Loss of soluble (released) protein at time of fertilization. At the time of fertilization the loss of radioactive valine is not limited to the

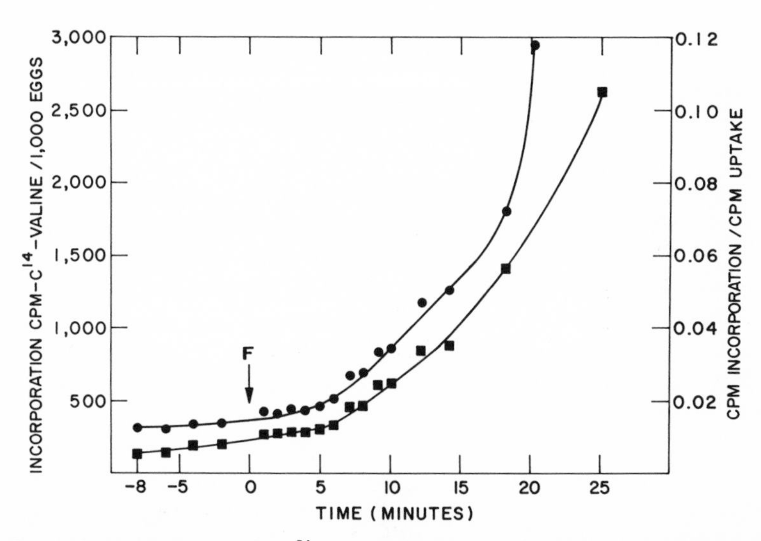

FIG. 4. Incorporation of valine-^{14}C into the protein fraction of eggs continually exposed to the isotope. ●——●, Total radioactivity (cpm) in protein fraction. ■——■, Ratio of incorporated radioactivity in protein fraction to uptake (radioactivity inside eggs). *F*, fertilization time.

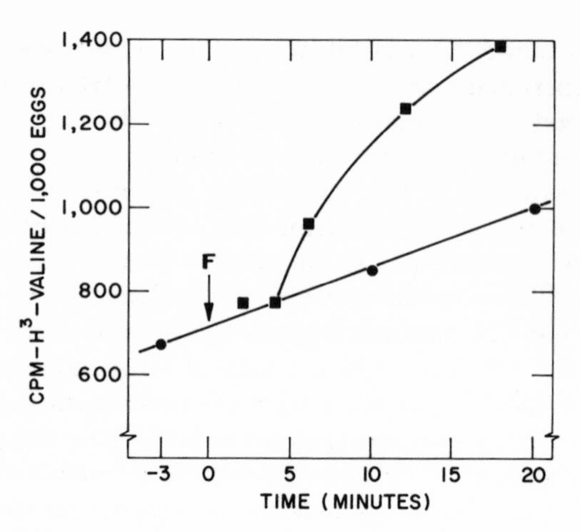

FIG. 5. Incorporation of valine-^{3}H into the protein fraction of unfertilized ●——● and fertilized eggs ■——■ preloaded for 60 minutes and then washed before fertilization.

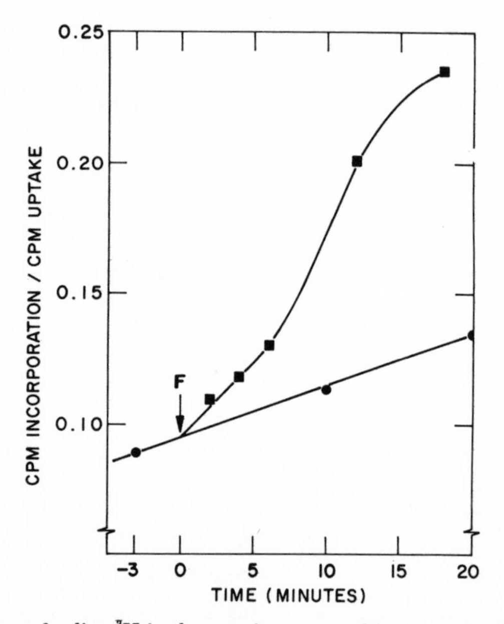

FIG. 6. Fraction of valine-^{3}H in the eggs incorporated into protein. Unfertilized eggs ●——●, fertilized eggs ■——■, were preloaded with valine-^{3}H for 60 minutes before fertilization and then washed to remove the external radioactivity.

free amino acid. Protein that is not bound to ribosomes is also lost. The soluble protein found in the supernatant after centrifugation of the total homogenate (at 40,000 rpm for 90 minutes) can be labeled before fertilization, and its loss can be detected at the time of fertilization. This is shown in Fig. 7, in which valine-^{3}H was incorporated by unfertilized eggs while the eggs were being preloaded. At fertilization, the amount of label in the soluble protein fraction immediately decreased. However, when valine-^{14}C was added at the time of fertilization, less ^{3}H was lost because the proteins that were being synthesized and released into the soluble fraction were also being diluted by ^{14}C-labeled proteins. By 5–10 minutes after fertilization, the increased rate of protein synthesis makes up for the lost protein. The incorporation of ^{14}C into the total protein fraction and the soluble (released from the ribosomes) protein fraction of the same eggs is shown in Fig. 8.

DISCUSSION

Fertilization is followed by changes in the metabolism, synthetic activities, and structure of the egg. These changes set the course for the formation of a new individual. Epel *et al.* (1969a) has reviewed and described the temporal sequence of many of the changes that occur in sea urchin eggs upon fertilization.

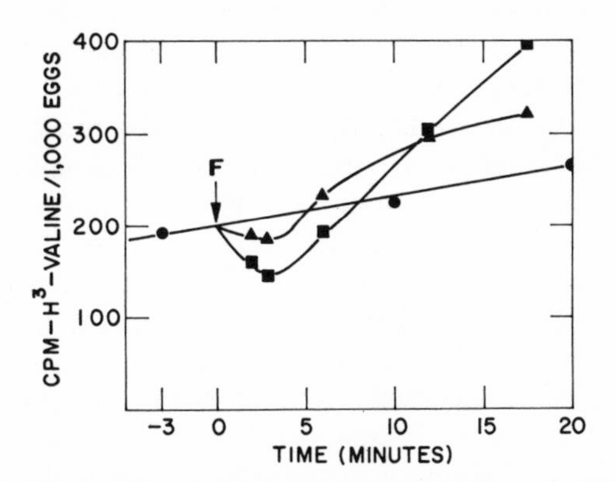

FIG. 7. Demonstration of loss of protein by eggs at time of fertilization. All eggs were preloaded with valine-^{3}H for 60 minutes before fertilization. At the time of fertilization one group was left unfertilized ●——●, one group was fertilized ■——■, and a third group was fertilized and incubated with valine^{14}C ▲——▲. *F*, fertilization time.

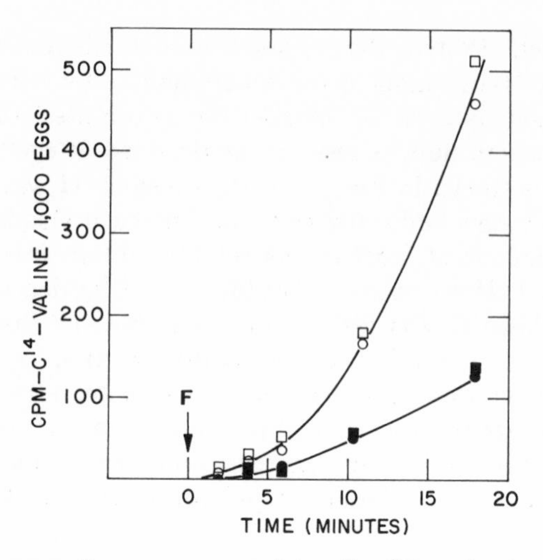

FIG. 8. Two groups of eggs were exposed to valine-^{14}C at the time of fertilization. One group was preloaded with valine-^{3}H for 60 minutes before fertilization and the other was not. Incorporation of valine-^{14}C into the total protein fraction by eggs preloaded with valine-^{3}H □ —— □ and eggs not preloaded ○ —— ○. Incorporation of valine-^{14}C into the soluble fraction by eggs preloaded with valine-^{3}H ■ —— ■ and eggs not preloaded ● —— ●.

After fertilization, the sea urchin eggs increase in their rates of uptake of amino acids from the surrounding seawater. The increased uptake has been measured for different amino acids by Tyler *et al.* (1966) and for valine by Timourian and Denny (1964) and Mitchison and Cummins (1966). Unlike the amino acids, nucleotides are not taken in by the unfertilized egg, and instead of stimulation there is an activation of their uptake (Mitchison and Cummins, 1966; Piatigorsky and Whiteley, 1965). Mitchison and Cummins (1966) showed that the increased uptake of amino acids after fertilization is not dependent upon the synthesis of new protein. However, it appears that specific carrier molecules mediate the uptake of amino acids and that their specificity is limited to the different types of amino acids (Tyler *et al.*, 1966). Epel *et al.* (1969a) described the increase in uptake of amino acids as taking place between 10 and 20 minutes after fertilization. Our present work shows increased rate of uptake of valine within 5–10 minutes after fertilization. However, we also find a decreased rate of uptake immediately after fertilization; although this decrease may not be an actual decrease in uptake, since valine inside the egg is lost at fertilization. Experiments with double-labeled valine showed that valine-^{3}H inside the egg is lost, while valine-^{14}C outside the egg goes in simultaneously.

The problem of measuring uptake is also complicated by the amount that can get in within a specific time. In earlier work (Timourian and Denny, 1964; Tyler *et al.*, 1966) the uptake was measured for 30 and 60 minutes. Mitchison and Cummins (1966) have shown that during long incubation periods the pool inside the egg may be filled up and initial uptake rates may not be accurately measured. In the present work, we measured the initial rate of uptake by sampling at 2, 4, 6, and 8 minutes after exposure to the labeled amino acid.

Measurements of the rate of protein synthesis are subject to variation if the specific activities of amino acids in their pools are shifted by changes in the uptake rate. For example, Fry and Gross (1970a) found that the changes in incorporation of amino acids into the protein fraction during the first cell cycle are paralleled by changes in the uptake by the egg. Different methods have been used to keep such fluctuations in uptake from obscuring the true picture of protein synthesis at the time of fertilization. Eggs have been preloaded with the radioactive isotope during oogenesis (Nakano and Monroy, 1958) or before fertilization (Epel, 1967); measurement of the radioactivity in the protein fraction then gives the amount of internal amino acid incorporated after fertilization. These workers found an increased rate of protein synthesis after fertilization. A second method of avoiding variability due to externally labeled amino acid is measurement of the absolute rates of incorporation based on consideration of the changes in specific activity of the amino acids inside the egg (Fry and Gross, 1970b). A third way to demonstrate an actual increase in the rate of protein synthesis after fertilization has been to show that the homogenates and subcellular fractions of fertilized sea urchin eggs have more capability for incorporating radioactive amino acid into protein *in vitro* than those of unfertilized eggs (Hultin and Bergstrand, 1960; Hultin, 1961; Timourian and Denny, 1964; Timourian, 1967; Stavy and Gross, 1967, 1969). However, the use of *in vitro* systems is subject to various problems, because we do not know whether the subcellular fractions are at their optimal activity. Although optimal activities for the *in vitro* systems have been described (cf. Hultin, 1961; Timourian and Denny, 1964; Timourian, 1967; Stavy and Gross, 1967, 1969), it is not known whether the optimal activities obtained *in vitro* are actually the same as occur *in vivo*, or, indeed, whether any such optimal conditions exist inside the egg. All that can be said is that under optimal conditions *in vitro* the components isolated from fertilized eggs are more active than those isolated from unfertilized eggs.

In our present work we can detect an increase in the rate of protein synthesis at 1–2 minutes after fertilization. Furthermore, we notice that during the first few minutes after fertilization not only amino acid is lost, but also soluble protein. Since these proteins are labeled with the radioactive amino acid and are found in the supernatant after centrifugation of the ribosomes, they are considered to be newly synthesized proteins released from the ribosomes. The decreased specific activity of the proteins lost at the time of fertilization in the double-labeled experiments indicates that proteins synthesized within the first 2 minutes of fertilization are released from the ribosomes and in turn may be lost out of the egg. Loss of materials from the egg noted in this report coincides with the loss of other soluble proteins, for example, the release of β-1,3-glucanase by Epel *et al.* (1969b); these proteins are lost at the time for breakdown of cortical granules (Runnström, 1966).

The fertilization membrane appears to be permeable to proteins (Epel *et al.*, 1969b), but if the labeled valine from the egg at the time of fertilization is trapped or slowed in the perivitelline space, our measured values of the loss of valine may be too low. Since we wished to sample soon after fertilization, we found it impractical to remove the fertilization membranes by mechanical means. We also avoided the chemical removal of membranes (i.e., treatments with trypsin, urea, or mercaptoethylgluconamide) since their effects on the surface mechanisms involved in amino acid uptake are unknown.

There is evidence that the increase in protein synthesis after fertilization does not require the concomitant synthesis of *m*RNA in the nucleus (Tyler, 1963, 1966; Denny and Tyler, 1964; Gross and Cousineau, 1963; Brachet *et al.*, 1963). It has therefore been assumed that most of the *m*RNA in the cytoplasm of the unfertilized egg is unable to participate in the process of translation either because the ribosomes are nonfunctional (Monroy *et al.*, 1965) or because the *m*RNA is unavailable or "masked" (Stavy and Gross, 1967; Gross, 1967; Tyler, 1967). On the other hand, MacKintosh and Bell (1969b), have been able to raise the incorporating capacity of unfertilized eggs by removing CO_2 from the seawater. MacKintosh and Bell

(1969b) proposed that the unfertilized egg may be metabolically inhibited and that the limiting factors are not the ribosomes or the *m*RNA, but factors required for peptide initiation. All these theories predict an increase in active polysomes after fertilization, as indeed seems to be the case (Monroy and Tyler, 1963; Stafford *et al.*, 1964; Hultin, 1964; Rinaldi and Monroy, 1969; MacKintosh and Bell, 1969; Denny and Reback, 1970).

The alternative hypotheses proposed to explain the post-fertilization increased rate of protein synthesis need not be mutually exclusive. However, it has been difficult to correlate them since the reported increased rate of incorporation 6–10 minutes after fertilization (Epel, 1967) is later than the following events that have been used to support the different alternatives. The following events have been reported within the first 2 minutes after fertilization: (1) An increased rate of O_2 consumption (Epel, 1969); (2) an increased number of polysomes (Rinaldi and Monroy, 1969) and (3) a burst of proteolitic activity (Lundbald, 1950), which may result in the activation of ribosomes (Monroy *et al.*, 1965). In this report we have presented evidence that the increased rate of protein synthesis takes place within the first 2 minutes after fertilization and therefore in closer agreement with the other events that are required for this increase.

SUMMARY

The uptake of valine into the sea urchin egg and its incorporation into the protein fraction were determined for the first few minutes after fertilization. By 10 minutes after fertilization, increased uptake can be measured. The rate of uptake immediately after fertilization is difficult to determine, because materials are lost from the egg at this time. Using eggs preloaded with valine-^{3}H and exposed to valine-^{14}C at the time of fertilization, it was shown that the loss and uptake are simultaneous. An increased rate of incorporation into the protein fraction can be detected within the first 2 minutes after fertilization if this loss of labeled valine is taken into consideration. There is also a loss of newly synthesized protein released from ribosomes at fertilization.

This work was performed under the auspices of the U.S. Atomic Energy Commission. Reference to a company or product name does not imply approval or recommendation of the product by the University of California or the U.S. Atomic Energy Commission to the exclusion of others that may be suitable.

REFERENCES

Brachet, J. A., Ficq, A., and Tencer, R. (1963). Amino acid incorporation into protein of nucleate and anucleate fragments of sea urchin eggs: Effect of parthenogenetic activation. *Exp. Cell Res.* **32,** 168–170.

Denny, P. C., and Reback, P. (1970). Active polysomes in sea urchin eggs and zygotes: Evidence for an increase in translatable RNA after fertilization. *J. Exp. Zool.* in press.

Denny, P., and Tyler, A. (1964). Activation of protein biosynthesis in non-nucleate fragments of sea urchin eggs. *Biochem. Biophys. Res. Commun.* **14,** 245–249.

Epel, D. (1967). Protein synthesis in sea urchin eggs: A "late" response to fertilization. *Proc. Nat. Acad. Sci. U. S.* **57,** 899–906.

Epel, D. (1969). Does ADP regulate respiration following fertilization of sea urchin eggs? *Exp. Cell Res.* **58,** 312–319.

Epel, D., Pressman, B. C., Elsaesser, S., and Weaver, A. M. (1969a). The program of structural and metabolic changes following fertilization of sea urchin eggs. *In* "The Cell Cycle" (G. M. Padilla, G. L. Whitson, and I. L. Cameron, eds.). Academic Press, New York.

Epel, D., Weaver, A. M., Muchmore, A. V., and Schimke, R. R. (1969b). β-1,3-Glucanase of sea urchin eggs: Release from particles at fertilization. *Science* **163,** 294–296.

Fry, B. J., and Gross, P. R. (1970a). Patterns and rates of protein synthesis in sea urchin embryos. I. Uptake and incorporation of amino acids during the first cleavage cycle. *Develop. Biol.* **21,** 105–124.

Fry, B. J., and Gross, P. R. (1970b). Patterns and rates of protein synthesis in sea urchin embryos. II. The calculation of absolute rates. *Develop. Biol.* **21,** 125–146.

Gross, P. R. (1967). The control of protein synthesis in embryonic development and differentiation. *Curr. Top. Develop. Biol.* **2,** 1–43.

Gross, P. R., and Cousineau, H. (1963). Effects of actinomycin D on macromolecular synthesis and early development in sea urchin eggs. *Biochem. Biophys. Res. Commun.* **4,** 321–326.

Hultin, T. (1961). Activation of ribosomes in sea urchin eggs in response to fertilization. *Exp. Cell Res.* **25,** 405–417.

Hultin, T. (1964). On the mechanism of ribosomal activation in newly fertilized sea urchin eggs. *Develop. Biol.* **10,** 305–328.

Hultin, T., and Bergstrand, A. (1960). Incorporation of C^{14}-L-leucine into protein by cell-free systems from sea urchin embryos at different stages of development. *Develop. Biol.* **2,** 61–75.

Lundbald, G. (1950). Proteolytic activity in the sea urchin gametes. *Exp. Cell Res.* **1,** 264–271.

MacKintosh, F. R., and Bell, E. (1969a). Proteins synthesized before and after fertilization in sea urchin eggs. *Science* **164,** 961–963.

MacKintosh, F. R., and Bell, E. (1969b). Regulation of protein synthesis in sea urchin eggs. *J. Mol. Biol.* **41,** 365–380.

Mans, R. J., and Novelli, G. D. (1961). Measurement of the incorporation of radioactive amino acids into protein by a filter-paper disk method. *Arch. Biochem. Biophys.* **94,** 48–53.

Mitchison, J. M., and Cummins, J. E. (1966). The uptake of valine and cytidine by sea urchin embryos and its relation to the cell surface. *J. Cell Sci.* **I,** 35–47.

MONROY, A., and TYLER, A. (1963). Formation of active ribosomal aggregates (polysomes) upon fertilization and development of sea urchin eggs. *Arch. Biochem. Biophys.* **103,** 431–435.

MONROY, A., MAGGIO, R., and RINALDI, A. M. (1965). Experimentally induced activation of the ribosomes of the unfertilized sea urchin egg. *Proc. Nat. Acad. Sci. U. S.* **54,** 107–111.

NAKANO, E., and MONROY, A. (1958). Incorporation of S^{35}-methionine in the cell fractions of sea urchin eggs and embryos. *Exp. Cell Res.* **14,** 236–243.

PIATIGORSKY, J., and WHITELEY, A. H. (1965). A change in permeability and uptake of [^{14}C] uridine in response to fertilization in *Strongylocentrotus purpuratus* eggs. *Biochim. Biophys. Acta* **108,** 404–418.

RINALDI, A. M., and MONROY, A. (1969). Polyribosome formation and RNA synthesis in the early post-fertilization stages of the sea urchin egg. *Develop. Biol.* **19,** 73–86.

RUNNSTRÖM, J. (1966). The vitelline membrane and cortical particles in sea urchin eggs and their function in maturation and fertilization. *Advan. Morphog.* **5,** 221–325.

STAFFORD, D. W., SOFER, W. H., and IVERSON, R. M. (1964). Demonstration of polyribosomes after fertilization of the sea urchin egg. *Proc. Nat. Acad. Sci. U. S.* **52,** 313–316.

STAVY, L., and GROSS, P. R. (1967). The protein-synthetic lesion in unfertilized eggs. *Proc. Nat. Acad. Sci. U. S.* **57,** 735–742.

STAVY, L., and GROSS, P. R. (1969). Protein synthesis *in vitro* with fractions of sea urchin eggs and embryos. *Biochim. Biophys. Acta* **182,** 193–202.

TIMOURIAN, H. (1967). Protein synthesis in sea urchin eggs. I. Fertilization-induced changes in subcellular fractions. *Develop. Biol.* **16,** 594–611.

TIMOURIAN, H., and DENNY, P. (1964). Activation of protein synthesis in sea urchin eggs upon fertilization in relation to magnesium and potassium ions. *J. Exp. Zool.* **155,** 57–70.

TYLER, A. (1963). The manipulation of macromolecular substances during fertilization and early development of animal eggs. *Am. Zoologist* **3,** 109–126.

TYLER, A. (1966). Incorporation of amino acids into protein by artificially activated non-nucleate fragments of sea urchin eggs. *Biol. Bull.* **130,** 450–461.

TYLER, A. (1967). Masked messenger RNA and cytoplasmic DNA in relation to protein synthesis and processes of fertilization and determination in embryonic development. *Develop. Biol.*, Suppl. 1, 170–226.

TYLER, A., and TYLER, B. S. (1966). The gametes: some procedures and properties. *In* "Physiology of Echinodermata" (R. A. Boolootian, ed.), pp. 639–682. Wiley, New York.

TYLER, A., PIATIGORSKY, J., and OZAKI, H. (1966). Influence of individual amino acids on uptake and incorporation of valine, glutamic acid, and arginine by unfertilized and fertilized sea urchin eggs. *Biol. Bull.* **131,** 204–217.

TYLER, A., TYLER, B. S., and PIATIGORSKY, J. (1968). Protein synthesis by unfertilized eggs of sea urchins. *Biol. Bull.* **134,** 209–219.

Active Polysomes in Sea Urchin Eggs and Zygotes: Evidence for an Increase in Translatable Messenger RNA after Fertilization [1,2]

PAUL C. DENNY AND PATRICIA REBACK

Protein synthesis increases severalfold in the sea urchin zygote within 30 minutes after fertilization (Epel, '67). There is evidence from several different approaches that this burst of synthesis can occur in the absence of *de novo* synthesis of the components of the protein synthesis system. (Tyler, '67; Gross, '68). It then appears likely that the low level of synthesis in eggs prior to fertilization is caused either by inefficient translation processes within the existing system or by the limited availability of a translational level material.

It was shown that the factors responsible for peptide elongation were available in eggs at activities adequate for *in vitro* synthesis rates which approached those of early embryos (Hultin, '61; Stavy and Gross, '67; Timourian, '67). It has also been shown that the availability of ATP and GTP for synthesis *in vivo* does not change after fertilization (MacKintosh and Bell, '69a). The time required for synthesis of an average size protein in eggs and zygotes appeared to be similar (Humphreys, '69; MacKintosh and Bell, '69b), suggesting that the factor(s) limiting synthesis in eggs was not directly related to peptide chain elongation, termination or release. Furthermore, the increase in polyribosomes observed at fertilization (Monroy and Tyler, '63; Stafford et al., '64) supplied evidence that the limiting condition for protein synthesis in eggs was localized at the site of peptide chain initiation. The two hypotheses which have been proposed to account for a block to synthesis at this site are either that the system is messenger RNA limited (Gross and Cousineau, '64; Stavy and Gross, '69) or that the peptide initiation step itself is inefficient possibly caused by initiation factors whch are limiting, suboptimal conditions or altered ribosomes (MacKintosh and Bell, '69b; Rinaldi and Monroy, '69). One approach to the decision between these two hypotheses is a comparison of the size distributions of the polysomes and the quantity of polysomal material in eggs and zygotes.

With messenger RNA as a limiting factor, it would be expected that the size distributions would be similar while the number of polysomes would be less in eggs than in zygotes. Messenger RNA becoming available for protein synthesis after fertilization would account for the increase in polysomes. On the other hand, an inefficient peptide initiation step which was limiting would tend to produce polysomes with only a partial load of ribosomes and would result in smaller polysomes in eggs than zygotes. In this case, the number of polysomes would be expected to be similar, and the increase in polysomal material would be proportional to the size difference of the polysomes before and after fertilization.

In this report, the relative amounts of polysomal material present in eggs and zygotes are given, and the size distribu-

[1] Supported by NIH grant GM 16090.

[2] One of us (PCD) would like to dedicate this report to the memory of the late Professor Albert Tyler.

tions of the polysomes are compared at levels sufficient to distinguish between the above hypotheses.

MATERIALS AND METHODS

Collection of gametes and pulse labeling

Eggs and semen were obtained from freshly collected *Lytechinus pictus* adults by the injection of 0.55 M KCl into the body cavity. The eggs from a single female were washed with sea water and divided in half. One group was left unfertilized and allowed to sit at 20°C for 30 minutes. The other aliquot was fertilized with a 1:5000 dilution of semen (95% fertilization acceptable) and allowed to sit at 20°C for 30 minutes. At this time both eggs and zygotes were washed two times with sea water and labeled for two minutes with 200 μc/ml of H^3 leucine (19.7 c/mmole; Amersham/Searle) at 20°C. The pulse was stopped by the addition of ten volumes of 0.55 M KCl at − 3°C, and two subsequent washes were given by gentle centrifugation. Previous to the labeling, a small aliquot of embryos was set aside for observation. Normal development through the fourth cleavage was observed in all cases.

Homogenization

In order to homogenize the eggs and zygotes at equal concentrations, the packed eggs were resuspended in an equivalent volume of 0.55 M KCl (to adjust for the fluid volume trapped by the fertilization membrane of the zygotes) plus two additional volumes of a buffered medium, SUHB (.0 66 M Tris-HCl pH 7.8, 0.0066 M Mg acetate, 0.132 M sucrose, 0.0001 M dithiothreitol). The zygotes and the associated 0.55 M KCl inside the fertilization membrane were resuspended in two volumes of SUHB. SUHB and 0.55 M KCl in this proportion gives Mg^{++} and K^+ concentrations which were optimal for *in vitro* amino acid incorporation in this system (Timourian and Denny, '64). Homogenization was accomplished by two passes through a 25 gauge needle, and the homogenates were then centrifuged at 15,000 × g for ten minutes. To solubilize the lipid containing fractions without damage to the ribosomes (Golub and Clegg, '69) or polysomes (Kuff and Roberts, '67), the supernatants were treated with non-ionic detergent, Triton X-100 (Rhom and Haas), at 1% and layered immediately onto the appropriate gradients.

Sucrose gradients

All gradients were made with sucrose (RNase-free; Mann) dissolved in SUDB (0.03 M Tris-HCl pH 7.8, 0.005 M Mg acetate, 0.06 M sucrose, 0.18 M KCl and 0.0001 M dithiothreitol).

Analysis of sucrose gradients

After centrifugation, the gradients were removed from the centrifuge tubes by pumping in 60% w/w sucrose containing a dye marker (Blue Dextran 2000; Pharmacia) at the bottom. Optical densities of the gradients were continuously monitored in a 2 mm path length flow cell (ISCO) and six drop fractions were collected directly on rectangles of filter paper. The papers were extracted by the method of Mans and Novelli ('61) and counted by liquid scintillation at an efficiency of 6%.

Ribonuclease treatment

A portion from both egg and embryo supernatants was subjected to ribonuclease treatment at a concentration of 5 μg/ml (5X cryst; Calbiochem) for 30 minutes at 4°C. Treatments of 0°C and 9°C gave similar results. The samples were layered over the gradients and centrifuged under the given conditions.

Radioautography

To test for possible parthenogenetic activation or immature oocytes in batches of eggs, aliquots of both eggs and embryos were withdrawn at the end of the pulse label period and fixed in Bouin's fixative for 24 hours. They were subsequently embedded, sectioned at 10 μ and stained with Harris' haemotoxylin and eosin blue, then dipped in NTB_2 nuclear tracking emulsion (Kodak) and allowed to sit at 4°C for one month before development. The resulting radioautographs showed no overlapping grain counts between eggs and zygotes. The egg sections had an average of three grains per unit area with 12 grains as an extreme. The zygote sections had an average in excess of 100 grains per unit area with the extreme minimum count of 50 grains.

RESULTS

Analysis of the sizes of the polysomes was carried out with a sucrose gradient which separated polysome classes up to seven ribosomes per polysome and concentrated the larger ones on a cushion. A rabbit reticulocyte fraction containing monoribosomes and polysomes primarily in the four to six ribosome class (Rich, '67) was used to standardize the gradient (fig. 1A). The rationale for using this type

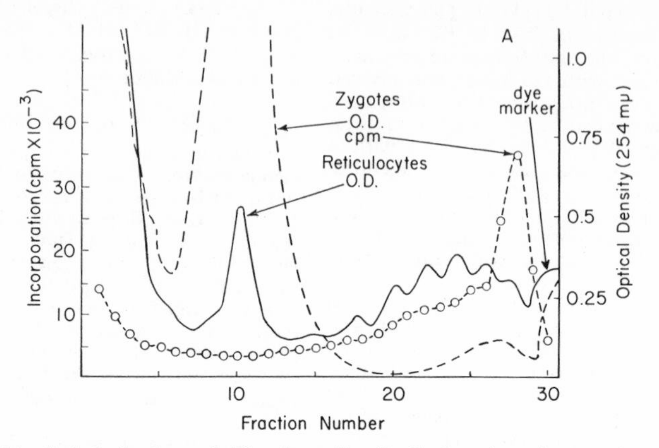

Fig. 1A Optical density and H^3 polypeptide distributions in polysomes from rabbit reticulocytes and sea urchin zygotes. Sedimentation was from left to right. Samples of 0.2 ml were layered over 15–30% w/w sucrose gradients (4 ml) with a 55% w/w cushion (0.5 ml) and centrifuged at 50,000 rpm in a SW 50.1 rotor (Spinco) for 70 minutes at 5°C. The monoribosomal peaks were in or near sample 10.

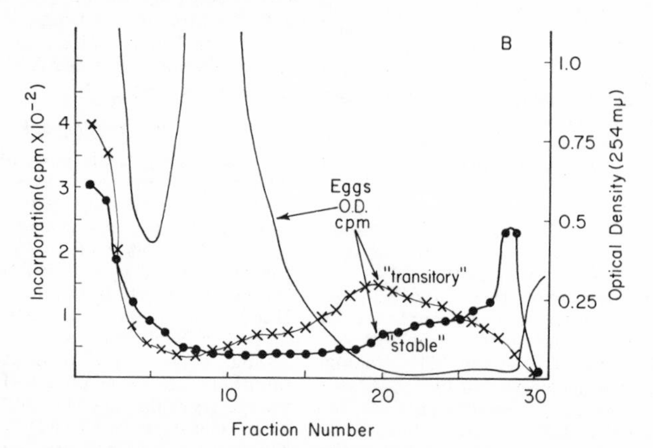

Fig. 1B Optical density and H^3 polypeptide distributions in polysomes from sea urchin eggs. Two representative types of distribution are shown. Additional studies indicated that the distribution containing the active small polysomes was transitory and eventually stabilized as the distribution with larger active polysomes.

of gradient is straightforward. It had been demonstrated previously by electron microscopy that a large portion of embryo polysomes contained approximately 23 ribosomes (Infante and Nemer, '67). A large increase of polysomal material at fertilization due solely to an addition of ribosomes to existing polysomes would require small sized egg polysomes to begin with. The accuracy of determining the magnitude of this type of increase would depend largely on how specifically this class of small polysomes could be identified. Although the gradient resolved polysomes with eight or more ribosomes collectively and not individually, an increase in size from eight ribosomes in the egg to 23 in the embryo would account for only a threefold increase in polysomal material. Preliminary experiments had indicated that the increase in material was considerably more than this.

As was expected, the polysomal fraction from sea urchin zygotes 30 minutes after fertilization contained many polysomes with more than seven ribosomes which were active in protein synthesis (fig. 1A). This time after fertilization was selected because the amount of polysomal material is no longer increasing rapidly (MacKintosh and Bell, '69b; Rinaldi and Monroy, '69). The polysomes from eggs were rarely detectable as U.V. absorbing material. There was considerable variation in the amounts and distributions of polysomal activity in the eggs of separate females as detected by amino acid incorporation. However, in more than 90% of the egg batches tested, the incorporation distributions fell into two general groups: those most active in the small polysome region of the gradients (3–5 ribosomes) and those with most of the activity concentrated in larger polysomes. Both distributions are shown in figure 1B.

Additional studies indicated that the observed variability was due to a transitory state existing in eggs just after shedding. Egg batches pulse-labeled immediately after the shedding period often showed the distribution primarily containing small active polysomes, while these same eggs, if allowed to sit at ambient temperature for an hour, showed an incorporation distribution containing the larger active polysomes with seven ribosomes or more in addition to the small polysomes (fig. 2). This incorporation distribution was stable and showed no further change after an additional hour. Moreover, many batches of eggs showed only the latter "stable" distribution regardless of the interval of time after shedding. Radioautographs of eggs containing this distribution of polysome incorporation gave no indication that the shift from the transitory to the stable pattern had been due to spontaneous par-

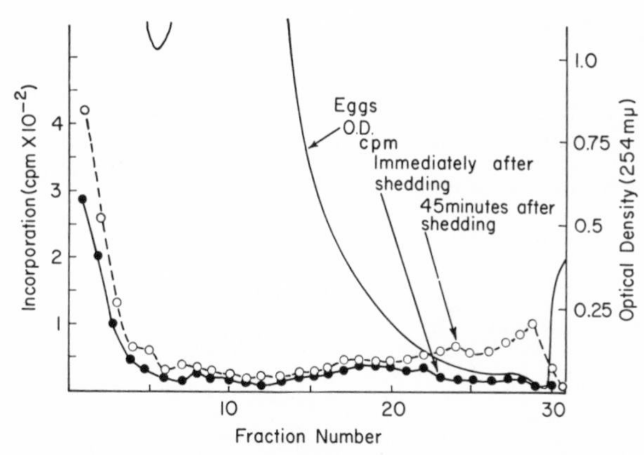

Fig. 2 Polysomal incorporation patterns from a single egg batch showing the change from the "transitory" to the "stable" distribution during the first hour after shedding. Gradients and sedimentation the same as in figure 1.

thenogenesis. These findings suggested that the variation in egg polysome distributions which were observed shortly after shedding may have been due to varying ovarian conditions and indicated that the eggs, once stabilized contained distributions of active polysomes which were similar to each other and similar to the distributions from zygotes.

The difficulty in visualizing the egg polysomes on the first gradient led to the use of a two part gradient for the quantitative measurements. The upper part, consisting of a linear gradient, separated out the monomers and dimers while the lower part, composed of a dense step gradient, concentrated the remaining polysomes into a single peak. Since the possibility existed that cell debris might also be concentrated in this peak, an aliquot from each sample was treated with ribonuclease. Sensitivity to the ribonuclease was then used as a measure of how much actual polysomal material was present. The result of a control experiment with rabbit reticulocyte ribosomal fractions is shown in figure 3A. All of the measurable U.V. absorbing material was removed from the polysome peak by ribonuclease treatment, and over 90% of this material was accounted for by the increase in size of the monomer peak and a small dimer peak.

When egg and embryo polysome fractions were placed on these gradients, the relative amount of polysomal material from zygotes 30 minutes after fertilization was about 77 times more than that from eggs. The absolute values for both were obtained by direct mechanical integration of optical density recordings of non-treated and ribonuclease treated polysome fractions from eggs and zygotes. The area of the ribonuclease sensitive material in each case, given in arbitrary units (figs. 3B, 3C), was determined by measuring the difference between the curves for the non-treated and the ribonuclease treated samples. Measurement of egg volume and total protein (Lowry et al., '51) were used to insure that equal numbers of eggs and embryos were being compared. All batches of eggs contained the "stable" distribution of polysome incorporation and gave no indication of the presence of parthenogenetic eggs or immature oocytes by radioautography.

Ribonuclease treatment of the zygote polysomes was effective in transferring 90% of the polypeptide label to the monomer peak. However, only about 60% of the label in the polysome peak from eggs was directly transferred to the monosome peak, and the remainder was concentrated in a smaller peak of slightly heavier material. While it was not evident from the above experiment, separate experiments which preceded and followed this one indicated that all of the labeled material in the polysome positions of the gradients was RNase sensitive and that the smaller secondary peak was probably partially degraded polysomes. Since this peak formed on the shoulder of the monoribosome peak, it did not interfere with the optical density measurements.

In these experiments 30 to 40% of the polysomal incorporation was discarded with the particulate fraction. This raised the possibility that the quantitative measurements reflected a shift from one fraction to another at fertilization. However, since the percentage of polysomal incorporation in the discarded fraction was the same in both cases, this possibility can be excluded.

DISCUSSION

The substantial rise in polysomal material at fertilization coupled with a similarity in distributions of polysomal activity in both eggs and zygotes indicate that there was an increase in the number of polysomes at fertilization. If it were assumed that the average sizes of the polysomes were the same in eggs and zygotes, then the increase in the number of polysomes would be proportional to the increase in the amount of polysomal material. Our observations do indicate a similar size distribution of polysomes in "stable" eggs and embryos and suggest that most of the increase in the polysomal material was due to an increase in the total population of polysomes.

A recent study by Humphreys ('69) also indicated that the size distribution of the polysomes were nearly identical in eggs and embryos of this species. Based upon this observation and the observation that the efficiency of translation was similar in both cases, he suggested that the number of polysomes increased by a factor of 10–15 times. The actual magnitude of increase in polysomes which is reported here (77-fold) should be regarded as a gross estimate because the amount of egg polysomes was very close to the minimum level for detection. On the other hand, it is unlikely that this value is greatly inflated since the detection system was quite reliable in picking up absorbances equivalent to more than ten arbitrary units of area, especially when this amount of ma-

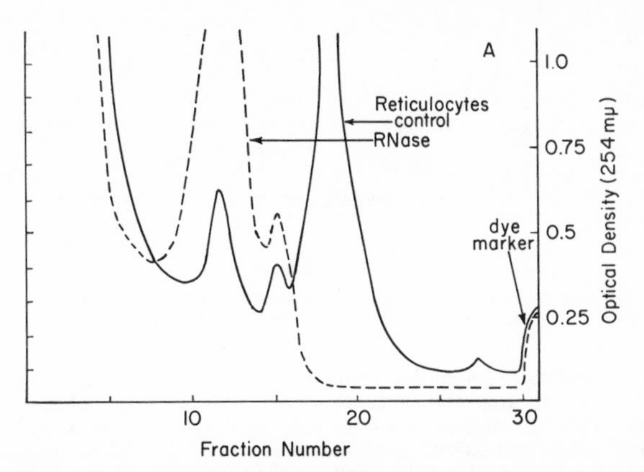

Fig. 3A Optical density distributions of rabbit reticulocyte ribosomal fractions before and after ribonuclease treatment. Samples of 0.2 ml were layered over 15–22% w/w sucrose gradients (2 ml) with steps of 45% (1 ml), 50% (1 ml) and 55% w/w sucrose (0.5 ml). The sedimentation forces were the same as in figure 1. The center of the monoribosomes was in sample 12.

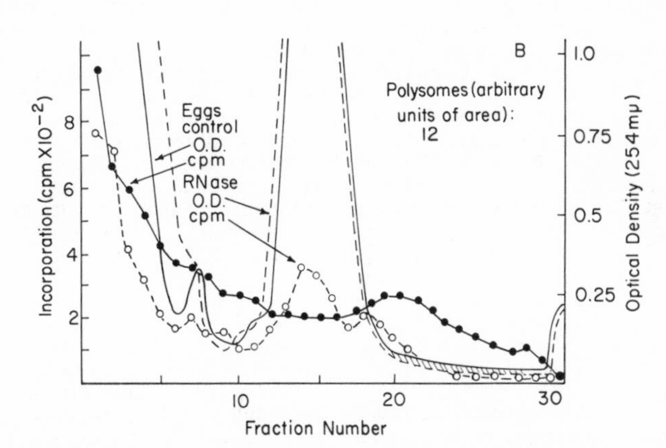

Fig. 3B Optical density and H^3 polypeptide distributions in polysomes from sea urchin eggs before and after ribonuclease treatment. The crosshatched area is given in arbitrary units and is proportional to the amount of polysomal material which was ribonuclease sensitive. The monoribosome peaks were in sample 15.

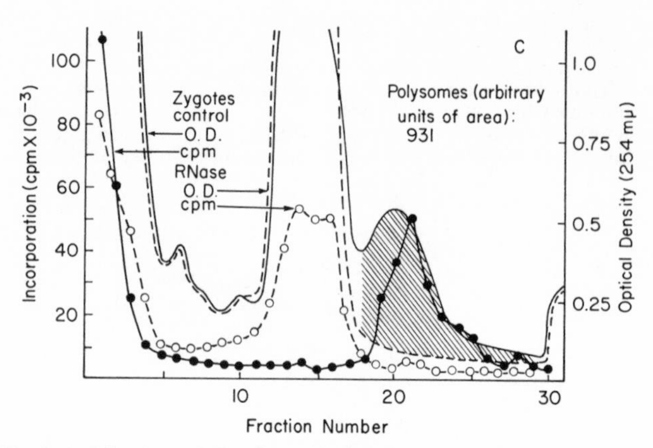

Fig. 3C Optical density and H^3 polypeptide distributions in polysomes from zygotes before and after ribonuclease treatment. The monoribosome peaks were in the same positions as in B.

terial was of a uniform size or localized at a gradient interface. It would be more likely that the reported magnitude of increase was an underestimation. Such a large difference in ribosomal involvement in protein synthesis before and after fertilization has not been reported previously. With the exception of the above observations by Humphreys ('69), the other studies dealing with this question have reported increases in polysomal material in the range of three to five times (Infante and Nemer, '67; MacKintosh and Bell, '69b; Rinaldi and Monroy, '69). The discrepancy in the results does not concern the number of ribosomes involved in synthesis after fertilization, as the above reports indicated that 20-30% of the total ribosomal pool was on the polysomes in zygotes. Reconstruction of the monoribosome peaks and comparisons with the amounts of polysomes formed at fertilization in our experiments also indicated that from 15 to 20% of the total ribosomes in zygotes were in polysomes. The disagreement appears to be based on the amount of polysomal material in eggs.

One explanation of the differing results is that of faulty technique in the isolation of the polysomes. However, our findings consistently demonstrated that polysomes could be isolated from zygotes and that there was a lack of polypeptide label in the monosomes. This suggests that there were no large losses of material or sources of breakdown in the technique. A possible difference in lability of egg and zygote polysomes was checked by testing the stability of the polysomes in the homogenates and by increasing the severity of homogenization. In neither instance was the lability of the polysomes different. On the other hand, parthenogenetic eggs or immature oocytes which can be a source of elevated synthesis rates in a batch of eggs (Piatigorsky, Ozaki and Tyler, '67) were not present in eggs used in this study.

Another mitigating factor may be that of species differences. However, until more is known about factors which can affect synthesis rates in sea urchin eggs, this possibility cannot be adequately tested. One study reports increases in rates at fertilization of 5 to 15 times in *S. purpuratus* eggs. (Epel, '67). An independent study using the same species and a similar technique reports a 30-fold increase (MacKintosh and Bell, '67). Furthermore, recent studies have indicated that levels of synthesis approaching those in zygotes can be induced in *S. purpuratus* eggs by removing the CO_2 from the sea water (MacKintosh and Bell, '69b). These eggs neither showed the cortical response associated with fertilization nor underwent development. They continued to resemble unfertilized eggs, and could subsequently be fertilized. A similar effect can be produced in *L. pictus* eggs by subjecting them to 0°C for ten minutes and then returning them to ambient temperature (Denny and Reback, unpublished). Analysis of these eggs with sucrose gradients and radioautography indicated that there had been an increase in polysomal material and that all of the eggs were equally stimulated to levels approaching that found in zygotes. These studies suggest that in the case of protein synthesis in sea urchins, the classification, unfertilized egg, does not denote a standard level of synthesis, nor does it allude to the natural or induced variations which can exist. They also suggest that unless care is taken to avoid stimulation of the system in eggs prior to fertilization, a study such as this could be quite misleading.

LITERATURE CITED

Epel, D. 1967 Protein synthesis in sea urchin eggs: a late response to fertilization. Proc. Natl. Acad. Sci. U.S., *57:* 899–906.

Golub, A. L., and J. S. Glegg 1969 Dissociation of oxidized 80- S ribosomes by sodium deoxycholate. Biochim. Biophys. Acta, *182:* 121–134.

Gross, P. R. 1968 Biochemistry of Differentiation. Ann. Rev. Biochem., *37:* 631–655.

Gross, P. R., and G. H. Cousineau 1964 Macromolecule synthesis and the influence of actinomycin on early development. Exptl. Cell Res., *33:* 368–395.

Hultin, T. 1961 Activation of ribosomes in sea urchin eggs in response to fertilization. Exptl. Cell Res., *25:* 405–417.

Humphreys, T. 1969 Efficiency of translation of messenger-RNA before and after fertilization in sea urchins. Develop. Biol., *20:* 435–458.

Infante, A. A., and N. Nemer 1967 Accumulation of newly synthesized RNA templates in a unique class of polyribosomes during embryogenesis. Proc. Natl. Acad. Sci. U.S., *58:* 681–688.

Kuff, E. L., and N. E. Roberts 1967 *In vivo* labeling patterns of free polyribosomes: relationship to tape theory of messenger ribonucleic acid function. J. Mol. Biol., *26:* 211–225.

Lowry, O. H., N. J. Rosebrough, A. L. Farr and R. J. Randall 1951 Protein measurement with the folin phenol reagent. J. Biol. Chem., *193:* 265–275.

MacKintosh, F. R., and E. Bell 1967 Stimulation of protein synthesis in unfertilized sea urchin eggs by prior metabolic inhibition. Biochem. Biophys. Res. Comm., *27:* 425–430.

——— 1969a Labelling of nucleotide pools in sea urchin eggs. Exptl. Cell Res., *57:* 71–73.

——— 1969b Regulation of protein synthesis in sea urchin eggs. J. Mol. Biol., *41:* 365–380.

Mans, R. J., and G. D. Novelli 1961 Measurement of the incorporation of radioactive amino acids into protein by a filter paper disk method. Arch. Biochem. Biophys., *94:* 48–53.

Monroy, A., and A. Tyler 1963 Formation of active ribosomal aggregates (polysomes) upon fertilization and development of sea urchin eggs. Arch. Biochem. and Biophys., *103:* 431–435.

Piatigorsky, J., H. Ozaki and A. Tyler 1967 RNA- and protein-synthesizing capacity of isolated oocytes of the sea urchin *Lytechinus pictus*. Develop. Biol., *15:* 1–22.

Rich, A. 1967 Preparation of polysomes from mammalian reticulocytes, lymph nodes, and cells grown in tissue culture. In: Methods in Enzymology, XIIA. L. Grossman and K. Moldave, eds. Academic, New York, p. 481.

Rinaldi, A. M., and A. Monroy 1969 Polyribosome formation and RNA synthesis in the early post-fertilization stages of the sea urchin egg. Develop. Biol., *19:* 73–86.

Stafford, D. W., W. H. Sofer and K. M. Iverson 1964 Demonstration of polyribosomes after fertilization of the sea urchin egg. Proc. Natl. Acad. Sci. U.S., *52:* 313–316.

Stavy, L., and P. R. Gross 1967 The protein synthetic lesion in unfertilized eggs. Proc. Natl. Acad. Sci. U.S., *57:* 735–742.

——— 1969 Protein synthesis *in vitro* with fractions of sea urchin eggs and embryos. Biochim. Biophys. Acta, *182:* 193–202.

Timourian, H. 1967 Protein Synthesis in Sea Urchin eggs. 1. Fertilization-induced changes in subcellular fractions. Develop. Biol., *16:* 594–611.

Timourian, H., and P. C. Denny 1964 Activation of protein synthesis in sea-urchin eggs upon fertilization in relation to magnesium and potassium ions. J. Exp. Zool., *155:* 57–70.

Tyler, A. 1967 Masked messenger RNA and cytoplasmic DNA in relation to protein synthesis and processes of fertilization and determination in embryonic development. Develop. Biol., Suppl. *1:* 170–226.

Effects of fertilization on the ribosomal subunit pool of sea urchin eggs

Ribosomal subunits appear to be intimately involved in peptide chain initiation in bacteria[1-3]. A similar situation is indicated in eukaryotic systems, but much of the evidence is indirect due to the difficulty in obtaining *in vitro* peptide chain initiation[4-6].

Homogenates of sea urchin eggs contain a pool of ribosomal subunits which represent approx. 3 % of the total ribosomal population (Fig. 1A). As these subunits were obtained by a gentle homogenization procedure, and under ionic conditions well within the limits of ribosomal stability[7,8], it is likely they represent native subunits rather than artifacts of the isolation procedure.

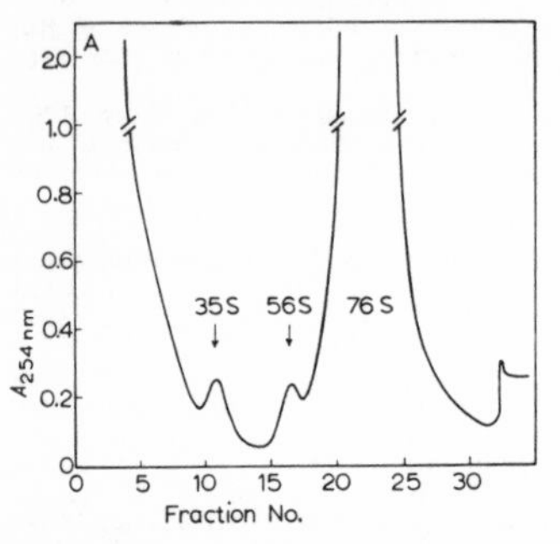

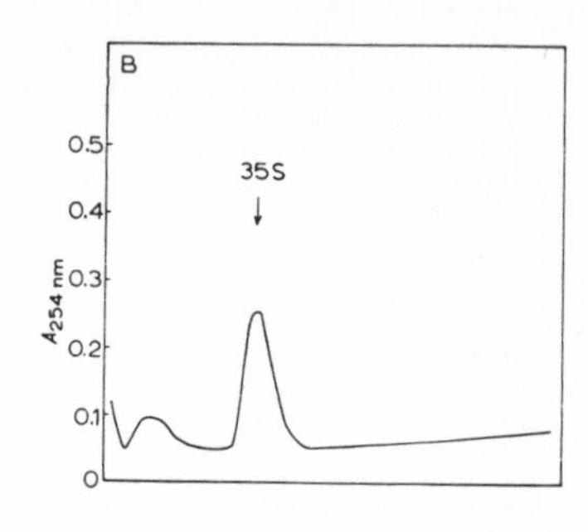

Fig. 1. Eggs and 20-min postfertilization zygotes of *Lytechinus pictus* were washed in 0.55 M KCl, suspended in 1 vol. 0.55 M KCl and 2 vol. buffer (6.6 mM magnesium acetate, 0.132 M sucrose, 66 mM Tris–HCl, pH 7.8) and homogenized by one pass through a 25-gauge needle. The homogenate was centrifuged at 15 000 × *g* for 10 min in an SW 50.1 rotor. The supernatant was retained, and the pellet was washed once and suspended in buffer to its original volume. Both pellet and supernatant were made 1 % with Triton X-100[15] and 2.5-ml samples were placed on linear gradients. Gradients were 35 ml, 15–30 % (w/w) sucrose, and were centrifuged at 20 000 rev./min for 16 h at 5° in a SW 27 rotor. Absorbances were continuously monitored at 254 nm (ISCO) and 1.0-ml fractions were collected. Due to the high background on the gradient from supernatant and monoribosome peaks, fractions containing the 35-S subunit were collected, diluted 3-fold in buffer, concentrated and recentrifuged as described above. The amount of subunit material was obtained by direct mechanical integration of the absorbance recording of the 35-S subunit peak. A. Sedimentation profile of zygote supernatant fraction; numbers indicate approximate *s* values of subunit and monoribosome peaks[16]. B. Recentrifugation of 35-S peak from zygote supernatant fraction.

Within 20 min after fertilization the sea urchin zygote undergoes an increase in the rate of protein synthesis of at least one order of magnitude[9,10], resulting in a shift of approx. 15 % of the monoribosomes to polysomes[11]. The effects of these postfertilization events on the ribosomal subunit pool are discussed below.

In initial experiments both increases and decreases in the 35-S ribosomal subunit pool were observed following fertilization (Table I, Expts. 1–3). Since these experiments dealt only with the 15 000 × *g* supernatant, the 15 000 × *g* pellet was examined and found to contain a substantial amount of subunits, representing 25–75 % of the total pool. These subunits remained with the pellet through successive washes in buffer and were only released after treatment with the detergent Triton X-100. In all further experiments both supernatant and pellet fractions were analyzed and the results combined to give the total subunit pool for a given sample. When these total subunit pools from eggs and zygotes were compared (Table I, Expts. 4–6), a small but consistent decrease was found by 20 min postfertilization. Apparently the wide variation observed initially represented shifts between the supernatant and pellet rather than changes in the total number of subunits. The lack of any consistent pattern of change from pellet to supernatant at fertilization would suggest that these shifts play no major role in the observed increase of protein synthesis.

TABLE I

A COMPARISON OF RIBOSOMAL SUBUNIT POOLS OF EGGS AND EMBRYOS

Representative experiments comparing 35-S ribosomal subunits from eggs and zygotes. Subunits were isolated and quantified as described in Fig. 1. Egg volume and total protein were measured to assure an equal number of eggs and zygotes. The amount of subunits is expressed in arbitrary units of area. Comparison of the 56-S subunits produced similar results but quantification was hampered by contamination with monoribosomes.

Expt.	*Subunits of supernatant*			*Subunits of pellet*			*Total subunits*		
	Egg	*Zygote*	*Zygote/egg*	*Egg*	*Zygote*	*Zygote/egg*	*Egg*	*Zygote*	*Zygote/egg*
1	900	480	0.53	—	—	—	—	—	—
2	900	450	0.50	—	—	—	—	—	—
3	370	500	1.50	—	—	—	—	—	—
4	990	720	0.73	2340	2100	0.91	3330	2820	0.85
5	1950	870	0.46	1260	2000	1.54	3210	2870	0.89
6	1220	2210	1.82	3900	1500	0.39	5120	3710	0.74

The small decrease in ribosomal subunits observed at fertilization is consistent with the work of HOGAN AND KORNER[12] who find only a small decrease in the subunit pool of ascites cells grown under conditions which cause a rapid increase in the rate of protein synthesis. Studies with bacterial systems[13] have also found subunit pools to be relatively stable while the rate of protein synthesis is altered substantially. As these particles act as an intermediate between monoribosomes and polyribosomes in bacteria, their rate of formation and utilization would appear to be linked. This is consistent with the model presented by KOHLER *et al.*[13] who propose a ribosomal dissociation factor which would also be required for initiation. A factor, DF, meeting these requirements, has since been isolated[14]. If the ribosomal subunits of the sea urchin are involved in peptide chain initiation, then a similar mechanism must be proposed to explain the relative stability of the subunit pool during fertilization.

It has been shown that the sea urchin egg contains a pool of native ribosomal subunits representing approx. 3 % of the total ribosomes. The number of subunits was found to be relatively stable following fertilization while the rate of protein synthesis increases rapidly. This observation is consistent with results obtained from other eukaryotes and from bacteria.

This work was supported by a National Institutes of Health research grant (GM-16090). One of us (K.D.J.) has held a National Science Foundation predoctoral traineeship during the course of this investigation.

KENNETH D. JENKINS
PAUL C. DENNY

1 J. M. EISENSTADT AND G. BRAWERMAN, *Proc. Natl. Acad. Sci. U.S.*, 58 (1967) 1560.
2 M. NOMURA AND C. V. LOWRY, *Proc. Natl. Acad. Sci. U.S.*, 58 (1967) 946.
3 M. NOMURA, C. V. LOWRY AND C. GUTHRIE, *Proc. Natl. Acad. Sci. U.S.*, 58 (1967) 1487.
4 B. COLOMBO, C. VESCO AND C. BAGLIONI, *Proc. Natl. Acad. Sci. U.S.*, 61 (1968) 651.
5 C. BAGLIONI, C. VESCO AND M. JACOBS-LORENA, *Cold Spring Harbor Symp. Quant. Biol.* 34 (1969) 555.
6 W. HOERZ AND K. S. MCCARTHY, *Proc. Natl. Acad. Sci. U.S.*, 63 (1969) 1206.
7 R. MAGGIO, M. L. VITTORELLI, I. CAFFARELLI-MORMINO AND A. MONROY, *J. Mol. Biol.*, 31 (1968) 621.
8 L. H. KEDES AND L. STAVY, *J. Mol. Biol.*, 43 (1969) 337.
9 T. HUMPHREYS, *Develop. Biol.* 20 (1969) 435.
10 P. C. DENNY AND P. REBACK, *J. Exptl. Zool.*, in the press.
11 A. M. RINALDI AND A. MONROY, *Develop. Biol.*, 19 (1969) 73.
12 B. L. M. HOGAN AND A. KORNER, *Biochim. Biophys. Acta*, 169 (1968) 129.
13 R. E. KOHLER, E. Z. RON AND B. D. DAVIS, *J. Mol. Biol.*, 36 (1968) 71.
14 A. R. SUBRAMANIAN, E. Z. RON AND B. D. DAVIS, *Proc. Natl. Acad. Sci. U.S.*, 61 (1968) 761.
15 A. L. GOLUB AND J. S. CLEGG, *Biochim. Biophys. Acta*, 182 (1969) 121.
16 A. A. INFANTE AND M. NEMER, *J. Mol. Biol.*, 32 (1968) 543.

PROTEIN SYNTHESIS *IN VITRO* WITH FRACTIONS OF SEA URCHIN EGGS AND EMBRYOS

LARY STAVY AND PAUL R. GROSS

INTRODUCTION

Within a few minutes following fertilization of sea urchin eggs, the rate of protein synthesis increases greatly. The increment is observed *in vitro* as well as *in vivo*. This fact has been available for a number of years[1], and the large literature concerned with its interpretation has been reviewed[2].

Though much has been learned about protein synthesis and its control in eggs and embryos, a detailed explanation of activation in the zygote remains elusive. There is good evidence demonstrating that eggs already contain a sufficient store of mRNA to account, at least quantitatively, for much of the protein synthesis that occurs between fertilization and the blastula stage—an interval of 5 to 24 h, depending upon the species and temperature[2–4,23].

Cell-free amino acid incorporating systems from eggs and embryos offer certain practical advantages over conditions for analysis of the fertilization-induced changes *in vivo*. The cell-free systems can, however, yield misleading results unless a broad range of the main experimental variables is first explored for a given species, and the optimal conditions chosen. Such a survey of conditions has not as yet been reported for any species, and it is evident from our own preliminary surveys[5] and from the data to be given here that some published results in this field are concerned with systems operating under far from favorable conditions. In such circumstances, the essential variables (*e.g.*, fertilized/unfertilized) have not been isolated, and quantitative and qualitative differences can be spurious.

The first step in our investigation was selection of a species, from among those readily available to us, with lowest ribonuclease activity. Among the three available, *Arbacia punctulata*, *Strongylocentrotus purpuratus*, and *Lytechinus pictus*, the last only yielded preparations of reproducibly long-term activity *in vitro*.

It was found early that Ca^{2+} inhibits cell-free amino acid incorporation very powerfully. Special procedures were therefore developed for removal of these ions from the cells before their rupture (details in MATERIALS AND METHODS).

The effect of each variation in the incubation conditions was examined simultaneously, with egg and embryo preparations from a single batch of eggs, in order to isolate so far as possible the changes that take place following fertilization. The results, like those reported by other investigators[5–9], indicate that incapacity of eggs for protein synthesis at the rates attained immediately after activation must be attributed to a lesion at the level of mRNA–ribosome interaction, *i.e.*, that neither the lack of co-factors nor the presence of inhibitors in any other cell compartment constitutes the lesion in the egg's translation machinery.

MATERIALS AND METHODS

The experiments were done with cell-free systems prepared from eggs and embryos of *L. pictus* (Pacific Bio-Marine Supply Co., Venice, Calif.). Gametes were obtained from animals stimulated to spawn by injection of 0.55 M KCl. Egg populations were selected for high fertilizability (>95 %) and for absence of significant numbers of oocytes and parthenogenetically activated cells.

Preparation of crude extracts

Egg extracts. Eggs were washed and filtered through cheesecloth, using Millipore-filtered sea water. They were freed of their jelly coats by washing in sea water with its pH reduced to 5.0, and then washed twice with Ca^{2+}- and Mg^{2+}-free sea water. A final wash was done in a solution containing 0.1 M KCl; 0.006 M $MgCl_2$; 0.035 M Tris buffer (pH 7.4); 0.25 M sucrose; 0.006 M β-mercaptoethanol. Eggs centrifuged through this medium were suspended in it once again. The ratio of packed cells to medium varied from 1/3 to 1/6, depending upon the product required. The suspension was homogenized gently in a Potter–Elvejhem homogenizer. A preliminary centrifugation of the homogenate was done at $5000 \times g$ for 5 min. The cell- and debris-free homogenate was then centrifuged at 21 000 $\times g$ for 15 min. The upper two-thirds of the supernatant from this centrifugation is the "crude extract".

Embryo extracts. Washed, jelly-free eggs were incubated and inseminated at 18°, following which harvesting was begun. Eggs or embryos were washed twice with Ca^{2+}- and Mg^{2+}-free sea water and the procedure continued as described above.

Preparation of fractions

Microsomes. This fraction was prepared by centrifugation of the crude extract for 90 min at 135 000 $\times g$, discarding the supernatant.

Ribosomes (deoxycholate-treated microsomes). Microsomes were suspended in KCl–$MgCl_2$–Tris–sucrose–β-mercaptoethanol solution containing 1/3 % sodium deoxycholate. The ribosomes were re-isolated by centrifugation for 90 min at 135 000 $\times g$.

Rat liver pH-5 fraction. Rat livers were washed and homogenized in KCl–$MgCl_2$–Tris–sucrose–β-mercaptoethanol solution. The homogenate was subjected to the same centrifugations as sea urchin microsomes. The pH of the upper two-thirds of the supernatant was brought to 5.0–5.1 with 1 M acetic acid and kept for 30 min with continuous stirring. The pH-5 fraction precipitate was collected by centrifugation at 12 000 $\times g$ for 10 min.

Cells and their products were kept at 4° or below.

Microsomes, ribosomes and pH-5 fraction were suspended in KCl–$MgCl_2$–Tris–sucrose–β-mercaptoethanol solution. The fractions were stored at −80 °.

Incubation conditions (additional information supplied in the legends)

A standard reaction mixture contained in μmoles/0.5 ml: 50 KCl, 2 $MgCl_2$, 17.5 Tris buffer (pH 7.4), 50 sucrose, 1.2 β-mercaptoethanol, 1 ATP, 0.2 GTP, 0.015 each of 20 L-amino acids *minus* L-leucine (in the experiments where L-[^{14}C]leucine was added), 0.0021 L-[^{14}C]leucine or 0.0016 L-[^{14}C]phenylalanine, 2 phosphoenolpyruvate, 6–8 μg phosphoenolpyruvate kinase, poly U, when used, 100 μg. Amounts of crude extract or microsomes and pH-5 fraction will be given in the legends. The reaction mixture was incubated for 90 min at 25°.

Assay of radioactivity

Incorporation was stopped by the addition of 3 ml of 6 % trichloroacetic acid containing an excess of either L-leucine or L-phenylalanine, unlabeled. The precipitates were centrifuged, resuspended in 3 ml 6 % trichloroacetic acid, kept for 30 min at 85°. After cooling, the samples were recentrifuged, resuspended in 6 % trichloroacetic acid, and the precipitates were finally collected on type HA Millipore filters. They were washed on the filter discs with 25 ml of 6 % trichloroacetic acid. Filters were dried and counted in a Nuclear–Chicago low background gas-flow counter (model C-115).

All data presented below are corrected for the values of the zero-time blanks. For L-[^{14}C]phenylalanine these varied between 50 and 95 counts/min and for L-[^{14}C]leucine between 2 and 10 counts/min.

Protein assay

Protein content was determined by the method of LOWRY *et al.*[10], with bovine serum albumin as standard.

Chemicals

Labeled amino acids were from New England Nuclear Corp. The specific activity of the L-[^{14}C]leucine was 275 mC/mmole and of the L-[^{14}C]phenylalanine 367 mC/mmole, both uniformly labeled. Phosphoenolpyruvate, phosphoenolpyruvate kinase, GTP, ATP, CTP, UTP were from Sigma Chemical Co.

Poly U was from Miles Chemical Co.

RESULTS

Effects of Mg^{2+}, Ca^{2+} and K^+ on the incorporation

Figs. 1–3 deal with the incorporation of leucine and phenylalanine into proteins of various preparations whose incubation media contained varying amounts of Mg^{2+} and K^+. Crude extracts were used for the experiments represented by Fig. 1, microsomes for those of Fig. 2, and ribosomes (deoxycholate-treated microsomes)

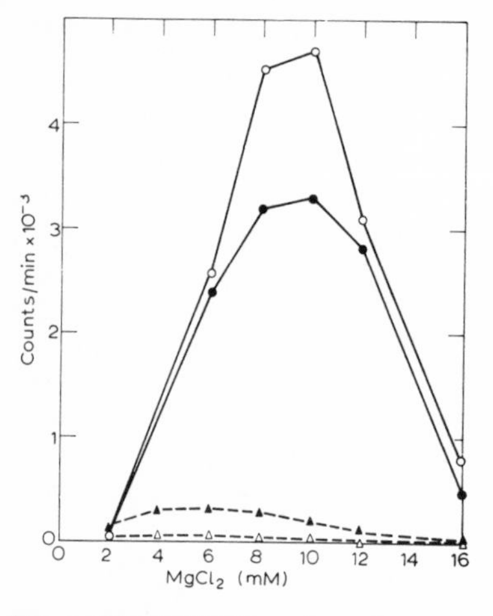

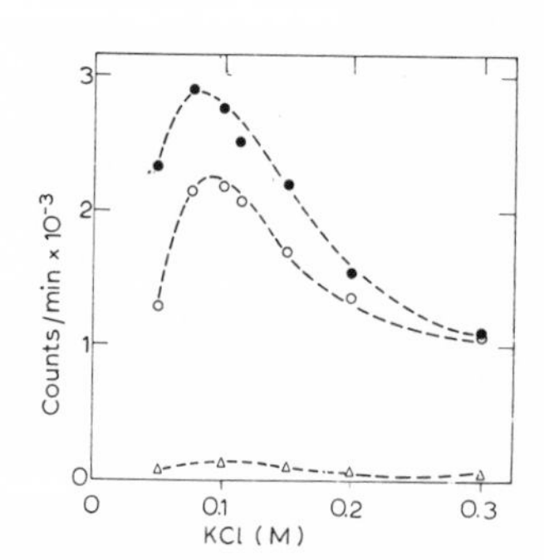

Fig. 1. Effect of Mg^{2+} on the endogenous and poly U-dependent incorporation of phenylalanine by crude extract. The reaction mixture contained 1.0 mg of crude extract protein from eggs and embryos. ○–○, egg, poly U-dependent incorporation; ●–●, embryo, poly U-dependent incorporation; ▲- - -▲, embryo, endogenous incorporation; △- - -△, egg, endogenous incorporation.

Fig. 2. Effect of K^+ on the incorporation of leucine by microsomes. The reaction mixture contained 0.4 mg protein of egg or embryo microsomes and 0.54 mg protein of rat liver pH-5 fraction. △- - -△, egg microsomes, 4 mM Mg^{2+}; ●- - -●, embryo microsomes, 4 mM Mg^{2+}; ○- - -○, embryo microsomes, 6 mM Mg^{2+}.

for those of Fig. 3. Microsomes and ribosomes were supplemented with pH-5 fraction prepared from rat liver.

As is shown, the level of incorporation is strongly dependent upon Mg^{2+} in the medium. Optimal Mg^{2+} concentrations are similar for incorporation systems prepared from eggs and embryos: for translation of endogenous templates, they lie between 4 and 6 mM $MgCl_2$, whereas for poly U-dependent phenylalanine incorporation, 8–10

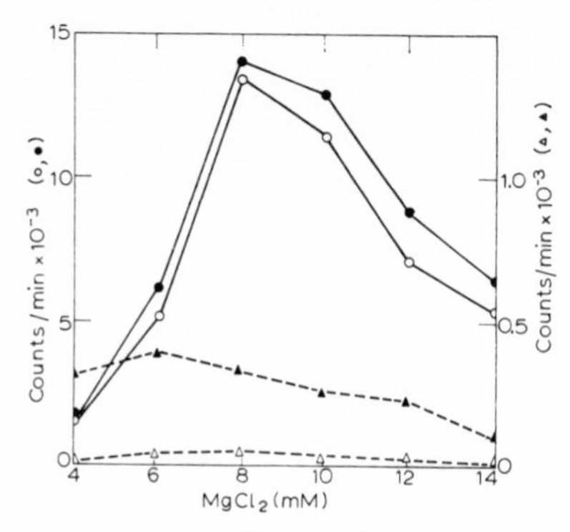

Fig. 3. Effect of Mg^{2+} on the endogenous and poly U-dependent incorporation of phenylalanine by ribosomes. The reaction mixture contained 0.2 mg protein of deoxycholate-treated microsomes from eggs or embryos and 0.54 mg protein of rat liver pH-5 fraction. Symbols as in Fig. 1.

mM $MgCl_2$ provide optimum activity. The K^+ concentration optimum is the same for all systems: it is at about 0.1 M (Fig. 2).

In all systems and with all the variations in Mg^{2+} and K^+ employed, fractions from embryos are much more active than those from unfertilized eggs. The fact that this difference exists unmodified in systems employing sea urchin microsomes or ribosomes proves that neither supernatant factors nor the membranous component of the microsomes can be responsible for the observed difference.

Table I exemplifies the effect of Ca^{2+} on the incorporation. Its inhibitory effect is probably due to competition with the essential Mg^{2+}.

TABLE I

EFFECT OF Ca^{2+} ON AMINO ACID INCORPORATION

The reaction mixture contained 1 mg crude extract of eggs or embryos, suspended in KCl–Tris–sucrose–β-mercaptoethanol (without $MgCl_2$). The labeled amino acid was L-leucine.

Concentration (mM)		*Counts/min*	
$CaCl_2$	*$MgCl^2$*	*Embryo*	*Egg*
3	—	11	5
4	—	5	1
2	1	40	3
3	1	47	4
4	1	21	3
1	2	27	—
1	3	298	21
1	4	346	37
—	3	1483	84
—	4	1288	103

Effect of temperature and pH variation

The incorporation was tested at 15°, 25°, and 35° (Figs. 4–6). The optimal temperature was found to be about 25° (Fig. 4) and the optimal pH about 7.4

(Figs. 5 and 6). (The temperature limits for normal, synchronous development in this species are 15–21°.)

The rates of incorporation under different experimental conditions are not necessarily matched by the amount of incorporation finally accomplished, *i.e.*, rate

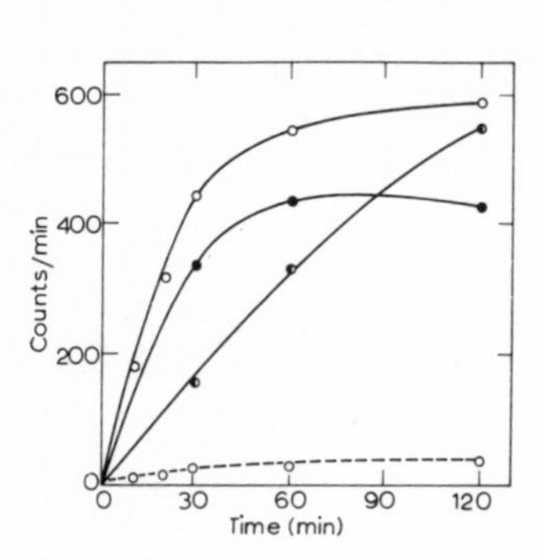

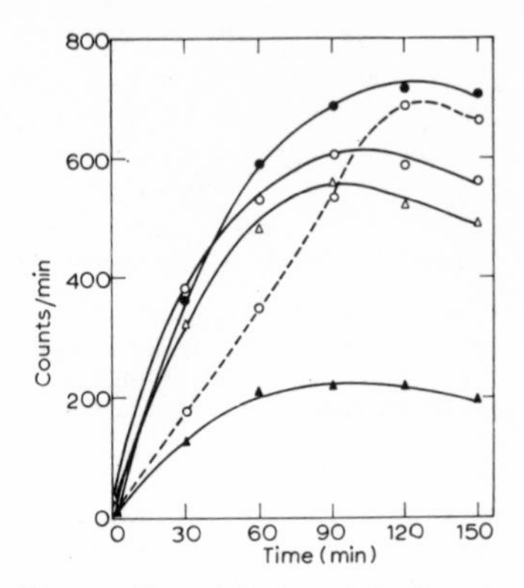

Fig. 4. Temperature dependence of leucine incorporation. The reaction mixture contained 1 mg protein of crude extract of eggs or embryos. Mixtures were incubated at 15°, 25° and 35°. ○–○, embryo, at 25°; ○- - -○, egg, at 25°; ●–●, embryo at 35°; ◐–◐, embryo, at 15°.

Fig. 5. The amount of incorporation of leucine at various pH values as a function of time. The reaction mixture contained 0.75 mg protein of crude embryo extract. ●–●, 25°, pH 7.4; ○–○, 25°, pH 7.2; ○- - -○, 15°, pH 7.2; △–△, 25°, pH 7.6; ▲–▲, 25°, pH 8.0.

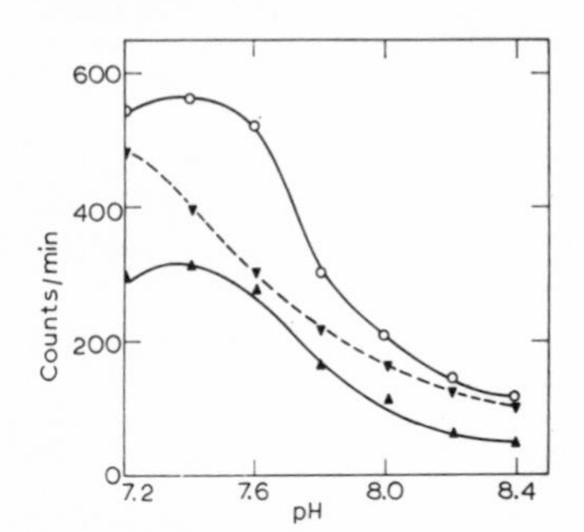

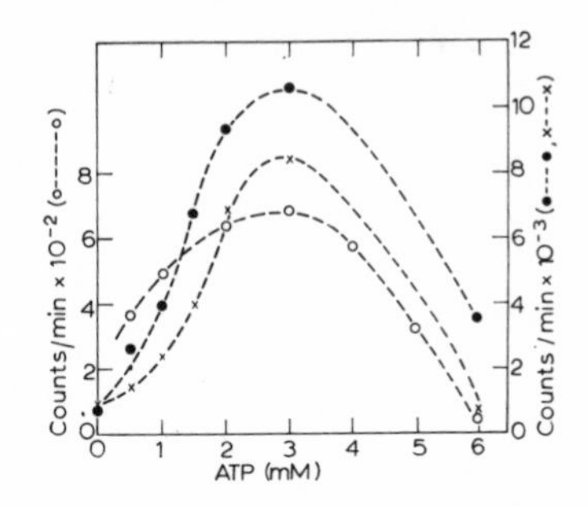

Fig. 6. Effect of pH on the incorporation of leucine at different temperatures. The reaction mixture contained 0.75 mg protein of crude extract of embryo. The reaction mixtures were incubated at 15°, 25° and 35°. ○–○, 25°; ▲–▲, 35°; ▼- - -▼, 15°.

Fig. 7. ATP dependence of leucine and phenylalanine incorporation by crude extract and microsomes. ○- - -○, leucine incorporation (endogenous) by crude embryo extract; ×- - -×, crude extract incorporation of phenylalanine in the presence of poly U; ●- - -●, microsomal incorporation of phenylalanine in the presence of poly U. (Ordinates: 10^{-2} scale for leucine incorporation; 10^{-3} scale for phenylalanine.)

and extent of reaction are not obligately coupled. Thus the rate of incorporation at 25° and pH 7.4 is higher during the first hour of incubation than is the corresponding rate for 15° and pH 7.2, but the extent of both is about the same (Fig. 5).

Optimal pH values appear to shift somewhat with changing temperature (Fig. 6).

This shift probably reflects a variation in pH with temperature of the Tris buffer. (The pH values of the buffer system were all measured at 25°.)

Response of the incorporation systems to added ribonucleoside triphosphates and to amino acids

Incorporation by either crude extract or microsomes is inhibited by ribonucleoside triphosphates when added as a mixture of ATP, GTP, UTP and CTP. Data are shown in Table II. The inhibition is actually accomplished by CTP or by UTP.

TABLE II

EFFECT OF NUCLEOSIDE TRIPHOSPHATES ON AMINO ACID INCORPORATION

The reaction mixture contained 0.75 mg protein of crude extract or 0.15 mg microsomal protein *plus* 0.35 mg rat liver pH-5 fraction. The labeled amino acid was L-leucine.

μmoles/tube					*Counts/min*			
ATP	*GTP*	*CTP*	*UTP*	*Phosphoenol-pyruvate**	*Crude extract*		*Microsomes*	
					Embryo	*Egg*	*Embryo*	*Egg*
—	—	—	—	—	8	5	6	11
1.5	—	—	—	—	223	6	302	32
1.5	0.25	—	—	—	259		507	
1.5		—	—	2.0	689		499	
1.5	0.25	—	—	2.0	675	20	820	47
1.5	0.25	0.125	0.125	2.0	494	18	544	39
1.5	0.25	—	0.25	2.0	517			
1.5	0.25	0.25	—	2.0	601			

* Phosphoenolpyruvate kinase was added with phosphoenolpyruvate.

In the crude extract, GTP is not limiting, since the full rate of incorporation is realized without it. This result stands in contrast to that obtained for microsomes. When supplemented by rat liver pH-5 fraction, the microsomes require added GTP for optimum amino acid incorporation.

TABLE III

EFFECT OF COMPLETE AMINO ACID MIXTURE ON THE INCORPORATION

The reaction mixture contained 0.75 mg of crude extract protein, or 0.3 mg of microsomal protein *plus* 0.35 mg of rat liver pH-5 fraction. The labeled amino acid was L-leucine. The amino acid mixture consisted of 0.015 μmole of each of 20 L-amino acids *minus* L-leucine. Other conditions as given under MATERIALS AND METHODS.

	Amino acid mixture	*Counts/min*
Egg crude extract	—	54
	+	59
Embryo crude extract	—	543
	+	600
Egg microsomes	—	81
	+	86
Embryo microsomes	—	1389
	+	1863

Incorporation by crude extract and by microsomes is very sensitive to added ATP. The optimum occurs at 1.5 mM ATP. Supra-optimal concentrations depress the incorporation severely. This effect is observed not only when endogenous mRNA serves as template, but also when poly U is employed as template for the synthesis of polyphenylalanine (Fig. 7).

As might be expected, ATP was found to determine the incorporation rate more critically in the purified systems (microsomes and ribosomes) than in the crude extract.

Table III shows that the amino acid pool is not limiting in the egg system. It does exert some small limitation upon the embryo system, which is stimulated to a certain extent by addition of an amino acid mixture to the medium.

Neither amino acid pool differences nor changes in the level of available ATP can account for the differences in efficiency between the egg and the embryo systems.

DISCUSSION

The information provided permits construction of optimal amino acid incorporating systems *in vitro* from Lytechinus eggs and embryos. It should be a guide in the use of other species as well. None of the variations in Mg^{2+}, K^{+}, temperature, pH, energy supply, nor in amino acids tested here, was able to bring the incorporation of egg fractions to the level observed in embryo preparations, so long as the templates for polypeptide chain assembly were of endogenous origin. The difference between the systems therefore resides either in the amount of available template RNA or in the changing competence of ribosomes to translate a constantly available amount.

Systems derived from embryos were found to incorporate amino acids linearly for more than 30 min at 25° and for more than 1 h at 15°. In the latter case, significant increase in accumulated polypeptide radioactivity continues for 90 min or more, provided that other environmental conditions are optimal. This behavior is somewhat different from that of systems derived from bacteria or from mammalian cells, in which linearity of incorporation at optimal temperatures is normally lost after a shorter time[11–13]. Some systems that permit RNA synthesis to proceed simultaneously with peptide synthesis allow amino acid incorporation to proceed longer[14–17], but RNA synthesis cannot be the explanation for the long life of sea urchin embryo systems. Were that so, amino acid incorporation after the first few minutes would be expected to depend upon a maintained supply of all four ribonucleoside triphosphates. Since such a dependency was not found in the sea urchin embryo systems, the above explanation is ruled out.

Potential for polypeptide synthesis is reflected by the poly U-directed synthesis of polyphenylalanine. There may be differences in the mechanism of chain initiation between poly U and normal mRNA[18–22]. Nevertheless, incorporation of phenylalanine in this system, under the influence of poly U, shows that the same fractions—crude extract, microsomes, ribosomes—whether from egg or embryo, have roughly the same ability to elongate polypeptide chains. In particular, eggs and very early embryos yield systems *in vitro* that have essentially identical activity with poly U, but are very different indeed with respect to endogenously directed incorporation. Since it is now certain that the endogenous mRNA responsible for most of the protein synthesis of the cleavage period resides in the egg cytoplasm[2], there is reason to assume that the difference between eggs and embryos is a difference in the availability of those messengers. If the difference involves some incompetence on the part of ribosomes from eggs, then that incompetence must be limited to chain initiation only.

Experiments parallel to those reported here for *L. pictus* were done with eggs and embryos of *A. punctulata* and *S. purpuratus*. The Lytechinus systems were by far the most reproducible in behavior. Some incorporation was always observed in cell-free systems made from eggs of all species, but it was generally higher for systems prepared from eggs of Arbacia and Strongylocentrotus. The level of incorporation provided by unfertilized egg systems varied with season (and hence with the relative maturity of the gametes). The techniques used for collection and washing of the eggs also influenced the incorporation rate ultimately obtained *in vitro*. Incorporation given by systems from populations of ripe ootids, in mid-season of spawning and carefully processed, is always very low. These results reinforce earlier suggestions[5] that variables in procedure of preparation and in the incorporation assay determine critically the relative rates and extents of polypeptide synthesis obtainable *in vitro* with subcellular particle systems derived from eggs and embryos.

Most of the incubation medium variables studied thus far show a single optimum, sufficiently sharp to demand careful technique in the study of other, and perhaps more biologically significant variables, *e.g.*, developmental stage and the presence of metabolic inhibitors.

ACKNOWLEDGEMENTS

The work reported in this paper was supported by research grants from the National Institutes of Health, U.S. Public Health Service (GM-13560), from the National Science Foundation (GB-5760), and from the American Cancer Society (E-285). Certain facilities were made available by an award of funds from a National Institutes of Health Institutional Grant (PR-07047) to the Massachusetts Institute of Technology. We thank Miss Michele Mastrolia for able technical assistance. One of us (L.S) has held an International Postdoctoral Fellowship from the National Institutes of Health during the time these investigations were carried out.

REFERENCES

1 T. Hultin, *Exptl. Cell Res.*, 25 (1961) 405.
2 P. R. Gross, in A. A. Moscona and A. Monroy, *Current Topics in Developmental Biology*, Vol. 2, Academic Press, New York, 1968, p. 1.
3 P. R. Gross, L. I. Malkin and M. Hubbard, *J. Mol. Biol.*, 13 (1965) 463.
4 P. R. Gross, *New Engl. J. Med.*, 276 (1967) 1230.
5 L. Stavy and P. R. Gross, *Proc. Natl. Acad. Sci. U.S.*, 51 (1967) 735.
6 F. H. Wilt and T. Hultin, *Biochem. Biophys. Res. Commun.*, 9 (1962) 313.
7 R. Maggio and C. Catalano, *Arch. Biochem. Biophys.*, 103 (1963) 164.
8 M. Nemer and S. G. Bard, *Science*, 140 (1963) 664.
9 C. Ceccarini, R. Maggio and G. Barbata, *Proc. Natl. Acad. Sci. U.S.*, 52 (1964) 140.
10 O. H. Lowry, N. J. Rosebrough, A. L. Farr and R. J. Randall, *J. Biol. Chem.*, 193 (1951) 265.
11 M. W. Nirenberg and P. Leder, *Science*, 145 (1963) 664.
12 G. Von Ehrenstein and F. Lippman, *Proc. Natl. Acad. Sci. U.S.*, 47 (1961) 41.
13 J. M. Fessenden and K. Moldave, *Biochemisty*, 1 (1962) 485.

14 R. Byrne, J. G. Levin, H. A. Bladen and M. W. Nirenberg, *Proc. Natl. Acad. Sci. U.S.*, 52 (1964) 140.
15 B. Nisman, H. Fukuhara, J. Demailly and C. Gerrin, *Biochim. Biophys. Acta*, 55 (1962) 704.
16 J. Bonner, R. C. Huang and R. V. Gilden, *Proc. Natl. Acad. Sci. U.S.*, 50 (1963) 893.
17 T. Kameyama and O. G. Novelli, *Proc. Natl. Acad. Sci. U.S.*, 48 (1962) 659.
18 M. Revel, M. Herzberg, A. Bascarevic and F. Gros, *J. Mol. Biol.*, 33 (1968) 231.
19 J. Lucas-Lenard and F. Lipmann, *Proc. Natl. Acad. Sci. U.S.*, 57 (1967) 1050.
20 W. Stanley, M. Salas, H. J. Wahba and S. Ochoa, *Proc. Natl. Acad. Sci. U.S.*, 56 (1966) 290.
21 G. Brawerman and J. Eisenstadt, *Biochemistry*, 5 (1966) 2784.
22 M. Revel and H. H. Hiatt, *J. Mol. Biol.*, 11 (1965) 467.
23 P. R. Gross, *New Engl. J. Med.*, 276 (1967) 1297.

Efficiency of Translation of Messenger-RNA before and after Fertilization in Sea Urchins

Tom Humphreys

INTRODUCTION

Upon fertilization the rate of protein synthesis in sea urchin eggs increases severalfold (Hultin, 1952; Nakano and Monroy, 1958; Epel, 1967). Because concurrent RNA synthesis is not required for this stimulation of protein synthesis, the messenger-RNA (mRNA) for the increased protein synthesis must be present in the eggs before fertilization (Gross *et al.*, 1964). Examination of protein synthetic machinery in the egg has suggested several possible conditions that may limit the translation of this mRNA stored in the unfertilized eggs. These range from inadequate aminoacyl-transfer-RNA (Glišin and Glišin, 1964; Ceccarini *et al.*, 1967), through missing cytoplasmic activators of ribosomes (Candelas and Iverson, 1966) and inhibited ribosomes (Hultin, 1961), inactive polysomes (Piatigorsky, 1968; Verhey *et al.*, 1965) or messenger-RNA-ribosome complexes (Monroy *et al.*, 1965) to inactive messenger-RNA-protein particles (Spirin, 1966; Mano and Nagano 1966). There is suggestive evidence supporting each of these possibilities, but existing information concerning protein synthesis before and after fertilization does not allow a definite choice among any of them at this time. It seemed appropriate, therefore, to compare quantitatively the basic characteristics of the protein synthetic system in eggs and embryos as a necessary step toward elucidating the mechanism for the translational regulation of protein synthesis which occurs at fertilization in sea urchins. These comparisons can substantially limit the current hypotheses and thus direct experimental approaches in fruitful directions.

In this paper the relative *in vivo* efficiency of translation, defined as the number of protein molecules produced per mRNA molecule per unit time, has been measured before and after fertilization

(Humphreys, 1968). Increased efficiency of translation represents one of the two possible modes by which the rate of protein synthesis can be increased at fertilization. The other is translation of additional mRNA molecules. The first possibility suggests that all mRNA molecules to be translated at a "normal" rate in the embryo are being read at a slow rate in the egg. The other suggests that there is a population of competely active mRNA molecules in the egg as well as a population of inactive mRNA molecules which are activated at fertilization.

Consideration of the details of the protein synthetic machinery generally found in cells allows the delineation of the changes expected in measurable parameters upon an increase in efficiency of translation. If each protein molecule is synthesized as the linear condensation of amino acids on a single ribosome as it moves down a messenger-RNA molecule, more ribosomes must move from one end to the other of each mRNA molecule per unit time in order to produce more protein molecules per mRNA molecule per unit time. Two ways for increasing the number of ribosomes moving the length of each mRNA molecule may be distinguished: either the number of ribosomes attached to each mRNA molecule at any one time must be increased or, if the number of ribosomes attached to the mRNA molecule remains constant, each ribosome must move more quickly from one end to the other of the mRNA molecule. These are not mutually exclusive and may occur simultaneously to increase efficiency.

Quantitative consideration of the available data on acceleration of protein synthesis at fertilization predicts the expected changes to be at least 10- to 15-fold in the species employed in this work, *Lytechinus pictus*, since protein synthesis accelerates by 10- to 15-fold (Epel, 1967; Humphreys, unpublished data and see below). Thus at least 10–15 times as many ribosomes would have to move down each mRNA molecule per unit time if the acceleration of protein synthesis is achieved by increasing the efficiency of translation. This increase could be achieved by stimulating the attachment of ribosomes to mRNA upon the initiation of new polypeptide chains. In this case, the size of the polysomes would increase 10- to 15-fold at fertilization. Available data suggesting that large physical aggregates of ribosomes appear only in quantity after fertilization (Infante and Nemer, 1967) conform to this possibility. Comparison of the size of *active* polysomes before and after fertilization is unavailable in the literature although active polysomes have been described in eggs

(Tyler *et al.*, 1968) and embryos (Monroy and Tyler, 1963; Stafford *et al.*, 1964; Malkin *et al.*, 1964; Cohen and Iverson, 1967). Therefore, the size distributions of active polysomes before and after fertilization were measured.

Increase in efficiency of translation could also be achieved by increasing the average rate of movement of the ribosomes. This would decrease the length of time required to synthesize a protein molecule. It must be assumed that the rate of attachment of ribosomes is concomitantly increased because the free end of the messenger-RNA is exposed more rapidly by the faster-moving ribosome. The time required for synthesis of a protein molecule is related to the average time an amino acid remains in the nascent protein and the relative distribution of the ribosomes along the mRNA molecule (Englander and Page, 1965). The distribution of the ribosomes is reflected in the size distribution of the nascent peptides (Hunt *et al.*, 1968). No data in the literature relate to the sizes of these parameters in eggs and embryos. Therefore the average time an amino acid remains in the nascent chains and the size distribution of the nascent chains were compared before and after fertilization.

Protein synthesis in eggs and embryos had to be examined to determine whether the assumptions in the experiments proposed to compare efficiency of translation were valid for the specific case of sea urchin eggs and embryos. Altogether three assumptions were recognized: (1) The rate of accumulation of newly synthesized protein during long labeling periods (20 minutes or more, Epel, 1967) was assumed to measure the rate of synthesis. If some of the proteins are rapidly turning over (Mackintosch and Bell, 1967), this assumption would not be true. (2) All protein was assumed to be synthesized on ribosomes. If there was significant enzymatic synthesis (Mach *et al.*, 1963), especially in the egg where protein synthesis is low, the apparent efficiency of synthesis on ribosomes would be greatly increased. (3) The size distribution of protein molecules being synthesized before and after fertilization were assumed to be similar. If the proteins are different sizes and one ribosome produces one polypeptide chain, the number of ribosomes producing equal weights of protein would be very different and greatly alter the predictions. Simple ways to test these three possibilities were devised.

MATERIALS AND METHODS

Eggs and embryos. Eggs from *Lytechinus pictus* collected at La Jolla were obtained by injection of about 0.5 ml of 0.5 *M* KCl into the perivisceral cavity and were then washed several times in artificial seawater. Fertilization (95% or greater was acceptable) was achieved with a 1:5000 final dilution of semen also obtained by KCl injection. Fertilization membranes were always removed 10 seconds after fertilization by addition of 0.005% dry trypsin powder (Nutritional Biochemicals, 1:300) followed by two washings in artificial seawater. Eggs or embryos were incubated at ratios of seawater to packed cell volume of 10 or greater at 18°C on a rotatory shaker. One milliliter of packed cells equals approximately 1.2×10^6 eggs or membraneless embryos in these early stages. Measurements before fertilization were made on eggs soon after they were shed by the female. Measurements after fertilization were made on embryos 6 hours after fertilization when protein synthesis on preformed mRNA is high and synthesis on newly formed mRNA has not yet begun (Gross *et al.*, 1964; Infante and Nemer, 1967).

Incubation with isotope. Preparatory to incubation with isotopes, eggs or embryos were washed in artificial seawater containing 30 μg of penicillin, G potassium, and 50 μg of streptomycin sulfate per milliliter and resuspended in the same solution at a ratio of 10 volumes of seawater to 1 volume of cells. Radioactive protein and nucleic acid precursors were added 30 minutes later in amounts and for times specified in the individual experiments.

Isolation of polysomes, ribosomes, and nascent proteins. Because the conclusions of this report depend completely on quantitative measurements of polysomes and nascent proteins, the techniques used to prepare these materials will be discussed in detail. Total polysomes and ribosomes with nascent protein were isolated from cells homogenized in two different buffers, selected after testing numerous other formulations. Both buffers were used with 0.5% of the detergent Nonidet P-40 (Shell Chemical Co.) added to dissolve all membranous components of the cells. They were a high salt buffer (HSB—0.24 *M* NH_4Cl, 0.01 *M* $MgCl_2$, 0.25 *M* sucrose, 0.01 *M* Tris, pH 7.8) developed by Hultin (1961) and used successfully with sea urchins by others (Stafford *et al.*, 1964; Spirin and Nemer, 1965) and a low ionic strength buffer (LSB—0.01 *M* KCl, 0.0015 *M* $MgCl_2$, 0.01 *M* Tris, pH 7.8) developed for reticulocytes (Warner *et al.*, 1963). Cells to be homogenized were washed once by centrifugation through the appropriate buffer. Sucrose (0.24 *M*) was added to LSB for this step and no internal cellular material, either trichloroacetic acid (TCA) soluble or precipitable, was lost in this wash in either buffer. The cells were then resuspended in 10 to 50 volumes of buffer plus detergent. They were homogenized (usually 3–6 strokes) with a Dounce homogenizer sufficiently to completely disintegrate all cells as determined during each homogenization with a microscope. An aliquot of the *total* homogenate, solubilized by detergent, was layered onto a sucrose gradient made up in the appropriate buffer and centrifuged. A second aliquot of the homogenate was dissolved in 1 N NaOH for Lowry protein determination. In some experiments, as noted, 1 μg of pancreatic ribonuclease (bovine, Sigma, 5$\times$ crystallized) per milliliter was added to a third aliquot which was held 30 minutes at 0°C before centrifugation was begun.

After centrifugation for various times, the gradients were analyzed for absorbance at 260 mμ and TCA-precipitable radioactivity. Nascent protein was defined as trichloroacetic acid-precipitable radioactivity (stable in 1 N NaOH for 30 minutes at 25°C) associated with ribosomes or polysomes after a short incubation of the eggs or embryos with exogenous radioactive amino acids. After homogenization in LSB plus detergent, virtually all ribosomal optical density and nascent protein were found in the 80 S monoribosome peak or in small polysomes (cf. Fig. 1). No significant amount of material sedimented faster than a 3-ribosome particle, and there was essentially no radioactivity in the pellet at the bottom of the gradient tube, even when the cells had been incubated up to 20 minutes in radioactive amino acids.

In homogenates prepared in 10 volumes of HSB with detergent, most ribosomes ($>$90%) again sedimented at 80 S and absorbance at 260 mμ was very low in the polysome region even though nascent protein sedimented with large polysomes ranging from 150 to 400 S rather than with the 80 S monoribosome peak (Fig. 3). The pellet at the bottom of the tube was of varying size ranging from 10 to 50% of the radioactivity found in the polysome region of the gradient. When the homogenate was treated with ribonuclease, all radioactivity in the polysome region sedimented to 80 S. Only a small proportion of the radioactivity in the pellet was caused to sediment at 80 S by ribonuclease. Thus, in these high salt buffers, most nascent protein sedimented in the gradient associated with polysomes of 150 to 400 S, and less than 5% was pelleted when the low concentrations of membraneless embryos employed in these experiments were used. If the cells were homogenized in 3 to 5 volumes of buffer, much more optical density and radioactivity not in nascent protein appeared both in the polysome region and the pellet of the gradient, apparently because considerable aggregation of material occurred in the homogenates (cf. Piatigorsky, 1968).

Although the different buffers yielded very different sizes of polysomes the total amounts of nascent protein associated with ribosomes were similar in the two buffers. Buffers with increasingly less monovalent salt, magnesium, or sucrose than HSB yielded increasingly smaller polysomes (Cohen and Iverson, 1967). Increasing monovalent salt or magnesium above the concentration in the HSB did not further increase the size of the polysomes in *L. pictus*.

There are three criteria which these methods must meet in order to provide a quantitative measure of the size of polysomes and the amount of nascent protein. (1) The yield must approximate 100%. (2) Binding of newly synthesized, nonnascent protein to ribosomes must be absent. (3) The polysomes must be isolated relatively intact. Criteria one and two appear to be met. Homogenization with detergent appears to solubilize the whole egg, since significant amounts of material were not found in the pellet when the total homogenate was layered onto the gradient. Because similar amounts of nascent protein were recovered after homogenization in the LSB and HSB, disintegration of ribosomes or loss of nascent protein from ribosomes in either of these buffers seem very unlikely. The sticking of nonnascent protein to ribosomes can be ruled out by data presented in Tables 1 and 2, which show that the amount of radioactivity associated with ribosomes decreases as the specific radioactivity of the amino acid pool decreases upon long incubations even though the radioactivity in the soluble protein has increased severalfold during the long incubation.

The large size of the polysomes in the optimal high salt buffer with the absence of nascent protein in the monoribosome peak indicate that the polysomes are intact. Attempts to extract larger polysomes were unsuccessful. Although at suboptimal concentrations of magnesium or potassium ions the inhibitors of ribonuclease, bentonite (1 mg/ml) (Infante and Nemer, 1967), or purified yeast RNA (5 mg/ml), were beneficial, they were not effective when optimal concentrations of these ions were present. Indeed they may have acted by contributing magnesium ions, since bentonite was less effective after it was washed in the chelating agent EDTA.

Breakdown of polysomes in homogenates has usually been attributed to endogenous ribonuclease. This is the case in sea urchin homogenates even though degradation is eliminated by increasing ions. If polysomes from rabbit reticulocytes, which are stable in LSB, are added to a homogenate of sea urchin eggs in LSB, the added polysomes are degraded. As the eggs are homogenized in buffers approaching the optimal concentrations of the various components, the degradation of reticulocyte polysomes decreases and finally cannot be detected in optimal homogenates.

There is, however, still detectable ribonuclease activity in the optimal homogenates. This was established by observing the degradation of radioactive RNA in optimal homogenates of embryos which had been incubated in uridine-^{3}H for 3 hours. The homogenate was allowed to stand at 0°C for 30 minutes, and the RNA was extracted (Scherrer *et al.*, 1963). The radioactive RNA extracted from these homogenates sedimented as heterogeneous population with a size distribution between 3 and 23 S. Radioactive RNA from intact embryos sedimented as a heterogeneous population distributed between 6 and 30 S. This change in size indicates each RNA molecule sustained one or two breaks. Similar experiments with suboptimal homogenates reveal greater degradation of the radioactive RNA. Increasing the concentration of monovalent and/or magnesium ions thus appears to act either by protecting the RNA from ribonuclease, by inhibiting the enzyme, or by preventing its release or activation during homogenization.

Analysis of sucrose gradients. Absorbance at 260 mμ in various levels of sucrose gradients was assayed by pumping the gradient through a continuous-flow cell of a Gilford Model 2000 recording spectrophotometer. Fractions were collected by counting equal numbers of drops from the effluent. TCA-precipitable radioactivity in each fraction was coprecipitated with a drop of 0.1% bovine serum albumen by adding TCA to a final concentration of 5% and cooling on ice for 10 minutes. The precipitate was collected on a Millipore filter, transferred to a vial on the filter, dissolved in 0.5 ml of 1 *N* NH_4OH, and counted in Beckman scintillation mix (10% napthalene and 0.6% PPO in dioxane) to 5% error. Counts were corrected for quench by an external standard.

Measuring uptake and incorporation of exogenous amino acids. Total radioactive amino acids entering eggs or embryos and radioactive amino acids incorporated into TCA-precipitable material were measured after various intervals of incubation of the cells in the precursor as follows: 0.2 ml of cell suspension was pipetted into 10 ml of ice cold seawater and the cells were pelleted by light centrifugation (15 seconds at 50–100 *g*). The supernatant was then aspirated. The cells were resuspended in 1 ml of cold seawater and layered over 10 ml of an ice cold 3:1 mixture of seawater and 1.1 *M* (isotonic) sucrose. They were again pelleted by centrifugation, and the supernatant was aspirated carefully from the top down. This washing by centrifuging the cells through sucrose could be repeated at least five times without further lowering the total radioactivity in the cells, indicating that all free, external isotope had been removed by the two washings. The twice-washed cells were then dissolved in 1 *N* NaOH at room temperature by passing the suspension through a 25-gauge needle several times during a period of 3 or more hours with a 1-ml disposable syringe. Total radioactivity was measured by adding 0.1 ml of the suspension to a vial with Beckman scintillation mix. Radioactivity incorporated into protein was measured by taking 0.7 ml of the solution in 1 *N* NaOH and precipitating it with an equal volume of 30% TCA at 0°. The precipitate was collected and counted just as were the fractions from sucrose gradients. Protein in the sample was determined by the Lowry procedure (1951).

The kinetics of uptake of amino acids by eggs and embryos were very different both quantitatively and qualitatively. The details of these kinetics do not directly affect the experiments, but they do elucidate the differences in the curves of incorporation of radioactive amino acids into protein (Fig. 4). Radioactive amino acids entered eggs at a linear rate for over an hour (Fig. 4b inset); thus, the specific radioactivity of the amino acid pool was constantly increasing with time after addition of label, resulting in a continuing increase in the rate of incorporation of radioactivity into protein (Fig. 4b). Exogenous radioactive amino acids entered embryos rapidly for several minutes until a maximum internal level was achieved (Fig. 4a). Within 20 minutes a significant amount of the isotope entering the cells had been incorporated. This was reflected in a slightly increasing or virtually linear incorporation of radioactivity into protein for 10 to 15 minutes followed by a gradual decrease in the rate of incorporation with time after embryos were given the exogenous radioactive amino acids. Because the actual molar concentrations of exogenous isotope were relatively low, there was no indication of the preferential use of exogenous isotope observed by Berg (1968) with high concentrations of valine-^{14}C.

Radioactive precursors. All isotopically labeled materials were purchased from Schwarz BioResearch. Reconstituted protein hydrolyzate-^{14}C (^{14}C-AA mix) contains a mixture of 13 1-amino acids with specific activities ranging from 50 to 250 mC/mmole. Leucine-1-^{3}H had a specific activity of 45 C/mmole; leucine-1-^{14}C 170 mC/mmole; uridine-^{3}H, 21 C/mmole.

Sedimentation of proteins. Eggs or embryos incubated with radioactive amino acids, were dissolved overnight at room temperature in a solution of 8 *M* urea and 0.5% sodium dodecyl sulfate (SDS). Nascent proteins were obtained from a sucrose gradient as illustrated in Fig. 2. The ribosomes and polysomes were collected as a pellet after a centrifugation for 4 hours 50,000 rpm in a Spinco Ti 50 rotor and dissolved in a freshly (less than 10 minutes) prepared solution of 0.5% SDS and 8 *M* urea in 0.3 *N* KOH. They were incubated at 37°C for 15 minutes to release all transfer-RNA from the nascent chains. The protein solutions were then immediately layered onto a 5–20% sucrose gradient made up in the 8 *M* urea and 0.5% SDS solution and centrifuged at 25°C for 15 hours at 50,000 rpm or 30 hours at 60,000 rpm in a Spinco SW 65 rotor. Fractions were collected and diluted 10 times with water before determination of TCA-precipitable radioactivity as described for gradients. Markers of light chain immunoglobin (3 S) added in these gradients sedimented slightly more rapidly then the peak of radioactive protein in both the total protein and nascent protein preparations. Greater than 99% of the radioactive protein was soluble in the buffer since less than 0.5% pelleted in the bottom of the gradient tubes.

RESULTS

Characteristics of Protein Synthesis in Eggs and Embryos

Three questions were raised which involved assumptions about protein synthesis in eggs and embryos made during the design of the tests for increased efficiency of translation. (1) Could the apparent slower rate of protein synthesis in eggs be due to rapid turnover of newly synthesized protein? (2) Does most protein synthesis occur on ribosomes in eggs as well as in embryos? (3) Do the protein molecules in eggs and embryos have the same size distribution such that each ribosome on a mRNA molecule synthesizes on the average the same amount of protein before and after fertilization?

Stabilization of protein. The acceleration of protein synthesis as measured by long-term incubation with isotopes has been accurately

TABLE 1

PERCENT OF FREE AMINO ACIDS INCORPORATED INTO TCA-PRECIPITABLE MATERIAL AFTER SHORT INCUBATIONS OF EGGS AND EMBRYOS WITH RADIOACTIVE AMINO ACIDS

Cells	Pulse[a] length (min)	Precipitable[b] radioactivity (cpm/mg protein)	Total[b] radioactivity (cpm/mg protein)	Percent incorporated	Average
Embryos	½	1042	7970	13.1	13.8
Embryos	1	1809	12540	14.4	
Eggs	½	19	1770	1.07	0.99
Eggs	1	26	2850	0.91	

[a] With 20 μC ^{14}C-labeled amino acid mixture.

[b] See Methods section for details of determinations.

determined (Epel, 1967). If very short-term incubations with isotopes give similar results, rapid decay of significant amounts of newly synthesized protein would be essentially ruled out. Eggs and embryos were incubated for 0.5 and 1 minute with radioactive amino acids, and total and precipitable radioactivity was determined (see Table 1 for typical experiment). Total uptake of radioactive amino acids in eggs is much lower than in embryos, thus, the ratio of total radioactivity incorporated into protein will not be a measure of the true relative rates of protein synthesis. However, the relative rates can be calculated because the entering isotope equilibrates quickly with the free amino acid pools as indicated by the similar results with 0.5 and 1 minute incubations and since the pool sizes do not change greatly at fertilization (Gustafson and Hjelte, 1951; Kavanau, 1954). This means that the ratio of the percentage of free isotope incorporated into protein should accurately reflect the factor by which protein synthesis is increased. The value of 14-fold increase in protein synthesis obtained from the figures in Table 1 is an average figure and agrees well with the values obtained by long-term incubations with the same species (Epel, 1967; Humphreys, unpublished). The accumulation of newly synthesized protein results from an acceleration of protein synthesis and is not an apparent result of decreasing protein turnover at fertilization.

Protein synthesis on ribosomes. Ribosomes active in *in vivo* protein synthesis have been demonstrated in unfertilized eggs (compare Tyler *et al.*, 1968, with Monroy and Tyler, 1963; Malkin *et al.*, 1964;

Cohen and Iverson, 1967). However, quantitative measurements were not made. Eggs that had been incubated for 1 minute with 30 μC ^{14}C-AA mix per milliliter were homogenized in LSB with 0.5% NP-40 and the total homogenate fractionated on a sucrose gradient (Fig. 1). As noted in the Methods section, this buffer yields all nascent protein attached to ribosomes and does not cause any nonspecific sticking of newly synthesized protein to ribosomes. Since polysomes are severely degraded by endogenous ribonuclease in such a buffer, the nascent protein is in the monoribosome peak or in small polysomes. In several such experiments 50–60% of the incorporated isotope was still in nascent protein. Because the labeling time is not long compared to the time required for synthesis of a polypeptide chain in other animal cells (Dintzis *et al.*, 1959; Conconi *et al.*, 1966; and see below), this is the appropriate percentage which might be expected on the ribosomes. It appears that a significant percentage of the protein synthesized in unfertilized eggs is polymerized on ribosomes.

Many previous studies have demonstrated synthesis of protein on

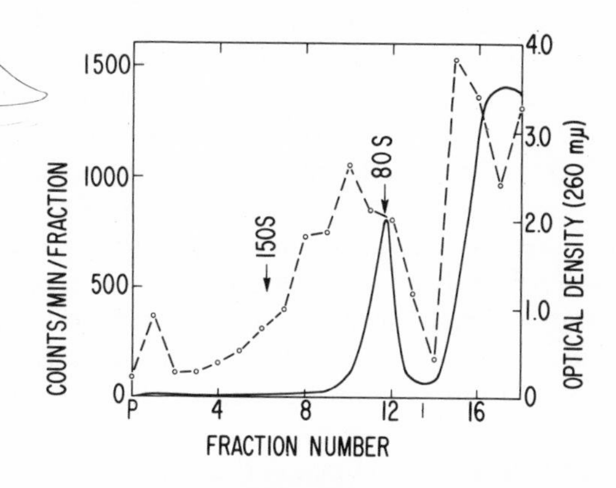

FIG. 1. Nascent protein and completed protein in eggs after a 1-minute incubation with a radioactive amino acid. Eggs (0.1 ml) were incubated 1 minute in 1 ml SW with 30 μC leucine-^{3}H washed, homogenized in 1 ml LSB with 0.5% NP-40, and layered onto a 15 to 30% sucrose gradient in LSB. The gradient was centrifuged 5 hours at 24,000 rpm in a Spinco SW 25.1 rotor at 7°C and analyzed for optical density at 260 mμ (——) and TCA-precipitable radioactivity (--○--). Polysomes are mostly degraded to monomers, dimers, or trimers, in this buffer. Radioactivity sedimenting at less than 60 S (fractions 14–18) was taken to be in completed protein, and radioactivity sedimenting more rapidly than 60 S (fractions 1–13) was taken to be nascent protein associated with ribosomes. *P* = pellet in gradient tube.

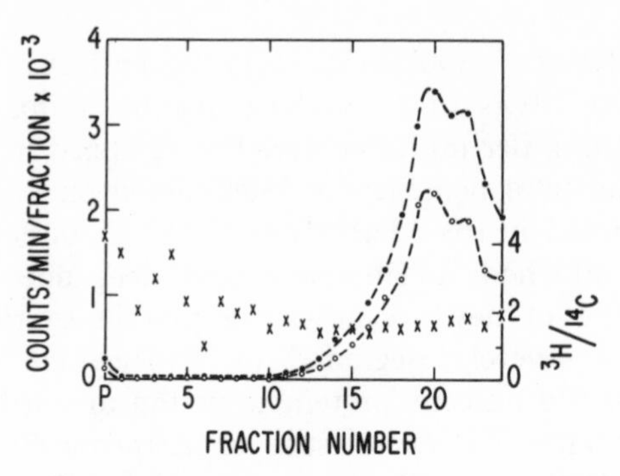

FIG. 2. Relative sedimentation rates of protein from eggs and 6-hour embryos. Eggs (0.2 ml) were incubated 10 minutes in seawater with 20 μC/ml leucine-H^3, and embryos (0.03 ml) were incubated 10 minutes in seawater with 60 μC/ml leucine-^{14}C. Both samples were combined and dissolved in 1 ml of 8 μ urea with 0.5% SDS; 0.2 ml was layered onto a 4.5 ml, 5 to 20% sucrose gradient in the urea-SDS solution and centrifuged 50,000 rpm for 15 hours at 25°C in a Spinco SW 65 rotor. Fractions were collected, and TCA-precipitable ^{14}C (--○--) and ^3H (--●--) were determined, and the ratio of ^3H to ^{14}C (×) was calculated for each. *P* = pelleted material in bottom of gradient tube.

the ribosomes of embryos. Quantitative experiments on embryos identical to those described above on eggs yielded very similar results. Similar percentages of protein must be synthesized on ribosomes in both eggs and embryos and most probably virtually all protein synthesis occurs on ribosomes in eggs as well as embryos.

Size distribution of newly synthesized protein. The size distributions of proteins from eggs and embryos were compared by cosedimentation of proteins from each cell type in a sucrose gradient. Eggs and embryos were incubated for 10 minutes with leucine-^3H and leucine-^{14}C, respectively. Both were dissolved in the same aliquot of SDS-urea buffer, a buffer which should disrupt all association of polypeptide chains not linked by covalent bonds. The solution was fractionated by zonal sedimentation on a sucrose gradient made up in the SDS-urea solution (Fig. 2). The TCA-precipitable ^3H and ^{14}C were exactly coincident in the gradient. Thus, the size distribution of the proteins being synthesized is very similar before and after fertilization.

Together the three results show that the acceleration of protein synthesis at fertilization is the synthesis of more protein molecules on more ribosomes and indicate the experiments designed to test efficiency are valid for sea urchin eggs and embryos.

Size Distribution of Polysomes

If the acceleration of protein synthesis at fertilization is achieved exclusively by attaching more ribosomes to the same population of messenger RNA, the size of the polysomes should increase 10- to 15-fold at fertilization. There are two implicit assumptions in this prediction. They are that the number of cistrons per mRNA does not change upon fertilization and that inactive ribosomes are not attached to active mRNA molecules. Because simple ways to test these unlikely possibilities could not be devised, they remain assumptions.

Polysomes from embryos. The size distribution of polysomes prepared from 6-hour embryos incubated with ^{14}C-labeled amino acids for 10 minutes to ensure that all nascent peptides were completely labeled is shown in Fig. 3a. As demonstrated by TCA-precipitable radioactivity, polysomes with sizes ranging up to 400 S with a broad peak at about 300 S were found. This size is comparable to the largest sizes of polysomes which have previously been isolated from developing embryos (Monroy and Tyler, 1963; Stafford *et al.*, 1964; Spirin and Nemer, 1965; Infante and Nemer, 1967; Cohen and Iverson, 1967). The radioactivity in the polysome region could be quantitatively shifted to the monoribosome peak by pretreatment of the homogenate with ribonuclease.

Polysomes from eggs. Polysomes in eggs active in protein synthesis are not well characterized (compare Tyler *et al.*, 1968 with Monroy and Tyler, 1963; Malkin *et al.*, 1964; Cohen and Iverson, 1967). This failure to characterize polysomes in eggs apparently reflects both the low permeability of the unfertilized eggs to exogenous precursors and the low activity of the polysomes in the egg. To overcome this, eggs were incubated in high concentrations of radioactive amino acids (20 μC/ml ^{14}C-AA mix or leucine-^{3}H). This increased the specific activity of the amino acid pool sufficiently so that considerable radioactivity was found on polysomes extracted from these eggs (Fig. 3b). The 200 S to 400 S size range of polysomes from unfertilized egg was very similar to the size of polysomes from embryos. The slightly sharper peak and possibly smaller size of the polysomes from unfertilized eggs, as shown in Fig. 3, was the greatest difference noted in any experiment and may indicate a slight difference between polysomes from eggs and embryos although in several experiments no difference was detectable. The radioactivity

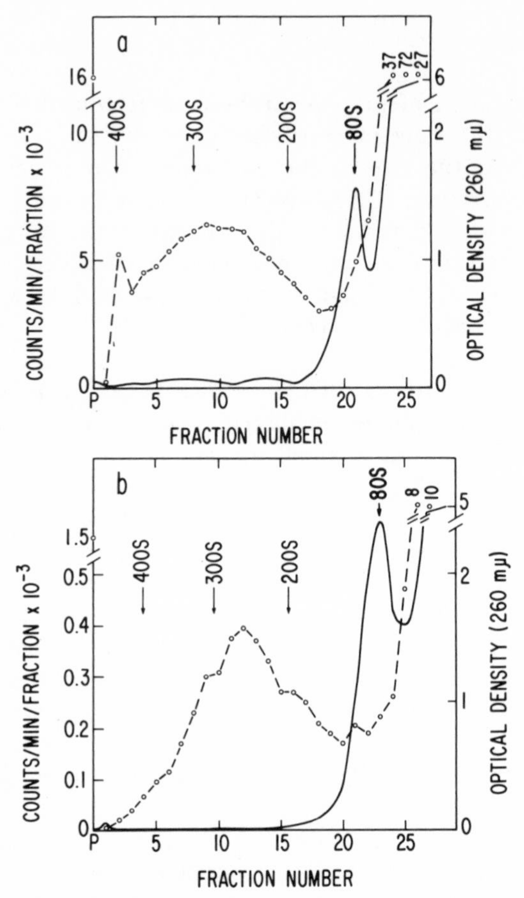

FIG. 3. Polysomes from 6-hour embryos (a) and from unfertilized eggs (b). Cells (0.1 ml) were incubated in SW with 20 μC/ml ^{14}C-AA mixture for 10 minutes, washed, homogenized in 1 ml of HSB with 0.5% NP-40, and layered onto a 15 to 30% sucrose gradient in HSB. Gradients at 7°C were centrifuged at 24,000 rpm for (a) 4.5 hours in a Spinco SW 25.3 rotor and (b) 3.5 hours in a Spinco SW 25.1 rotor. Gradients were analyzed for optical density at 260 mμ (——) and TCA-precipitable radioactivity (--○--). *P* = pellet in bottom of gradient tube.

in the polysome region of unfertilized eggs was also quantitatively shifted to the monosomes by ribonuclease.

These results suggest that the size distribution of polysomes is essentially the same in eggs and embryos. There certainly are not 10–15 times more ribosomes per polysome in embryos than in eggs.

Therefore, the increased protein synthesis at fertilization cannot be attributed to a more active attachment of ribosomes to messenger-RNA molecules.

Time for Synthesis of a Polypeptide Chain

Efficiency of translation could be increased upon fertilization by increasing the average rate of movement of the ribosomes along the mRNA molecules. This would decrease the average interval required to synthesize a complete polypeptide chain. The relationship between the magnitudes of this interval before and after fertilization could be obtained by comparing the average time an amino acid remains in the nascent chains and the size distribution of the nascent chains before and after fertilization (Englander and Page, 1965; Hunt *et al.*, 1968).

Amino acid flow through nascent chains. During synthesis of protein molecules, free amino acids from the pool are incorporated into a growing nascent chain and remain in the nascent chain until the protein molecule is completed and released from the ribosome. The amino acid residues then move from the nascent proteins into the completed proteins. The average time which an amino acid residue remains in the nascent proteins can be calculated as the amount of amino acid in the nascent chains divided by the amount of amino acid which flows through the nascent protein per unit time. Both quantities can be easily measured during a pulse of radioactive amino acids. The first is the radioactivity in polysomes since, as discussed in the Methods section, the extraction and fractionation procedures used for polysomes gave a 100% yield of nascent protein uncontaminated by other newly synthesized protein. The second is the rate of incorporation of amino acids into protein as defined by standard TCA precipitation, since the amino acids being incorporated are the ones flowing through the nascent chains. The final numbers obtained have dimensions of time only, thus the specific activities of the amino acid pools do not affect the correctness of the values.

Typical curves of incorporation of radioactivity into TCA-precipitable material at various intervals during incubation of eggs and embryos with radioactive amino acids appear in Fig. 4. The rate of incorporation expressed as radioactivity per milligram of egg or embryo protein per minute incubation was determined from these curves for any time point during the incubation by measuring the slope of

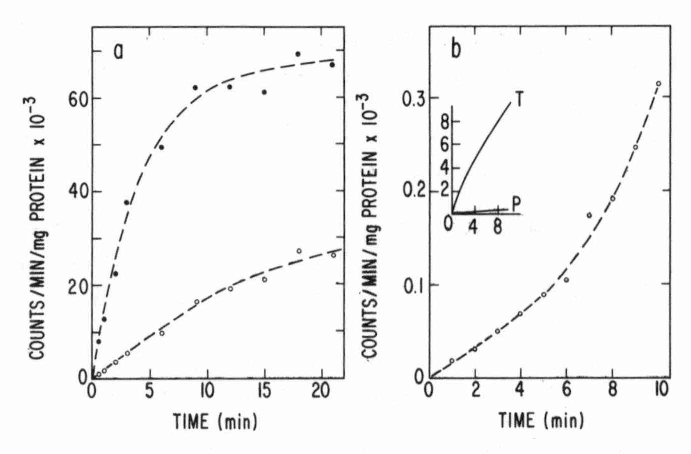

FIG. 4. Uptake and incorporation of exogenous radioactive amino acids by embryos 6 hours after fertilization (a) and by unfertilized eggs (b). Cells were incubated in 20 μC/ml ^{14}C-AA mixture, and samples were taken at intervals for determination of total radioactivity (--○--) and TCA-precipitable radioactivity (--●--) per milligram of cell protein. Inset in (b) shows total (*T*) and precipitable (*P*) radioactivity from eggs on a reduced scale (units in inset are the same as the larger graph).

the incorporation curve at that time point. The nascent protein at this time point was determined by removing eggs or embryos from the incubation flask at that time and determining radioactivity in the polysomes per milligram of total protein layered onto the gradient in which the polysomes were isolated. The rate of incorporation was then divided into the amount of nascent protein to determine the average time an amino acid remains in the nascent chains. Values for nascent protein, rates of incorporation, and the calculated average time that an amino acid remains in the nascent protein in embryos and eggs, respectively, are shown in Tables 2 and 3. As controls on technique, the experiment was repeated with different buffers, isotopes, and incubation times and with and without ribonuclease in the polysome extract.

The resulting value for the average time an amino acid remains in the nascent protein as obtained by averaging the data in Tables 2 and 3, respectively, is 1.16 minutes for embryos and 0.61 minute for eggs. These averages indicate, contrary to possible expectations, that ribosomes actually move more slowly after fertilization. However, the individual determinations in both Tables 2 and 3 show sufficient variation and overlap to make a real difference questionable. Basically, the data suggest that the average time an amino acid remains in the nascent protein does not change significantly at fertilization. There certainly is no indication that ribosomes might be moving 10- to 15-fold faster after fertilization. These values for average time an amino acid remains in nascent chains obtained in sea urchins at 18°C

TABLE 2

AVERAGE LENGTH OF TIME THAT AMINO ACIDS INCORPORATED INTO PROTEIN REMAIN IN THE NASCENT PEPTIDES ATTACHED TO RIBOSOMES IN SEA URCHIN EMBRYOS[a]

Expt. No.	Homogenization buffer	Source of isotope	Incubation time (min)	RNase	Rate of incorporation[b]	Amount of nascent protein (cpm/mg cell protein)	Average time on ribosomes (min)
1.	HSB	^{14}C-AA[c]	18	−	643	399	0.62
1.	LSB	^{14}C-AA	10	E[d]	1125	796	0.71
1.	HSB	^{14}C-AA	10	−	1125	1177	1.05
2.	HSB	^{14}C-AA	3	−	1050	1231	1.17
3.	HSB	Leu-^{3}H	10	−	2700	3220	1.19
1.	LSB	^{14}C-AA	18	E[d]	643	845	1.31
2.	HSB	^{14}C-AA	3	+	1050	2174	2.04

[a] See text for details of measurements and calculations.

[b] Counts per minute per milligram of cell protein per minute of incubation.

[c] ^{14}C-labeled amino acid mixture.

[d] Endogenous Rnase is active in LSB homogenates.

TABLE 3

AVERAGE LENGTH OF TIME THAT AMINO ACIDS INCORPORATED INTO PROTEIN REMAIN IN THE NASCENT PEPTIDES ATTACHED TO RIBOSOMES IN SEA URCHIN EGGS[a]

Expt. No.	Homogenization buffer	Source of isotope	Incubation time (min)	RNase	Rate of incorporation[b]	Amount of nascent protein (cpm/mg cell protein)	Average time on ribosomes (min)
1.	LSB	Leu-^{3}H	10	E[d]	175	90	0.51
2.	HSB	^{14}C-AA[c]	20	−	63	32	0.51
3.	HSB	^{14}C-AA	10	−	36	20	0.56
3.	LSB	^{14}C-AA	10	E[d]	36	21	0.58
4.	HSB	^{14}C-AA	10	−	75	50	0.67
2.	LSB	^{14}C-AA	20	E[d]	63	45	0.71
4.	HSB	^{14}C-AA	10	+	75	57	0.76

[a] See text for details of measurements and calculations.

[b] Counts per minute per milligram of cell protein per minute of incubation.

[c] ^{14}C-labeled amino acid mixture.

[d] Endogenous RNase is active in LSB homogenates.

are similar to the times obtained in rabbit reticulocytes at similar temperatures (1.6 minutes at 17°C, 0.9 minutes at 27°C; Conconi *et al.*, 1966) and indicate that ribosomes move at similar rates in cells of these two species.

Size distribution of nascent chains. Although the average time an amino acid remains in the nascent protein does not change appreciably at fertilization, the average rate of movement of the ribosomes could change because the pattern of movement of the ribosome along the mRNA molecule determines the actual relationship between these two parameters (Englander and Page, 1965). The size distribution of nascent protein changes as the pattern of movement of the ribosomes and thus provides a marker for these changes (Hunt *et al.*, 1968).

The size distributions of nascent proteins before and after fertilization were compared by cosedimentation of these proteins extracted after incubation of eggs and embryos, respectively, with leucine-^{3}H and leucine-^{14}C. The nascent proteins were obtained from ribosomes and small polysomes in a sucrose gradient after homogenization of the cells in low salt buffer (cf. Fig. 1). They were collected from the sucrose gradient fractions as a centrifuged pellet, solubilized and cleaved from the transfer-RNA in a freshly prepared alkaline SDS-urea solution, layered onto a sucrose gradient in the SDS-urea solution, and centrifuged (Fig. 5). A broad distribution of protein was spread across the gradient indicating a very heterogenous preparation. The constant ratio of ^{3}H to ^{14}C across the gradient shows that the size distribution of nascent proteins was the same in eggs and embryos. Therefore, the sizes of the nascent protein and, thus, the pattern of movement of the ribosomes appear to be similar before and after fertilization.

There is no indication that the rate of movement of ribosomes increases at fertilization. It clearly does not change the 10- to 15-fold that would be necessary to account for the acceleration of protein synthesis.

DISCUSSION

These results establish that there is little change in the efficiency of translation of mRNA molecules upon fertilization of sea urchin eggs. By elimination, the increase in rate of protein synthesis which occurs at fertilization must be a result of initiation of translation of additional mRNA molecules.

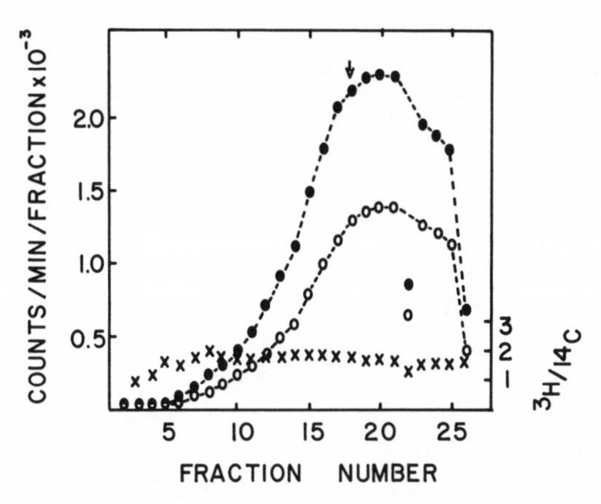

FIG. 5. Relative sizes of nascent protein in eggs and embryos. Ribosomes with nascent protein were obtained from appropriate fractions of sucrose gradients in low salt buffer as shown in Fig. 1 from 0.1 ml of eggs and 0.1 ml of embryos incubated for 20 minutes in 50 μC/ml leucine-^{3}H and 30 μC/ml leucine-^{14}C, respectively. The ribosomes and nascent protein from both eggs and embryos were pooled and precipitated by centrifugation for 4 hours at 50,000 rpm in a Spinco Ti 50 rotor at 5°C. The pellet was dissolved in 0.4 ml of alkaline urea-SDS solution and 0.2 ml was layered onto a 5–20% gradient in urea-SDS solution. The gradient was centrifuged at 60,000 rpm for 30 hours at 25°C in a Spinco SW 65 rotor. Fractions were collected from the gradient and total ^{3}H(-●-) and ^{14}C (-○-) were determined and ^{3}H/^{14}C (×) was calculated for each fraction. The arrow indicates the position of a 3 S ^{14}C-light chain of an immunoglobin molecule in a companion gradient with only ^{3}H nascent protein.

The present data could adequately be explained by newly synthesized mRNA entering the polysomes at fertilization. However, experiments with actinomycin D suggest that the additional mRNA is not synthesized at fertilization but is present in the egg before fertilization (Gross *et al.*, 1963), and many other observations are consistent with this conclusion (see Tyler, 1967). These results indicate that this mRNA must be completely inactive until fertilization. Since there is also a population of mRNA molecules in the egg which is active, the mechanism which inhibits the translation of the inactive mRNA must be specific and not inhibit the active population. Because the mechanism is specific for certain mRNA molecules, any hypothesis of a general inhibition of the protein synthetic machinery cannot be valid. Thus, a general inhibition of ribosomes (Hultin, 1961), even if only partial, cannot occur in the egg. If some charged transfer-RNA is missing (Glišin and Glišin, 1964; Ceccarini *et al.*,

1967), its codon must not appear in the mRNA active in the egg, but must occur in all the inactive mRNA molecules. A number of the possible mechanisms which have been suggested can specifically inhibit the translation of certain mRNA molecules in cells which are translating others (see introduction; see also Spirin, 1966; Tyler, 1967).

Translation of additional mRNA molecules requires an increase in the number of *active* polysomes. Increased activity of the polysomes at fertilization has been suggested repeatedly after pulse incubations with radioactive amino acids (Monroy and Tyler, 1963; Malkin *et al.*, 1964; Stafford *et al.*, 1964; Cohen and Iverson, 1967) although no correction for permeability or yields were made in any of these studies. If it is accepted that the size of the amino acid pools does not change at fertilization (Kavanau, 1954) and that entering isotope equilibrates with these pools, data from the experiments presented in this paper confirm these suggestions quantitatively. The amount of nascent protein can be expressed as a percentage of the amino acid pool by determining the ratio of radioactivity in the nascent proteins to radioactivity in the pool. In our experiments this ratio increases about 10- to 15-fold upon fertilization (Humphreys, 1968). Because the size distribution of the nascent protein chains does not change, the number of nascent chains and thus the number of active ribosomes must increase by this factor. Since the active polysomes remain the same size, their number must also increase by 10- to 15-fold to accommodate the extra ribosomes.

The number of ribosomes physically present in the free polysomes contained in supernatants of centrifuged homogenates also increases at fertilization (Infante and Nemer, 1967). Comparison of the optical density in the polysome regions of the graphs of total homogenates (Fig. 3) where both free and bound polysomes are displayed indicate that total polysomes increase also. However, the differences in Fig. 3 are very small and did not appear in all experiments, probably because the optical density in the gradients is barely above background levels. These results support the idea that additional mRNA molecules are activated and enter the polysomes at fertilization. However, the techniques used in this study were not evaluated for their yield of polysomes defined as physical aggregates of ribosomes, and thus the data are not conclusive. Other data have suggested that the mRNA might be stored in the polysomes before fertilization (Stavy and Gross, 1967; Piatigorsky, 1968). These differences are based on different methods of assaying for polysomes. More careful attention to development of adequate techniques is needed to demonstrate directly that additional mRNA becomes active at fertilization.

The techniques used in this study could elucidate other problems. For example, a severalfold inhibition of translation occurs during mitosis (Prescott and Bender, 1962; Scharff and Robbins, 1966; Salb and Marcus, 1965). Although the ribosomes seem to be involved in the inhibition (Salb and Marcus, 1965), it is not clear whether it is achieved by a general decrease in the efficiency of translation or by the inhibition of translation of a specific population of mRNA molecules. An appropriate analysis could easily distinguish these alternatives.

SUMMARY

The efficiency of translation, defined as the number of protein molecules produced per mRNA molecule per unit time, was compared before and after the acceleration of protein synthesis which occurs upon fertilization of sea urchin eggs. Examination of protein turnover, extent of protein synthesis on ribosomes, and the size distribution of newly synthesized protein before and after fertilization showed that this acceleration of protein synthesis involved the synthesis of 10 to 15 times more protein molecules on 10 to 15 times as many active ribosomes. To determine whether these increased numbers of active ribosomes were interacting more efficiently with an unchanged number of mRNA molecules or at the same efficiency with an increased number of mRNA molecules, the size of the polysomes and the time a ribosome stayed on the polysomes was compared before and after fertilization. The latter time interval was determined by measuring the average time a newly incorporated amino acid remained in the nascent chains and the size distribution of the nascent chains. New methods giving essentially 100% yield of active polysomes from sea urchin cells made these measurements possible.

The results of the experiments show that efficiency of translation is similar in egg and embryos. Protein synthesis, thus, is accelerated at fertilization by the translation of additional mRNA molecules. This translation level control of protein synthesis therefore cannot be a general change in activity of some component in the cellular synthetic machinery such as ribosomes, but must specifically control the activity of a defined population of mRNA molecules.

It has been a pleasure to have the invaluable assistance of Mrs. Pamela Tyler Lindstrom and Mrs. Hazel Schubert during the course of these experiments.

This work has been supported by Grant No. HD03480 from the National Institute of Child Health and Development.

REFERENCES

BERG, W. E. (1968). Kinetics of uptake and incorporation of valine in the sea urchin embryo. *Exptl. Cell Res.* **49,** 379-395.

CANDELAS, G. C., and IVERSON, R. M. (1966). Evidence for translational level control of protein synthesis in the development of sea urchin eggs. *Biochem. Biophys. Res. Commun.* **24,** 867-871.

CECCARINI, C., MAGGIO, R., and BARBATA, G. (1967). Amino acyl-sRNA synthesis as possible regulators of protein synthesis in the embryo of the sea urchin *Paracentrotus lividus. Proc. Natl. Acad. Sci. U.S.* **58,** 2235-2239.

COHEN, G. H., and IVERSON, R. M. (1967). High resolution density-gradient analysis of sea urchin polysomes. *Biochem. Biophys. Res. Commun.* **29,** 349-355.

CONCONI, F. M., BANK, A., and MARKS, P. A. (1966). Polyribosomes and control of protein synthesis: Effects of sodium fluoride and temperature in reticulocytes. *J. Mol. Biol.* **19,** 525-540.

DENNY, P. C., and TYLER, A. (1964). Activation of protein biosynthesis in non-nucleate fragments of sea urchin eggs. *Biochem. Biophys. Res. Commun.* **14,** 245-249.

DINTZIS, H. M. BORSOOK, H., and VINOGRAD, J. (1959). Microsomal structure and hemoglobin synthesis in the rabbit reticulocyte. *In* "Microsomal Particles and Protein Synthesis" (R. B. Roberts, ed.), pp 95-99. Pergamon Press, New York.

ENGLANDER, S. W. , and PAGE L. A. (1965). Interpretation of data on sequential labeling of growing polypeptides. *Biochem. Biophys. Res. Commun.* **19,** 565-570.

EPEL, D. (1967). Protein synthesis in sea urchin eggs: A "late" response to fertilization. *Proc. Natl. Acad. Sci. U.S.* **57,** 899-906.

GILBERT, W. (1963). Polypeptide synthesis in *Escherichia coli*. II. The polypeptic chain and s-RNA. *J. Mol. Biol.* **6,** 389-403.

GLIŠIN, V. R., and GLIŠIN, M. V. (1964). Ribonucleic acid metabolism following fertilization in sea urchin eggs. *Proc. Natl. Acad. Sci. U.S.* **52,** 1548-1553.

GROSS, P. R., MALKIN, L. I., and MOYER, W. A. (1964). Templates for the first proteins of embryonic development. *Proc. Natl. Acad. Sci. U.S.* **51,** 407-414.

GUSTAFSON, T., and HJELTE, M. (1951). The amino acid metabolism of the developing sea urchin egg. *Exptl. Cell Res.* **2,** 474-490.

HULTIN, T. (1952). Incorporation of ^{15}N-labeled glycine and alanine into the proteins of developing sea urchin eggs. *Exptl. Cell Res.* **3,** 494-496.

HULTIN, T. (1961). Activation of ribosomes in sea urchin eggs in response to fertilization. *Exptl. Cell Res.* **25,** 405-417.

HUMPHREYS, T. (1968). Efficiency of translation before and after fertilization of sea urchin eggs. *Ann. Embryol. Morphol., Proc. 6th Intern. Embryol. Congr.*, in press.

HUMPHREYS, T. (1968). Translation of additional mRNA upon fertilization of sea urchin eggs. *J. Cell Biol.* **39,** 63a-64a.

HUNT, T., HUNTER, T., and MUNRO, A. (1968). Control of haemoglobin synthesis: Distribution of ribosomes on the messenger RNA for α and β chains. *J. Mol. Biol.* **36,** 31-45.

INFANTE, A. A., and NEMER, M. (1967). Accumulation of newly synthesized RNA templates in a unique class of polyribosomes during embryogenesis. *Proc. Natl. Acad. Sci. U.S.* **58,** 681-688.

KAVANAU, J. L. (1954). Amino acid metabolism in the early development of the sea urchin *Paracentrotus lividus*. *Exptl. Cell Res.* **7,** 530–557.

LOWRY, O. H., ROSEBROUGH, N. J., FARR, A. L., and RANDALL, R. J. (1951). Protein measurement with the Folin phenol reagent. *J. Biol. Chem.* **193,** 265–275.

MACH, B., REICH, E., and TATUM, E. L. (1963). Separation of the biosynthesis of the antibiotic polypeptide tyrocidine from protein biosynthesis. *Proc. Natl. Acad. Sci. U.S.* **50,** 175–181.

MACKINTOSH, F. R., and BELL, E. (1967). Stimulation of protein synthesis in unfertilized sea urchin eggs by prior metabolic inhibition. *Biochem. Biophys. Res. Commun.* **27,** 425–430.

MALKIN, L. I., GROSS, P. R., and ROMANOFF, P. (1964). Polyribosomal protein synthesis in fertilized sea urchin eggs: The effect of actinomycin treatment. *Develop. Biol.* **10,** 378–394.

MANO, Y., and NAGANO, H. (1966). Release of maternal RNA from some particles as a mechanism of activation of protein synthesis by fertilization in sea urchin eggs. *Biochem. Biophys. Res. Comm.* **25,** 210–215.

MONROY, A., and TYLER, A. (1963). Formation of active ribosomal aggregates (polysomes) upon fertilization and development of sea urchin eggs. *Arch. Biochem. Biophys.* **103,** 431–435.

MONROY, A., MAGGIO, R., and RINALDI, A. M. (1965). Experimentally induced activation of the ribosomes of the unfertilized sea urchin egg. *Proc. Natl. Acad. Sci. U.S.* **54,** 107–111.

NAKANO, E., and MONROY, A. (1958). Incorporation of ^{35}S-methionine in the cell fractions of sea urchin eggs and embryos. *Exptl. Cell Res.* **14,** 236–243.

NEMER, M., and INFANTE, A. A. (1965). Messenger RNA in early sea urchin embryos: Size classes. *Science* **150,** 217–221.

PIATIGORSKY, J. (1968). Ribonuclease and trypsin treatment of ribosomes and polyribosomes from sea urchin eggs. *Biochim. Biophys. Acta* **166,** 142–155.

PRESCOTT, D. M., and BENDER, M. A. (1962). Synthesis of RNA and protein during mitosis in mammalian tissue culture cells. *Exptl. Cell Res.* **26,** 260–268.

SALB, J. M., and MARCUS, P. I. (1965). Translational inhibition in mitotic HeLa cells. *Proc. Natl. Acad. Sci. U.S.* **54,** 1353–1358.

SCHARFF, M. D., and ROBBINS, E. (1966). Polyribosome disaggregation during metaphase. *Science* **151,** 992–995.

SCHERRER, K., LATHAM, H., and DARNELL, J. E. (1963). Demonstration of an unstable RNA and of a precursor to ribosomal RNA in HeLa cells. *Proc. Natl. Acad. Sci. U.S.* **49,** 240–248.

SPIRIN, A. S. (1966). On "masked" forms of messenger RNA in early embryogensis and in other differentiating systems. *Current Topics Develop. Biol.* **1,** 1–38.

SPIRIN, A. S., and NEMER, M. (1965). Messenger RNA in early sea urchin embryos: Cytoplasmic particles. *Science* **150,** 214–217.

STAFFORD, D. W., SOFER, W. H., and IVERSON, R. M. (1964). Demonstration of polyribosomes after fertilization of the sea urchin egg. *Proc. Natl. Acad. Sci. U.S.* **52, 313—316.**

STAVY, L., and GROSS, P. R. (1967). The protein synthetic lesion in unfertilized eggs. *Proc. Natl. Acad. Sci. U.S.* **57,** 735–742.

STEWARD, D. L., SHAEFER, J. R., and HUMPHREY, R. M. (1968). Breakdown and assembly of polysomes in synchronized Chinese hamster cells. *Science* **161,** 791–793.

Tyler, A. (1967). Masked messenger RNA and cytoplasmic DNA in relation to protein synthesis and processes of fertilization and determination in embryonic development. *Develop. Biol., Suppl.* **1,** 170–226.

Tyler, A., Tyler, B. S., and Piatigorsky, J. (1968). Protein synthesis by unfertilized eggs of sea urchins. *Biol. Bull.* **134,** 209–219.

Verhey, C. A., Moyer, F. H., and Iverson, R. M. (1965). Differences in the microsome complexes of unfertilized and fertilized sea urchin eggs revealed by chemical dissection followed by microscopy. *Am. Zoologist* **5,** 637.

Warner, J., Knopf, P., and Rich, A. (1963). A multiple ribosomal structure in protein synthesis. *Proc. Natl. Acad. Sci. U.S.* **49,** 122–129.

Synthesis of Nuclear and Mitotic Proteins During Cleavage

Histone Synthesis. Assignment to a Special Class of Polyribosomes in Sea Urchin Embryos

Boaz Moav and Martin Nemer

A class of polyribosomes containing 3–7 ribosomes/mRNA plays a prominent role in the early development of the sea urchin embryo (Infante and Nemer, 1967). These polyribosomes, designated "s-polysomes," increase in concentration from barely perceptible amounts in the early cleaving embryo to over one-third of the total ribosomes in the 10-hr, 200-cell early blastula of *Stronglyocentrotus purpuratus*. The formation of s-polysomes is largely dependent on RNA newly synthesized by the early stage embryo, rather than upon RNA preexisting in the egg (Infante and Nemer, 1967). This new polysomal RNA is predominantly of a 9–10S class (Nemer and Infante, 1965; Kedes and Gross, 1969). The significance of the very extensive synthesis of this narrow class of mRNAs and the accumulation of these polyribosomes in early sea urchin blastulae has been the subject of several recent studies. Nemer and Lindsay (1969) have reported that the incorporation ratio of tryptophan/arginine in nascent protein was substantially less in the s-polysomes than in the rest of the polysomal population. The absence of tryptophan in chromosomal histones (Mirsky and Pollister, 1946; Hnilica, 1967) suggests that the tryptophan-deficient nascent proteins of the s-polysomes may include nascent histones, in accordance with the rationale of Borun *et al.* (1967), bearing on their observations with mammalian tissue culture cells. A similar tryptophan/lysine asymmetry in s-polysomes has been noted recently by Kedes *et al.* (1969), together with the observation through autoradiography that preponderantly nuclear protein is synthesized in the early sea urchin embryo.

In order to assign a specific function to the newly accumulating class of s-polysomes of the early stage embryo, we have attempted to analyze the polysomal protein products directly. Our approach has not been to characterize the nascent proteins *in toto*, whose vastly heterogeneous population might present a morass for analysis, but instead we have submitted this material to purification procedures that might reasonably be applied to completed histones. The result was indeed the selection from this heterogeneous population of a small fraction resembling completed histones. Purification of histone-like protein was conducted by first obtaining the material freed from polyribosomes that was soluble in 0.4 N H_2SO_4. This acid-soluble protein was fractionated by cation-exchange chromatography. A striking result was that approximately 70% of this nascent protein became irreversibly bound to the cation-exchange resin. The rest could be fractionated into nonhistone and histone-like proteins. The histone-like proteins were then submitted to acrylamide gel electrophoresis, as a final step in purification. We conclude that a major portion of the proteins of the s-polysomes that are susceptible to fractionation closely resemble histones. The rest of the polysomal population yields considerably less histone-like protein.

Materials and Methods

(*a*) *Incubation of Embryos and Extraction of Polyribosomes.* Sources of sea urchins, conditions for fertilization and development, and incubation have been documented previously (Nemer and Infante, 1967). Embryos were incubated with 5 μM [^{3}H]L-leucine (750 Ci/mole), 0.26 μM [^{3}H]DL-tryptophan (24 Ci/mmole), 0.7 μM [^{14}C]L-arginine (312 Ci/mole), 3 μM [^{14}C]L-lysine (312 Ci/mole), and 0.016 μM 2,3-[^{3}H]L-alanine (50 Ci/mmole) Schwarz BioResearch, Orangeburg, N. Y. Also, a reconstituted protein hydrolysate (40 Ci/atom of carbon, Schwarz BioResearch), consisting of Ala, Arg, Asp, Glu, Ilu, Leu, Lys, Phe, Pro, Ser, Thr, Tyr, and Val, was incubated at 2 μCi/ml.

The standard medium in which cells were disrupted and cell-free components were examined consisted of 240 mM KCl, 5 mM $MgCl_2$, and 50 mM triethylamine·HCl, at pH 7.8. Cell disruption was effected by a modification of the method of Hinegardner (1962), involving a single passage of embryos through a narrow gauge hypodermic needle (No. 20 or 25), with bentonite added at a concentration of 1 mg/ml. Embryos of *S. purpuratus* were demembranized by treatment with hatching enzyme in sea water from just-hatched blastulae. Embryos of *Lytechinus pictus* could be demembranized at any stage by rapid passage through bolting silk (No. 16 standard). They were washed twice in ice-cold 1 M dextrose and once in standard medium, suspended in medium (1:5 volumes) in a hypodermic syringe, and then passed through a hypodermic needle. Only a single passage could be made without some nuclear lysis also occurring (Fromson and Nemer, 1970). The use of this method insures that the system under study is present in the cytoplasm and not leaked from the nucleus.

The embryo lysate was centrifuged for 10 min at 15,000*g*, and the supernatant fluid (S15) was layered onto 15–30% (w/w) sucrose gradients, prepared in standard medium. Centrifugation was at 50,000 rpm for 27 min in the Spinco rotor SW50 or SW50.1. Gradient fractions were collected after passage through a recording spectrophotometer. Incorporation was assayed after plating fractions on membrane filters (Infante and Nemer, 1968).

(*b*) *Characterization of Nascent Protein by Gel Filtration.* Polysomes were prepared from embryos incubated for 15 min with [^{3}H]L-arginine and [^{3}H]L-lysine or [^{14}C]amino acid mixture. The polyribosomes, collected from sucrose gradients as distinct sedimentation classes, were incubated with previously boiled pancreatic ribonuclease (Worthington Biochemical Corp., Freehold, N. J.) at 0.1 mg/ml for 30 min at 37°. The nuclease digest of polysomes was made 0.5 M with urea, 1% with sodium dodecyl sulfate, and 0.1% with 2-mercaptoethanol and dialyzed against phosphate buffer (0.01 M, pH 7.4) containing 0.1% sodium dodecyl sulfate and 0.1% 2-mercaptoethanol. The dialyzed extracts (1–2 ml) were mixed and applied to a Sephadex G-100 column (100 cm × 2 cm), previously equilibrated with the same phosphate buffer with sodium dodecyl sulfate and 2-mercaptoethanol. Fractions were collected at room temperature and a portion of each (0.2–0.5 ml) was used for radioactive assay in scintillation fluid (Bray, 1960).

(*c*) *Incubation of Polysomes in Vitro.* Ten-hour embryos were incubated for 20 min with tritiated amino acids and their polyribosomes fractionated on sucrose gradients. The s-polysomes and r-polysomes were pooled from the gradients and incubated separately *in vitro* for 1 hr. The incubation mixture at 25° contained the following substituents in a total of 15 ml: s-polysomes (2 mg) with nascent proteins labeled *in vivo* with [^{3}H]arginine, [^{3}H]lysine, [^{3}H]alanine, and [^{3}H]-leucine (or similarly labeled r-polysomes); 100,000*g* supernatant fluid (15 mg of protein); ATP (1 mM); phosphoenolpyruvate (5 mM), GTP (0.06 mM); 2-mercaptoethanol (5 mM), phosphoenolpyruvate kinase (5 μg/ml); a mixture of 20 unlabeled amino acids (1 mM each); KCl (240 mM), $MgCl_2$ (5 mM); and triethanolamine·HCl buffer (50 mM at pH 7.8). After this incubation the polyribosomes were pelleted by ultracentrifugation. The resuspended polyribosomes were incubated for 1 hr at 37° with pancreatic ribonuclease (100 μg/ml), then sonicated 30 sec (Bronson Sonifier, setting 4). The solution was made 0.4 N with H_2SO_4, and after 2 hr at 4°, an acid-insoluble residue was removed by centrifugation at 20,000*g* for 20 min. The acid-soluble extract was dialyzed against 8% guanidine phosphate buffer, pH 6.8, then concentrated by ultrafiltration, and layered on a column for ion-exchange chromatography. Fractions were collected and precipitated with 20% trichloroacetic acid. The precipitates were plated on glass fiber filter disks and assayed for radioactivity by scintillation spectrometry.

(*d*) *Extraction of Nuclear Histones.* Demembranized embryos were washed twice with 1 M dextrose, then diluted with 5 volumes of 2 mM $MgCl_2$ and passed through a No. 20 needle 2–4 times. The cell lysate was layered on a discontinuous sucrose gradient, consisting of 6 ml of 2.0 M; 6 ml of 1.75 M, 4 ml of 1.5 M, and 4 ml of 1.25 M sucrose in standard medium. Gradients were centrifuged for 45 min in the Spinco SW25.1 rotor at 20,000 rpm. Nuclei were pelleted under these conditions and were then washed twice with diluted (1:5) standard buffer. Purified nuclei were extracted with 0.25 N H_2SO_4. In several preparations 0.05 M $NaHSO_3$ was present both in the wash buffer and in the sulfuric acid, which could also be used at 0.4 N. Extraction was complete after 2 hr of constant stirring at 4°. A second extraction yielded only 5% more material. The extract was centrifuged at 42,000 rpm in the Spinco rotor 50 for 17 hr, to remove residual nucleoprotein and nucleic acid. The supernatant fluid was dialyzed against 8% guanidine phosphate, pH 6.8, in preparation for chromatography. In some cases the acid extract was dialyzed against 5% acetic acid, lyophilized, and kept at −70°. Comparative studies were done with preparations of calf thymus total histones from Sigma Chemical Company, St. Louis, Mo.

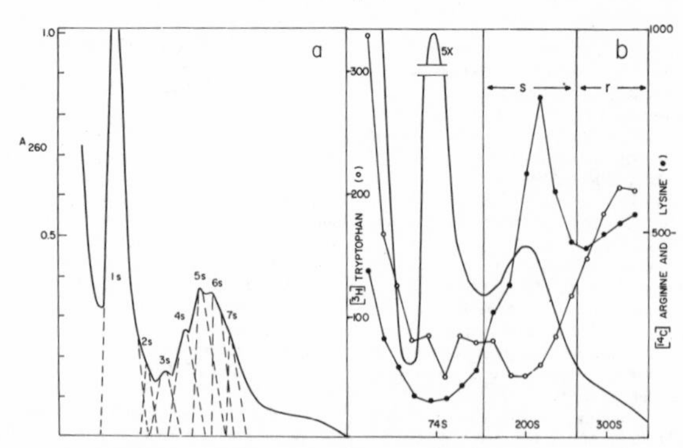

FIGURE 1: Polysomal classes in early sea urchin blastulae. (a) Eight-hour embryos of *S. purpuratus* were used without labeling. (b) Embryos of the same stage were incubated for 10 min (Methods) with [^{3}H]DL-tryptophan, [^{14}C]L-lysine, and [^{14}C]L-arginine. In both cases cells were disrupted by a single passage through a No. 20 gauge hypodermic needle and S15 supernatant was prepared. The S15 preparation was centrifuged through a 15–30% (w/w) sucrose gradient in the Spinco rotor SW50.1 at 50,000 rpm for 27 min.

(*e*) *Ion-Exchange Chromatography and Acrylamide Gel Electrophoresis.* The cation-exchange resin, Amberlite CG-50 (200–400 mesh, Mallinckrodt, New York, N. Y.) was regenerated by treatment successviely with 2 N HCl, water, 2 N NaOH, water, 2 N HCl, water, and then 2 N NaCl, before finally equilibrating it with 8% (w/w) guanidine chloride (Mann Research Laboratories, New York, N. Y.), containing 0.1 M Na_2HPO_4, pH 6.8. A 17 cm × 0.78 cm^2 column was packed with the resin by air pressure and washed with the 8% guanidine phosphate buffer. The sample protein solution was applied to the column and developed with 6 ml of 8% guanidine phosphate buffer, followed by a linear gradient of 8–13% (21 ml of each). The strongly bound material (histone fractions III and IV) was eluted by addition of 12 ml of 40% guanidine phosphate buffer. This was followed by 12 ml of 8% buffer solution. Radioactivity of fractions was assayed either after precipitation with 20% trichloroacetic acid in the presence of 200 μg of albumin carrier, followed by plating on glass fiber filter disks (grade 934AH, Reeve Angel, Clifton, N. J.) which were dried and counted in toluene scintillation fluid, or the fractions from the column were added directly to Bray's solution that was acidified by the addition of 0.1 ml of 2 N H_2SO_4 to 15 ml, and counted by scintillation spectrometry.

Acrylamide gel electrophoresis (Panyim and Chalkley, 1969) was preformed on preparations that had been lyophilized, then dissolved in 0.9 N acetic acid.

Results

(*a*) *Polysomal Classes and Their Associated Nascent Proteins.* In tracing the changes in the polysomal population during early sea urchin development, Infante and Nemer (1967) noted that the slowly sedimenting s-polysomes reached a maximum concentration at the 200-cell, 10-hr blastula stage. Sedimentation diagrams of good resolution, including Figure 1a, reveal that the approximate proportions of the component polyribosomes of this class do not change substantially during development from 7 to 10 hr. The weighted average size of this class holds constant at 4.5 ribosomes/polyribosome during this period, while the concentration of this class increases as a whole. The average size of protein produced by such a size class of polyribosomes might be expected to be 17,000, from the assumption that a ribosome occupies template equivalent to 30 amino acids (Warner *et al.*, 1963; Becker and Rich, 1966; Williamson and Askonas, 1967). Chromosomal histones might be expected to be synthesized on such a size class of polyribosomes, since the molecular weights of histones from several species range between 11,000 and 25,000 (Haydon and Peacocke, 1968; Butler *et al.*, 1968; Fambrough and Bonner, 1968; DeLange *et al.*, 1968).

Various lines of evidence (Robbins and Borun, 1967; Borun *et al.*, 1967; D. T. Lindsay, submitted for publication) have led us to postulate that the s-polysomes synthesize a special class of protein, namely the chromosomal histones. Nemer and Lindsay (1969) have reported that relatively less tryptophan compared to arginine was incorporated into the s-polysomes than into the more rapidly sedimenting "r-polysomes," and have suggested that the trytophan-deficient nascent protein population of the s-polysomes might include histones. The same asymmetric distribution was noted by Kedes *et al.* (1969) with lysine and tryptophan. The results with arginine and lysine, in combination (Figure 1b), are not readily distinguishable from the results obtained with these amino acids incubated separately. Such information as delineated in Figure 1a,b indicates only the possibility that histones may be synthesized by s-polysomes. A more direct analysis of the nascent proteins associated with these polyribosomes will be needed in order to identify the proteins they are synthesizing.

Nascent proteins derived from s-polysomes of 10-hr embryos were resolved into size classes by passage through Sephadex G-100 in the presence of 0.1% sodium dodecyl sulfate and 0.1% mercaptoethanol. Proteins of known molecular weights were included and used as markers (Figure 2). As shown, the elution position of the markers were linearly related to the logarithms of their molecular weights, so that

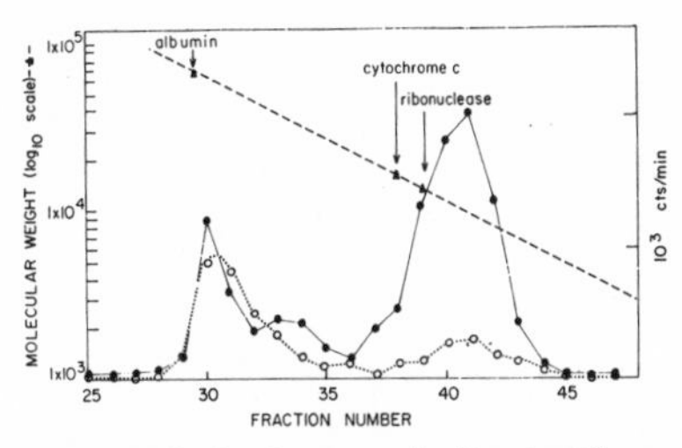

FIGURE 2: Gel filtration of nascent proteins. Nascent proteins were extracted from the s-polysomes of 10-hr early blastulae, labeled with [³H]arginine and [³H]lysine and from the r-polysomes of the same stage labeled with a [¹⁴C]amino acid mixture. The preparations were mixed and passed through a column of Sephadex G-100, as described in Methods. Nascent protein of s-polysomes (●) and r-polysomes (○).

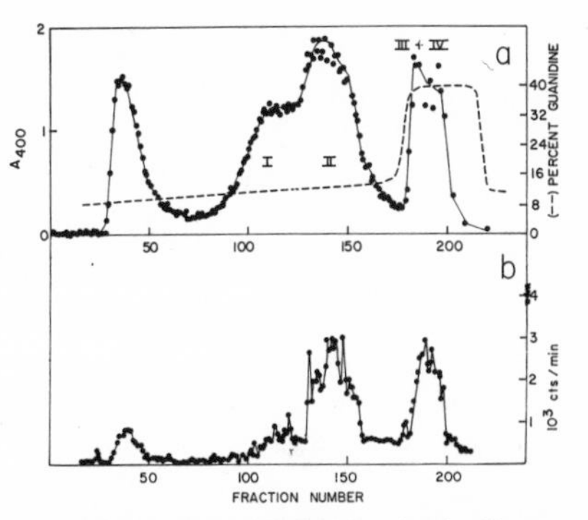

FIGURE 3: Cation-exchange chromatography of nuclear histones. Histone preparations were applied to columns of Amberlite CG-50 and eluted differentially with guanidine chloride in phosphate buffer. The per cent guanidine was monitored by measuring the refractive indices of fractions. (a) Commercial preparation of calf thymus histones. Aliquots of fractions were adjusted to 1.1 M trichloroacetic acid and the quantity of protein indicated by readings of turbidity at 400 mμ. Numerals I, II, III, and IV refer to the approximate locations of the histone classes according to Rasmussen *et al.* (1962). (b) Ten-hour embryos of *S. purpuratus* were incubated with [³H]L-arginine and [³H]L-lysine for 1 hr, then submitted to cellular fractionation and extraction of nuclear histones. Aliquots were taken for assay of radioactivity (Methods).

estimation of the molecular weights of the nascent proteins could be made. The major portion of r-polysomal nascent protein resided in a component of 57,500 average molecular weight. The major portion of the s-polysomal nascent protein (70%) was distributed in a peak with an average molecular weight of 9350. The remaining 30% of radioactivity appeared in material of higher molecular weight, which could have arisen from trailing of the r-polysomes into the segment taken for s-polysome analysis or from the formation of aggregates. The value for the average molecular weight for the nascent protein agrees favorably with the average nascent protein molecular weight (8500) derived from the average ribosome number in the s-polysomes given in Figure 1a.

The sizes of the polypeptides synthesized by the two polysomal classes differ by a factor of 6.15. If the rates of synthesis of these polypeptides were equivalent and their amino acid compositions similar, the incorporations of amino acid in nascent polypeptide per ribosome (counts/min per A_{260}) would reflect these size differences. Using the data of Figure 1b, we may test this relationship for tryptophan and for the combined arginine and lysine incorporation. The specific activity for tryptophan in r-polysomes is 8.6 times that in s-polysomes. This ratio is 1.4 times the ratio of nascent protein sizes, and may be indicative of a relatively greater content of tryptophan in r-polysomal nascent protein. The specific activity for arginine and lysine in r-polysomes is 3.2 times that in s-polysomes. This ratio, being less than the ratio of nascent protein sizes, indicates that the combined arginine and lysine content in s-polysomal nascent protein is 1.9 times that of the r-polysomes. We do not know whether in either case the deviation from the ratio of nascent protein sizes is due, at least partly, to a difference in translation rate between the two polysomal classes. Irrespective of such a difference in rates, the relative amino acid compositions of s- and r-polysomal nascent proteins can be quantitated on a per residue of tryptophan basis. Thus the ratio of the amounts of arginine and lysine per tryptophan in s-polysomes compared to r-polysomes is obtained from $(Arg,Lys)_s/(Trp)_s/(Arg,Lys)_r/(Trp)_r$, where each term refers to the specific incorporation in the respective polysomal classes, and is equivalent to the product $1.9 \times 1.4 = 2.6$. This value indicates the extent to which the s-polysomes are enriched by a class of proteins of relatively high arginine and lysine and low tryptophan, as compared to the r-polysomes. The extent of this asymmetry will be shown later to be a function of developmental stage.

(*b*) *Extraction of Histone-Like Proteins from Nascent Proteins Associated with the s-Polysomes.* The broad array of polypeptides among the nascent proteins associated with polyribosomes may not be characterized in any precise manner. One has two approaches toward an examination of the function of the polyribosomes in question. (i) Employ a purification applicable to known proteins which these polyribosomes supposedly synthesize; (ii) allow completion of proteins in reconstituted *in vitro* systems, preferably under conditions which afford release of finished proteins. We have pursued both approaches with informative results.

The purification of histone-like proteins from nascent proteins was performed in a manner similar to the extraction and characterization of chromosomal histones from purified sea urchin nuclei. Nuclei of early blastulae were acid extracted [Methods (d)], and the acid-soluble proteins were characterized by successive chromatography and electrophoresis. Ion-exchange chromatography was used to separate the nuclear histones into the major classes, similar to those reported for calf thymus histones (Rasmussen *et al.*, 1962). A commercial preparation of calf thymus histones has been chromatographed in Figure 3a. In Figure 3b, labeled chromosomal histones, derived from 10-hr early blastuale of *S. purpuratus*, were chromatographed under the same conditions. The histones from the sea urchin embryos display the same general classes as that of the calf thymus histones, but apparently in different proportions. In this early blastula class I, corresponding to lysine-rich histones, is comparatively under-

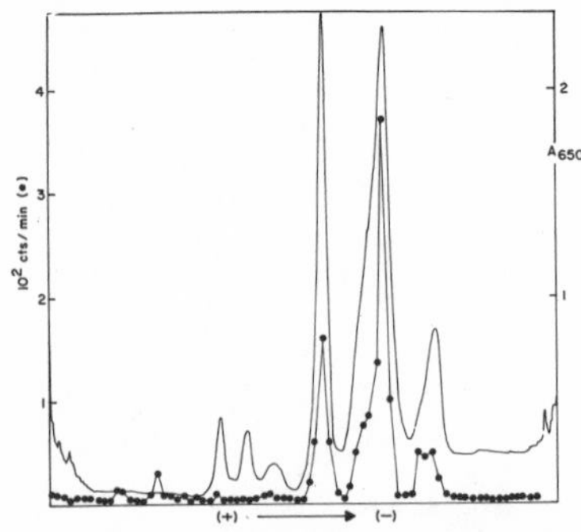

FIGURE 4: Acrylamide gel electrophoresis of nuclear histones of sea urchin embryos. A nuclear histone preparation from 7-hr embryos of *L. pictus* together with a trace amount of a similar preparation from 10-hr embryos of *S. purpuratus*, labeled as in Figure 3, was applied to a 15% gel (4.5 × 85 mm) and run for 2.5 hr at 2 mA/gel. Protein bands were stained with Amido Black. All procedures were according to Panyim and Chalkley (1969). Bands of stained protein were detected by scanning of gels with a recording spectrophotometer at 650 mμ (—).

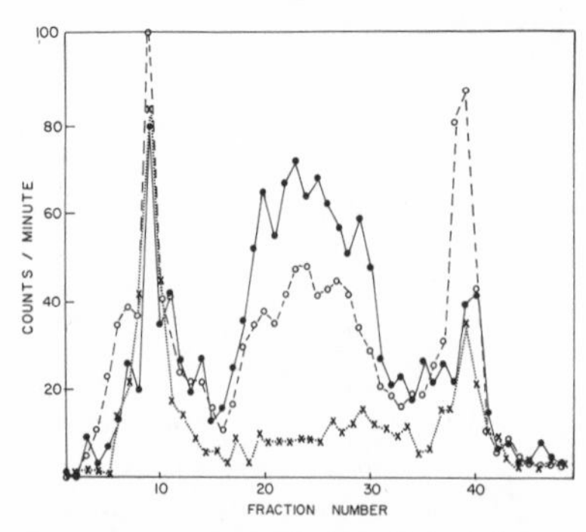

FIGURE 5: Cation-exchange chromatography of acid-soluble nascent proteins from s-polysomes and r-polysomes. Ten-hour embryos of *S. purpuratus* were incubated for 20 min either with a [^{14}C]amino acid mixture or a mixture of tritiated amino acids [as in Methods (c)]. The s-polysomes, labeled with ^{14}C, were isolated on sucrose gradients and their nascent proteins extracted by RNase treatment and subsequent acidification. The s- and r-polysomes, labeled with tritium, were separated and each incubated *in vitro* [Methods (c)]. After pelleting the polyribosomes, the labeled proteins were extracted as above. The ^{14}C-labeled nascent proteins of the s-polysomes (○) were mixed with the tritiated proteins of the s-polysomes, that had been subjected to *in vitro* incubation (●), and these preparations were then cochromatographed through a column of CG-50 resin. In a parallel column the tritiated proteins of the r-polysomes that had been similarly incubated *in vitro* were chromatographed through CG-50 resin (×).

represented. This observation is in agreement with that of Thaler *et al.* (1970). A subsequent characterization (Figure 4) was by means of high-resolution acrylamide gel electrophoresis (Panyim and Chalkley, 1969). Unlabeled nuclear histones of *L. pictus* blastulae were submitted to gel electrophoresis together with the labeled histones of blastulae of *S. purpuratus*, analyzed in Figure 3b. Six major bands were revealed by the dye binding assay. An exact coincidence of labeled histones with unlabeled bands was evident, except that little label was associated with the less-pronounced bands, which correspond to lysine-rich histones. It appears then that the histones of the blastulae of these two species are not readily distinguishable.

Nascent proteins were extracted from polyribosomes [Methods (b)] of early blastulae labeled *in vivo* with amino acids. The fraction of this material that was soluble in 0.40 N sulfuric acid (approximately 20%) was submitted to cation-exchange chromatography (Figure 5). Whereas all of the acid-soluble chromosomal protein so treated (Figure 3) was recovered, approximately 70% of the acid-soluble nascent proteins became irreversibly bound to the cation-exchange resin, defying elution even with 60% guanidine and with 1 N sulfuric acid. In other experiments, nascent proteins obtained from total polyribosomes of several embryonic stages (5, 10, 15, and 20 hr) were examined in the same way, and found to be irreversibly bound to the resin in the same high proportion. We cannot offer an explanation for this effect, but may suggest that it is the result of both the charge and the peculiarities of nascent polypeptide configuration. The phenomenon allows, at least, the isolation of those proteins that resemble histones from the large and diverse population of nascent proteins. It is likely that only those nascent histones that are comparatively near completion exhibit histone-like behavior under these circumstances. The proteins derived from the s-polysomes, that can be eluted and thus fractionated, fall into two categories (Figure 5). The class which does not adhere to the resin can be designated as nonhistone. It comprises approximately 30% of the eluted material. The next proteins are eluted at the positions of the histone classes (see Figure 3). These histone-like nascent proteins constitute 70% of the total eluted material. Ribosomal proteins would have been eluted with the initial, nonhistone-like proteins (Bonner *et al.*, 1968). However, the synthesis of ribosomal protein in this early stage embryo would not be expected, since ribosomal RNA synthesis is not detectable (Nemer, 1963), and since the two syntheses are apparently linked (Hallberg and Brown, 1969).

In vivo labeled polyribosomes were incubated *in vitro* with unlabeled amino acids, 100,000*g* supernatant fluid, and energy sources, in an attempt to promote completion of nascent polypeptides. At the end of the incubation the polyribosomes were pelleted by centrifugation, and the soluble and sedimented proteins were analyzed separately. There was negligible release of labeled protein during this incubation. The labeled proteins associated with the s-polysomes were analyzed by ion-exchange chromatography, just as the unincubated nascent proteins had been, and the results were similar, except for slight quantitative shifts (Figure 5). Upon incubation the proportion of nonhistone protein decreased slightly, and the relative proportions of histones shifted with a decrease in arginine-rich classes and an increase in lysine-rich classes. Although these shifts proceeded in the direction of more nearly completed histones, a large proportion of acid-soluble nascent proteins remained again irreversibly bound to the resin. Thus the apparent shift is in the

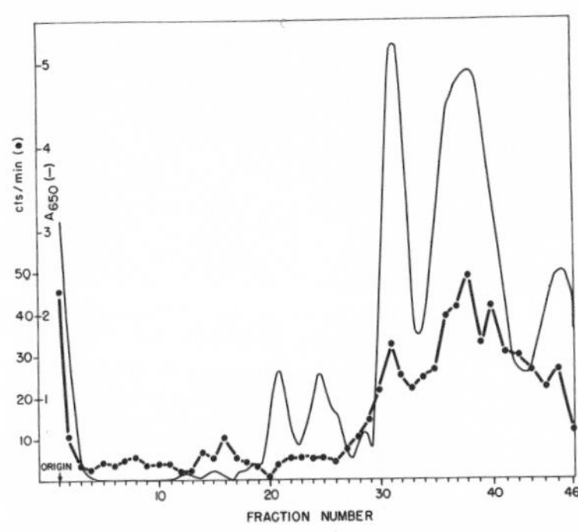

FIGURE 6: Acrylamide gel electrophoresis of nascent proteins of s-polysomes allowed to proceed toward completion of polypeptide chains *in vitro*. The s-polysomes were incubated *in vitro*, extracted with acid, and chromatographed as in Figure 4. The histone-like protein (fractions 18–45 of Figure 4) was dialyzed against 0.9 N acetic acid, lyophilized, and dissolved in 25 μl of 15% sucrose in 0.9 N acetic acid. This sample was mixed with 25 μl of the same unlabeled nuclear histones from *S. purpuratus* and electrophoresed as in Figure 5.

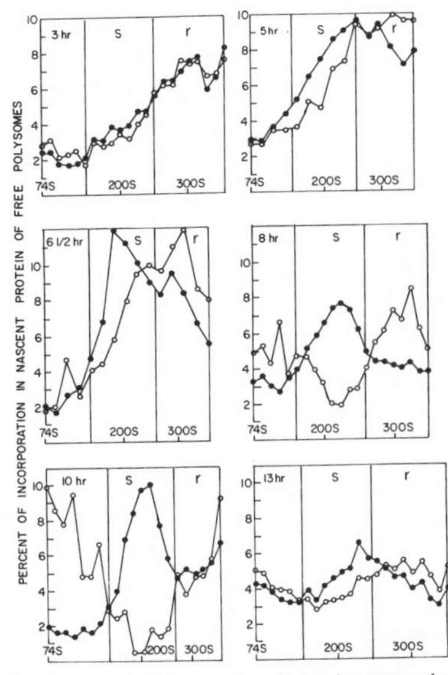

FIGURE 7: Differential incorporation of tryptophan compared to arginine and lysine in nascent proteins of polysomes at different embryonic stages. Embryos of *S. purpuratus* at the indicated stages of development were incubated as in Figure 1b. Incorporation in gradient fractions of sucrose gradients are indicated as the per cent of the total incorporation in the free polysomes, *i.e.*, material sedimenting more rapidly than the 74S monoribosomes: (○) [^{3}H]DL-tryptophan, (●) [^{14}C]L-arginine and lysine. The s-polysomal and r-polysomal regions are demarcated as indicated in Figure 1b.

nature of the isolated histone-like material. A comparison between the s-polysomes and r-polysomes was undertaken by incubating *in vivo* labeled r-polysomes under the same *in vitro* conditions applied to the labeled s-polysomes. The labeled protein associated with pelleted r-polysomes (again release was not detected) was analyzed in parallel with similar material from s-polysomes (Figure 5). There was a striking difference: considerably more chromatographed protein was nonhistone and very little corresponded to histone classes I and II. We may conclude that the nascent proteins of the s-polysomes have a much higher representation of histone-like proteins than those of the r-polysomes.

The size distribution of acid-soluble s-polysomal nascent protein was compared to that which was subsequently submitted to cation-exchange chromatography and eluted as histone-like protein. Gel filtration of these preparations was performed with Sephadex G-50, equilibrated with 8% guanidine phosphate buffer. The histone-like proteins displayed an appreciably higher average size (by an increment of 4000) than the unfractionated material (B. Moav and M. Nemer, unpublished). We may conclude that a selection of larger, presumably more complete, proteins occurs during fractionation of these proteins on the cation exchanger. The larger proportion of proteins that become irreversibly stuck to the resin probably represent the major portion of small, incomplete proteins.

The histone-like proteins eluted from the cation-exchange resin were concentrated and then submitted to acrylamide gel electrophoresis (Figure 6). The labeled nascent proteins were run together with unlabeled chromosomal histones from the same embryonic stage and found to coincide with the major chromosomal histone bands. The resolution of lysine-rich classes (the components in fractions 20–30) is superior to that obtained by chromatography. The ersult indicates that little radioactivity is associated with these histone classes, which are present in the least abundance (Thaler *et al.*, 1970). The chromatography and electrophoresis represent two independent characterizations. By either test the purified s-polysomal proteins under question were not distinguishable from chromosomal histones. This evidence greatly supports the proposal that the s-polysomes synthesize histones. However, more detailed resemblances would have to be shown before a firm conclusion could be drawn.

(*c*) *Developmental Changes.* The amino acid labeling pattern of polyribosomes, exemplified by the early blastula of Figure 1b, changes markedly during the course of embryonic development. In the early cleaving embryo the distribution of arginine and lysine and tryptophan incorporation appear to be uniform in the s- and r-polysomes (demarcated by vertical lines in Figure 7). However, a period of development (morula and early blastula) is reached during which the incorporation of arginine and lysine in s-polysomes, $(Arg,Lys)_s$, is proportionately high, and that of tryptophan, $(Trp)_s$, is relatively low, particularly when these are compared to their counterparts in the r-polysomes, $(Arg,Lys)_r$ and $(Trp)_r$. The ratio $[(Arg,Lys)_s/(Trp)_s]/[(Arg,Lys)_r/(Trp)_r]$ changes from 1.5 at 5 hr to 2.6 at 8 hr and 4.0 at 10 hr. The significance of the developmental increase in this value resides in the probability that the s-polysomes contain increasing amounts of nascent protein enriched in arginine and lysine and deficient in tryptophan. Such a situation would be consistent with an increased representation of nascent histones.

If we assume that there does indeed exist a class of nascent proteins in the s-polysomes lacking tryptophan, and by this criterion "histone-like," we may estimate the relative output of these histone-like proteins for each developmental stage. We may subtract from $(Arg,Lys)_s$ that portion of the arginine and lysine incorporation in the s-polysomes which is present in the same proportion to $(Trp)_s$ as arginine and lysine is to tryptophan in the r-polysomes. Thus $(Trp)_s$ $(Arg,Lys)_r/(Trp)_r$ is equivalent to arginine and lysine incorporation in those s-polysomal proteins that resemble the proteins of the r-polysomes in their arginine, lysine, and tryptophan composition. The difference resulting from this subtraction would yield the incorporation due to tryptophan-deficient, histone-like, s-polysomal proteins. The relative fractional output of these proteins would be represented by $[(Arg,Lys)_s - (Trp)_s\,(Arg,Lys)_r/(Trp)_r]/[(Arg,Lys)_r + (Arg,Lys)_s]$. This relative output by the polysomes has been plotted during the course of development up to the 20-hr late blastula (Figure 8). A striking feature of these developmental changes is that a maximum is reached at the 10-hr, 200-cell blastula stage. The value at this point is ten times that for the earlier, 3-hr cleaving embryo or the later, 16-hr swimming blastula. If we assume that the combined arginine and lysine content of the histone-like proteins is twice that of the rest of the protein population, then the values for relative output in Figure 8 may be divided by 2, to obtain values approximating the actual fraction of the total polysomal production representing histone synthesis. A positive correlation is afforded here by the inclusion in Figure 8 of the rate of cell formation per embryo, which may be regarded as indicative of the rate of DNA synthesis. In the 3-hr, 4-cell embryo the rate of cell formation is 4 cells/hr. This rate reaches 55 cells/hr at the 200-cell blastula stage, but declines thereafter. The changes in amino acid labeling pattern and in rate of cell formation occur in approximate parallel. A similar correlation between histone synthesis and DNA synthesis would be expected (Bonner, 1965).

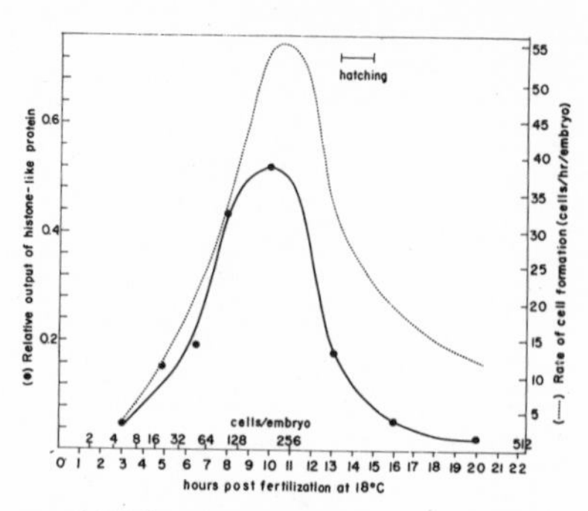

FIGURE 8: Relative output of histone-like protein by free polysomes during embryonic development. The data of Figure 7 are used together with similar sedimentation diagrams at other embryonic stages to estimate by the method described in the text the relative output of tryptophan-less proteins by the s-polysomes. The rates of cell formation were calculated from the data of Hinegardner (1967).

Discussion

The principal objective of this study has been to assign a function to the predominant class of polyribosomes in the early sea urchin blastula. These "s-polysomes" in the 200-cell blastula constitute over 70% of the polysomal ribosomes (Figure 1) and over 90% of the mRNA molecules, as estimated from the ribosomal content and the relative cistron sizes indicated by Figure 2. The newly synthesized RNA associated with these polyribosomes displays discrete size classes, the most prominant of which is 9–10 S (Nemer and Infante, 1965), recently confirmed by Kedes and Gross (1969). Another, intriguing property of this class of mRNA is its rapid annealing with DNA: Nemer and Infante (1965) observed greater than 40% hybridization of purified 9–10S RNA with 112 μg of DNA/ml in 24 hr. Thus the hybridized RNA transcripts could be estimated (Britten and Kohne, 1968) to correspond to highly "redundant" DNA. It is possible that the mRNAs of the s-polysomes are derived from genes present in multiple copies. The significance of the present study may therefore be viewed from several standpoints. The proteins synthesized by the s-polysomes are a sizeable portion of the total protein output during a brief and circumscribed period of early embryonic development. Their synthesis appears to be promoted almost entirely by newly synthesized mRNA (Infante and Nemer, 1967), as opposed to preexisting egg mRNA. And, a repository of multiple genes for their templates may exist, to serve, perhaps, some important developmental and evolutionary functions.

Our approach to the s-polysomal proteins involved separate qualitative and quantitative evaluations. Qualitatively, we analyzed the nascent proteins associated with these polyribosomes. An apparent limitation of this analysis is that the vast portion of the nascent polypeptides are not susceptible to meaningful characterization and cannot be expected to resemble the proteins to which they give rise. We have thus limited ourselves to the task of purifying from this large array of protein fragments those proteins resembling a narrow group of proteins, suspected of being synthesized by these polyribosomes. We were lead to suspect this group of proteins to be the chromosomal histones, because the relatively high arginine and low tryptophan incorporations in the nascent proteins associated with these polyribosomes tended to mimic the peculiar amino acid composition of histones (Nemer and Lindsay, 1969).

The purification of histone-like proteins involved three simple steps: acid extraction of labeled proteins removed from polyribosomes, cation-exchange chromatography, and, finally, acrylamide gel electrophoresis. Whereas preparations of chromosomal histones were recovered completely after passage through the cation-exchange resin, a considerable portion of the nascent polypeptides were irreversibly bound to the resin. The s-polysomes furnished a high proportion of the eluted proteins that were indistinguishable from histones, both according to their characteristic elution pattern from the resin and a coincidence with histone bands in acrylamide gel electrophoretograms. The recovered proteins can be presumed from our results to be complete or nearly completed. We have thus defined the function of the polyribosomes in question by virtue of the properties of the most nearly completed members of their nascent protein population. This rationale may be implicit in the several studies directed at the assignment of specific enzyme synthesis to given classes of polyribosomes (Kiho and Rich, 1964, 1965; Bagdasarian *et al.* 1970). It is likely that only the completed or most nearly completed enzyme molecules are detected. Our conclusions are that the s-polysomes are responsible for the synthesis of histone-like protein and that such synthesis is restricted to these polyribosomes, since histone-like proteins are not significantly demonstrable in larger polyribosomes.

Developmental changes were examined on the basis of shifts in the incorporation in nascent protein of tryptophan relative to arginine and lysine. The output of histone-like protein (lacking tryptophan) by the s-polysomes changed in a manner similar to the changes both in the relative concentration of s-polysomes (Infante and Nemer, 1967) and in the rate of DNA synthesis, indicated by Figure 8. For all three a maximal value was reached in the 200-cell early blastula. The actual fraction of the total polysomal production representing histone synthesis at this embryonic stage could be calculated on the basis of assumptions that all of the tryptophan-less proteins are histones and as such contain twice as much arginine and lysine as the rest of the protein population. The value for this output might be estimated at 26% of the total. An absolute rate of protein synthesis of 4.7 to 8.7 pg per embryo per min at this embryonic stage has been measured by Fry and Gross (1970). According to this measurement the ouput of histone would be 1.2–2.3 pg/embryo per min. The formation of 50 diploid nuclei per hr in the 10-hr blastula (Figure 8) entails an average synthesis of 90 pg of DNA/hr or 1.5 pg/min per embryo. It appears then that equivalent weights of DNA and histone may be synthesized, in agreement with conclusions for other organisms (Marushige and Ozaki, 1967; Chalkley and Jensen, 1968). A correlation between the activity of the s-polysomes and DNA synthesis was shown previously through an inhibition of DNA synthesis (Boron *et al.*, 1967; Kedes and Gross, 1969). The present study indicates a quantitative coincidence between DNA synthesis and the production of histone-like proteins by the s-polysomes. We can conclude from several lines of evidence that these polyribosomes are responsible for the synthesis of chromosomal histones.

Acknowledgments

The authors wish to acknowledge the excellent technical assistance of Mrs. Doreen McMurry, Mrs. Eva Skrenta, and Miss Catherine Foley. We wish to thank Mr. Seth Finkelstein for performing the analyses involving gel electrophoresis.

References

Bagdasarian, M., Ciesla, Z., and Sendecki, W. (1970), *J. Mol. Biol. 48*, 53.

Becker, M. J., and Rich, A. (1966), *Nature* (*London*) *212*, 142.

Bonner, J., Chalkley, G. R., Dahmus, M., Fambrough, D., Fujimura, F., Huang, R. C., Huberman, J., Jensen, R., Marushige, K., Ohlenbusch, H., Olivera, B., and Widholm, J. (1968), *Methods Enzymol. 12*, 3.

Bonner, J. F. (1965), The Molecular Biology of Development, New York, N. Y., Oxford University Press.

Borun, T. W., Scharff, M. D., and Robbins, E. (1967), *Proc. Nat. Acad. Sci. U. S. 58*, 681.

Bray, G. (1960), *Anal. Biochem. 1*, 279.

Britten, R. J., and Kohne, D. E. (1968), *Science 161*, 529.

Butler, J. A. V., Johns, E. W., and Phillips, D. M. P. (1968), *Progr. Biophys. Mol. Biol. 18*, 211.

Chalkley, R., and Jensen, R. H. (1968), *Biochemistry 7*, 4380.

DeLange, R. J., Smith, E. L., Fambrough, D. M., and Bonner, J. (1968), *Proc. Nat. Acad. Sci. U. S. 61*, 1145.

Fambrough, D. M., and Bonner, J. (1968), *J. Biol. Chem. 243*, 4434.

Fromson, D., and Nemer, M. (1970), *Science 168*, 266.

Fry, B. J., and Gross, P. R. (1970), *Develop. Biol. 21*, 125.

Hallberg, R. L., and Brown, D. D. (1969), *J. Mol. Biol. 46*, 393.

Haydon, A. J., and Peacocke, A. R. (1968), *Biochem. J. 110*, 243.

Hinegardner, R. T. (1962), *J. Cell Biol. 15*, 503.

Hinegardner, R. T. (1967), *in* Methods in Developmental Biology, Wilt, F. H., and Wessells, N. K., New York, N. Y., Thomas Crowell Co., p 139.

Hnilica, L. A. (1967), *Progr. Nucl. Acid Res. Mol. Biol. 7*, 25.

Infante, A. A., and Nemer, M. (1967), *Proc. Nat. Acad. Sci. U. S. 58*, 681.

Infante, A. A., and Nemer, M. (1968), *J. Mol. Biol. 32*, 559.

Kedes, L. H., and Gross, P. R. (1969), *Nature* (*London*) *223*, 1335.

Kedes, L. H., Gross, P. R. Cognetti, G., and Hunter, A. L. (1969), *J. Mol. Biol. 45*, *337*.

Kiho, Y., and Rich, A. (1964), *Proc. Nat. Acad. Sci. U. S. 51*, 111.

Kiko, Y., and Rich, A. (1965), *Proc. Nat. Acad. Sci. U. S. 54*, 1751.

Marushige, K., and Ozaki, H. (1967), *Develop. Biol. 16*, 474.

Mirsky, A. E., and Pollister, A. W. (1946), *J. Gen. Physiol. 30*, 117.

Nemer, M. (1963), *Proc. Nat. Acad. U. S. 50*, 230.

Nemer, M., and Infante, A. A. (1965), *Science 150*, 217.

Nemer, M., and Infante, A. A. (1967), *J. Mol. Biol. 27*, 73.

Nemer, M., and Lindsay, D. T. (1969), *Biochem. Biophys. Res. Commun. 35*, 156.

Panyim, S., and Chalkley, R. (1969), *Arch. Biochem. 130*, 337.

Rasmussen, P. S., Murray, K., and Luck, J. M. (1962), *Biochemistry 1*, 74.

Robbins, E., and Borun, T. W. (1967), *Proc. Nat. Acad. Sci. U. S. 57*, 409.

Smith, E. L., DeLange, R., J. and Bonner, J. (1970), *Physiol. Rev. 50*, 159.

Thaler, M. M., Cox, M. C. L., and Villee, C. A. (1970), *J. Biol. Chem. 245*, 1479.

Warner, J. B., Knopf, P. M., and Rich, A. (1963), *Proc. Nat. Acad. Sci. U. S. 49*, 122.

Williamson, A. R., and Askonas, B. A. (1967), *J. Mol. Biol. 23*, 201.

Synthesis of Nuclear and Chromosomal Proteins on Light Polyribosomes during Cleavage in the Sea Urchin Embryo

Laurence H. Kedes, Paul R. Gross,

Goffredo Cognetti and Anne L. Hunter

1. Introduction

The early development of animal embryos is characterized by rapid cell division, with little or no change in embryonic mass. The sea urchin zygote, for example, begins to cleave soon after fertilization and undergoes nine to ten division cycles within a few hours†. Dry mass change is negligible. The resulting multicellularity is essential for succeeding morphogenetic movements, which lead to the production of a motile, feeding and eventually growing larva. Nuclear, and not cytoplasmic, replication is therefore a major process of development during the period of cleavage.

Exponential synthesis of DNA is a necessary part of nuclear replication (Hinegardner, Rao & Feldman, 1964). During the cleavage stage of development of *Arbacia punctulata*, the cell cycle (28 to 32 min at 23°C) is characterized by almost continuous DNA synthesis (S) followed by mitosis (M). Gap phases (G_1 and G_2), characteristic of growing cells, are brief or missing entirely. Cultures of sea urchin embryos are synchronous in respect to S and M for the first three cycles. Later divisions become asynchronous as the rates of fission for micromeres and macromeres diverge. During any short period of time (20 to 30 min), however, virtually all the cells can be found in some part of S. Such cultures, then, in addition to being developmentally synchronous, are effectively synchronized in S phase.

Concomitant synthesis of nuclear proteins would appear necessary to keep pace with nuclear numbers and with the growing nuclear mass. Experiments described in this paper take advantage of the cleavage situation in order to examine in detail the biosynthesis of nuclear proteins, especially those whose production is coupled to DNA replication.

Robbins & Borun (1967) were the first to demonstrate synthesis of histones on small polyribosomes of cells in culture and the clear coupling between ongoing DNA synthesis and the continued synthesis of histones. We take advantage of a number of their experimental designs in the work to be described here. The evidence suggests that nuclear proteins are synthesized in the cytoplasm and accumulate in the nucleus after a brief delay. Lysine and tryptophan content of the nascent proteins at the site of synthesis and their later attachment to chromatin suggest that some of these proteins are histones. At the 64- to 128-cell (morula) stage about half the protein synthetic activity of the embryos is devoted to production of molecules that find their way into nuclear structures. The templates for these proteins are the subject of a separate report (Kedes & Gross, 1969*b*).

2. Materials and Methods

(a) *Fertilization and culturing of embryos*

Adult sea urchins of the species *Arbacia punctulata* were from the Florida Gulf Coast. Collection of gametes, fertilization and incubation were performed as described previously (Kedes & Gross, 1969*a*).

(b) *Pulse and chase labeling with radioactive amino acids*

When development was to take place in the presence of actinomycin D, the eggs were first washed for 30 min in sea water containing the drug at a concentration of 20 to 25 μg/ml. and then fertilized. Embyros were concentrated by centrifugation for exposure to labeled amino acids at an appropriate stage of development. After an appropriate period of exposure to the labeled amino acid(s), the embryos were collected and washed by centrifugation, after which they were either processed or resuspended in sea water

† The rates of cleavage of the sea urchin species used in these studies (*Arbacia punctulata*) have been well documented by E. B. Harvey (1956). At 23°C the first cleavage occurs 50 min after fertilization and subsequent cell divisions at fairly regular intervals of 22 to 32 min thereafter. The blastula stage is reached by about 7 hr and the embryo has about 1000 cells, i.e. there were 2^{10} cleavages and 10 division cycles. The intermitotic interval increases drastically from about this point with only a doubling in cell number over the next 24 hr of development.

containing a large excess of the appropriate unlabeled amino acid(s). Details of timing and labeling conditions are included in the legends for the Plates, Figures and Table. At the end of a chase, the embryos were concentrated by centrifugation for final processing.

(c) *Electron microscope autoradiography*

Fixation for electron microscopy was in 3% glutaraldehyde prepared in 0·1 M-phosphate buffer, pH 7·2. The fixative also contained 4% dextrose or NaCl (2%) plus $MgCl_2$ (0·5%). The percentage concentrations given are all w/v. Fixation was for 2 hr (Plate I) or 48 hr (Plates II to IV). The glutaraldehyde solution was then replaced for 1 hr by 4% NaCl. The material was treated with 1% OsO_4 in 0·1 M-phosphate buffer–4% NaCl for 1 hr. After several washes in 4% NaCl, embryos were dehydrated and embedded in Epon by routine methods.

Blocks were sectioned on a Porter-Blum MT-2 ultramicrotome employing a diamond knife. The sections were mounted on copper grids and coated with a thin carbon film. These grids were attached to glass slides preparatory to application of the emulsion.

For autoradiography, Ilford L4 nuclear track emulsion was applied in the form of a film formed on a copper wire loop. The filmed slides were stored in dry, lightproof boxes in the refrigerator. Exposure periods varied with the expected protein radioactivity in the sections, and are given in the legends. At termination of the photographic exposure, the specimens were developed at 20°C in Kodak D19 for 4 min. Fixation, washing and drying followed. Finally, the grids were removed and stained with lead acetate (Millonig, 1961) before examination in the electron microscope.

All specimens were studied and photographed with a Hitachi HU-11A electron microscope, operated at 50 kv. Instrumental magnifications for micrographs of the type shown in the plates ranged from 2500 to 5750 diameters. The silver grain distribution over many whole embryos was examined by means of photomontages reconstructed from 3 to 14 micrographs at the magnifications stated above. The montages are not shown here, but micrographs in the Plates are representative selections from montage sets.

(d) *Preparation of nuclear and cytoplasmic fractions*

Nuclei and cytoplasmic fractions were prepared by methods derived in this laboratory (B. Hogan & G. Cognetti, unpublished observations) but based in part on isolation procedures already available (Penman, 1966; Giudice & Mutolo, 1969). Embryos were labeled with [^{3}H]leucine for 30 min at the 64- to 128-cell stage in the presence or absence of actinomycin D. They were then chased in 0·01 M-leucine until late blastula or prism stage. The embryos were collected by centrifugation, washed 3 times with calcium–magnesium free artificial sea water and twice in RSB (0·01 M-NaCl, 0·0015 M-$MgCl_2$, 0·01 M-Tris–HCl pH 7·4). The pellet was resuspended in 5 to 10 vol. of RSB and made 1% in Triton X100 (Rohme & Haas, Inc., Philadelphia, Pa). The embryos were disaggregated with two strokes of a Dounce homogenizer (Kontes Glass Co., type B), allowed to stand at 0°C for 10 min and homogenized with 2 or 3 more strokes of the pestle. The degree of cell breakage was monitored by light microscopy. The homogenate was diluted with an equal volume of RSB and centrifuged at 600 ***g*** for 1 min/ml. The pellet was resuspended in concentrated RSB (0·1 M-NaCl, 0·02 M-$MgCl_2$, 0·01 M-Tris–HCl pH 7·4), centrifuged as above and resuspended in a small volume of the same buffer. To each ml. of resuspended nuclei was added 0·15 ml. of a 2:1 mixture of 10% Tween 40 (Mann Research Laboratory, New York, N.Y.) and 10% sodium deoxycholate. The mixture was agitated on a vortex mixer for 1 min and centrifuged at 600 ***g*** for 1 min/ml. All supernatant fractions were combined and considered cytoplasm. The nuclear pellet was suspended in 0·25 N-HCl for 30 min at 0°C and centrifuged at 20,000 ***g*** for 15 min. The pellet was resuspended in acid and deposited by centrifugation. These combined supernatants were considered acid soluble nuclear proteins (histones) and the washed pellet was considered to contain non-acid soluble nuclear proteins. The pellet was dissolved in 0·5% sodium dodecyl sulfate and samples of each fraction were analyzed for radioactivity and protein by the Lowry method (Lowry, Rosebrough, Farr & Randall, 1951). Standards were prepared with bovine serum albumin for determination of cytoplasmic and acid precipitable nuclear proteins and with calf thymus histones for the analysis of acid soluble nuclear proteins.

(e) *Preparation of polyribosomes*

The methods of homogenization and polyribosome preparation, collection of sucrose gradients and hot trichloroacetic acid precipitation of proteins were described previously (Kedes & Gross, 1969*a*). After heating the 6% trichloroacetic acid precipitates to 80°C for 30 min, the trichloroacetic acid concentration was increased to 10%, and after cooling the precipitates were collected on Millipore filters. Radioactivity was assayed in a Beckman liquid scintillation counter.

(f) *Chemicals*

Radioisotopes were obtained from New England Nuclear Corporation at the following specific activities:

L-[4,5-^{3}H]leucine, 5·0 c/m-mole;
L-[^{14}C]leucine, 251 mc/m-mole;
L-[^{3}H]tryptophan, 5·01 c/m-mole;
L-[3-^{14}C]tryptophan, 22·8 mc/m-mole;
L-[^{3}H]lysine, 3·96 c/m-mole;
L-[^{14}C]lysine, 271 mc/m-mole;
[5-^{3}H]uridine, 26·6 c/m-mole;
[*Methyl*-^{3}H]thymidine, 19·2 c/m-mole.

Actinomycin D as Cosmagen was obtained from Merck, Sharpe & Dohme, Inc. Hydroxyurea was obtained from Aldrich Chemical Corporation.

3. Results

(a) *Localization by autoradiography of proteins synthesized during cleavage*

The synthesis of all, or practically all, proteins of sea urchin embryo cells takes place in the cytoplasm. This point is established by the two electron microscope autoradiograms of Plate I. Fertilized eggs were labeled for periods of 5 and 15 minutes with [^{3}H]leucine. Immediately following the pulse exposures, the zygotes were chilled to 0°C and centrifuged on a sucrose gradient, also at 0°C, in order to stratify them. Both the fact that incorporation of amino acids stops at 0°C and the conditions required for stratification of the egg contents, have been established in other investigations (Gross, Hunter & Millonig, unpublished observations; Harvey, 1956). Stratification of subcellular particles removes large cytoplasmic objects from the immediate vicinity of the nuclei and makes eventual detection of silver grains easier. The result to be described is invariant, whether or not stratification is done, and, when it is done, whether the centrifugation is done before or after the pulse labeling.

After brief exposures to labeled amino acids radioactive proteins are easily detected in the cytoplasm, but none or very few appear in the nuclei. Within a few minutes, however, the nuclei begin to show internal labeling, and eventually (see below), they accumulate radioactive proteins relative to the cytoplasm. In Plate I(a) is shown the perinuclear region of a centrifuged zygote labeled with leucine for five minutes. This low magnification micrograph permits inspection of a large area (about 600 μ^2, after fixation). The area shown contains 12 silver grains (after an autoradiographic exposure of three weeks), and these are a very small fraction of the total found over the entire embryo section. No grains at all are found within the nucleus. Absolute zeros like this one are not common, but many observations suggest that the shorter the exposure to labeled amino acid, the less likely it is to find any radioactive protein in the nuclei of embryos at any stage of cleavage. As the pulses are lengthened (even at 15 min) radioactivity is localized within the nuclear envelope (Plate I(b)).

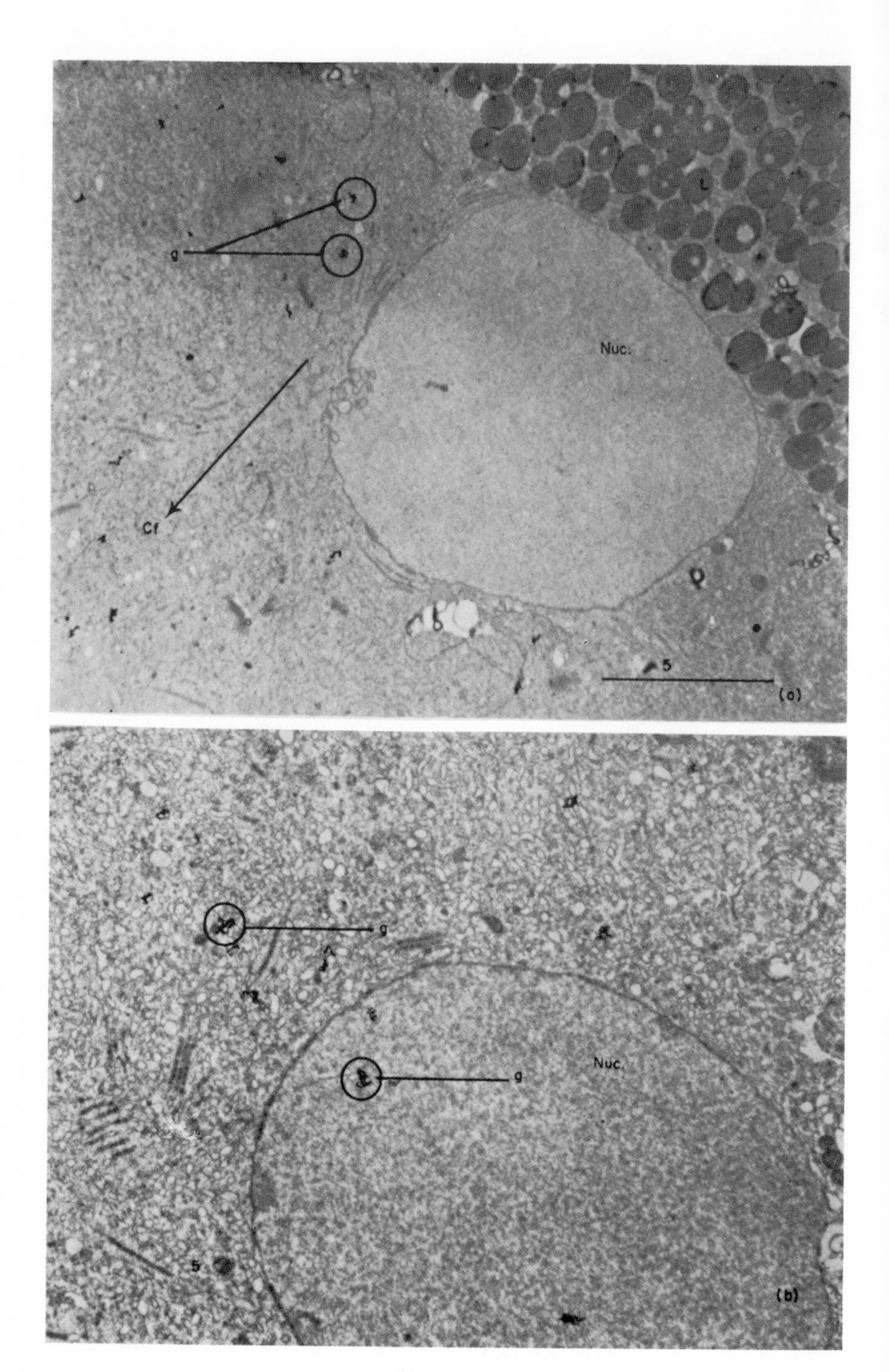

PLATE I.

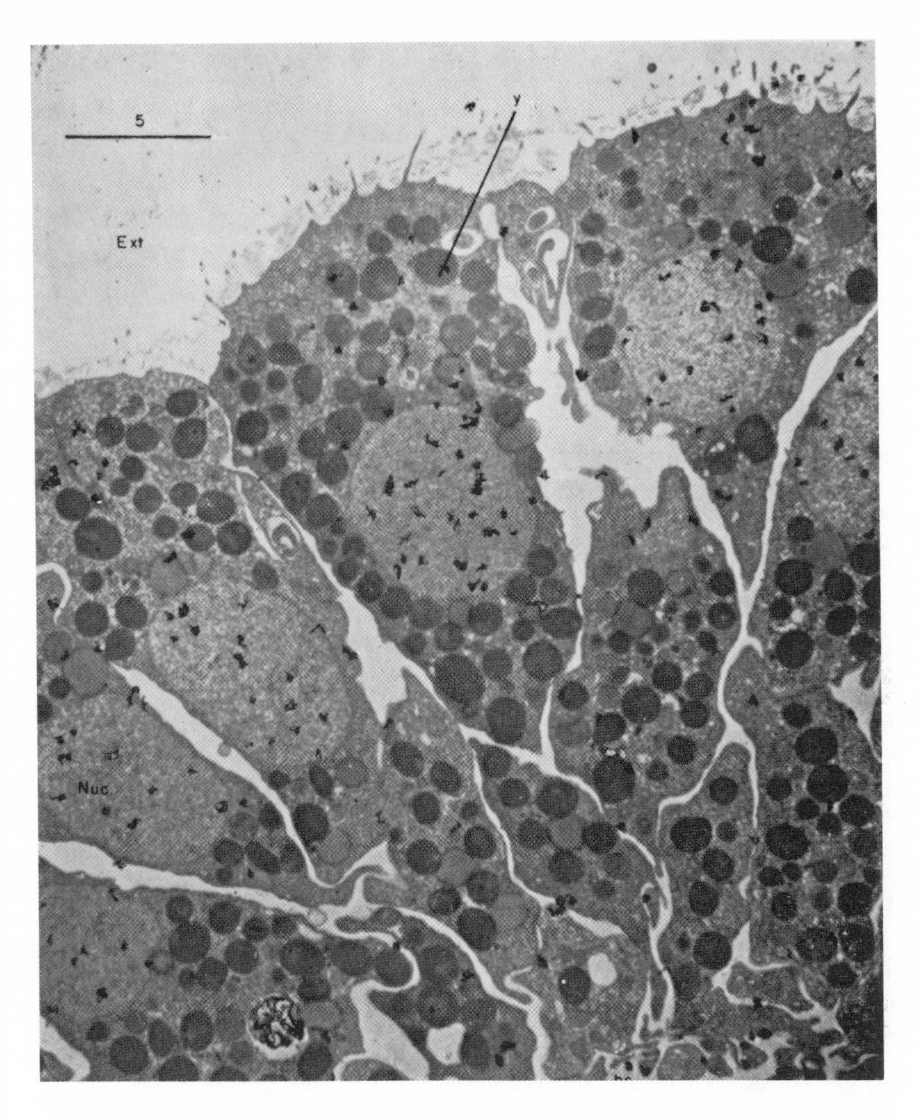

Plate II.

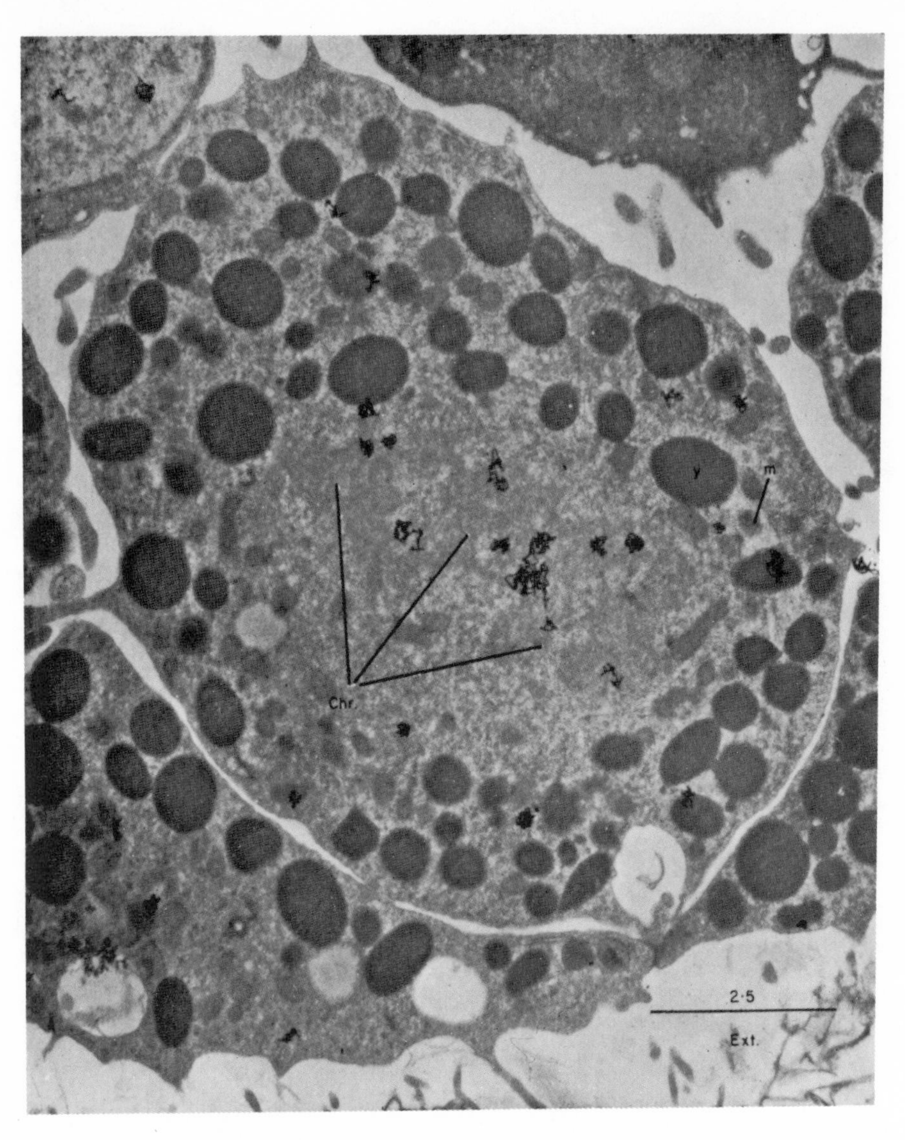

PLATE III.

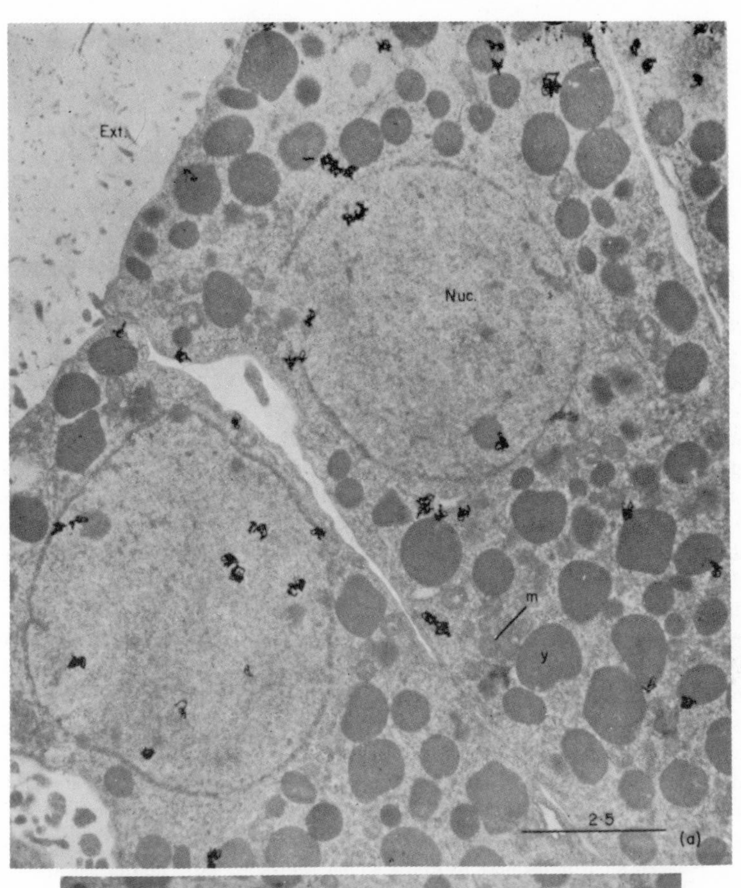

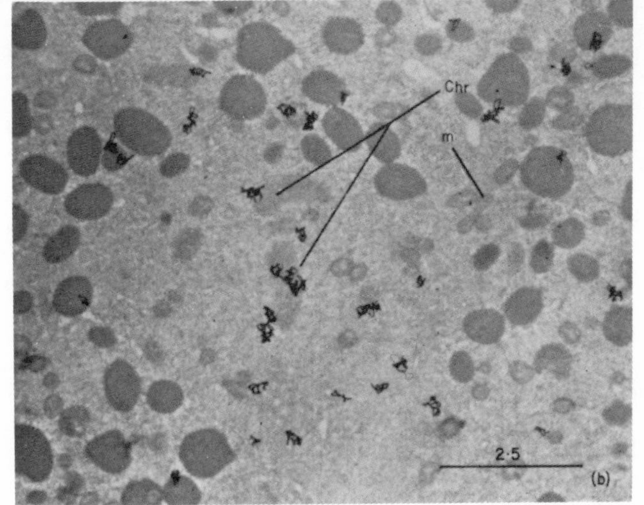

PLATE IV.

Longer exposures followed by cold amino acids produce a distinct concentration of radioactive proteins within the nuclei. There seems to be little doubt, therefore, that protein synthesis occurs in the cytoplasm, whether or not the destination of the products is cytoplasmic.

The results of pulse–chase experiments and autoradiography in the period between mid-cleavage and blastulation are represented by Plates II and III. In Plate II is shown an autoradiogram of part of a blastula exposed to [^{3}H]leucine according to the following schedule: four-hour embryos (23°C), i.e. embryos in mid-cleavage, were incubated for one hour with the labeled amino acid. Exposure was terminated by washing with sea water, after which further development took place in a large excess of unlabeled leucine. Hatching occurred 3·5 hours later, at which time the embryos were harvested, fixed, embedded and sectioned. The distribution of radioactivity in Plate II shows a striking concentration of newly synthesized proteins in the nuclei: the number of silver grains per unit section is much higher for nuclei than it is for cytoplasm. This is true even if a correction is made for the presence of impenetrable yolk particles in the cytoplasm. When grain counts were made on a typical photomontage, 207 grains of a total of 339 were nuclear (61% nuclear). This distribution was identical for labeling with either leucine or lysine. At the blastula stage (i.e. about 1000 cells), radioactivity incorporated two or more hours earlier is distributed to all nuclei in the sections, and neither the ratio of grain density to nuclei nor distribution of grains between cytoplasm and nuclei fluctuates greatly from one cell to another.

Most new proteins of the nuclear compartment are found associated with chromosomes when a metaphase cell is examined. The cell in Plate III is from a blastula section like that in Plate II, except that in the experiment represented the labeling was done with [^{3}H]lysine. The same result is obtained, however, with other amino acids, such as leucine. The radioactivity is found localized in the condensed metaphase chromosomes. Hence a significant fraction of proteins made during mid-cleavage find their stable positions close to or on the chromosomes. The production of nuclear and chromosomal proteins must therefore be a major commitment of mid-cleavage macromolecule synthesis.

Experiments like those just described were done with embryos developing in the presence of actinomycin D, at dose levels sufficient to stop transcription or to reduce it by 90% or more. The synthesis of proteins continues, and nuclei accumulate radioactivity, just as they do in the normal case. This is shown in Plate IV(a), which represents an experiment like that of Plate II, except that the embryos were grown in 20 μg actinomycin/ml. and were labeled with [^{3}H]lysine. The concentration of nuclear relative to cytoplasmic radioactive proteins is distinctly lower than it is in the normal case, however. When cells from many embryos developed in the presence of actinomycin were examined, only 134 of 395 grains appeared over the nuclei (34% nuclear). The metaphase chromosomes of actinomycin-treated embryos also seem to become radioactive (Plate IV(b)), as do normal ones, but once again, there is a tendency for a smaller fraction of the total silver grains on any section of a cell to be found *directly* over the chromosomes. The embryo makes chromosomal proteins, whether or not its normal transcriptional activity takes place, but there seem to be at least quantitative differences between the two situations as regards the fractional commitment to biosynthesis of nuclear proteins. This latter point is examined by completely independent methods in experiments now to be described.

(b) *The relative distribution of radioactive proteins between cytoplasm and nuclei*

Pulse-and-chase experiments were performed according to schedules like those employed for the autoradiographic preparations. Each cell fraction (see Materials and Methods) was examined for radioactivity, and the protein content determined by the Lowry method. In a typical experiment (Table 1), about half of the proteins synthesized at morula stage are recovered in the combined nuclear fractions.

TABLE 1

Distribution of protein radioactivity between nuclei and cytoplasm in control and actinomycin D-treated morulae

	Protein (mg)	% of total protein	Cts/min ($\times 10^{-3}$)	% of total cts/min	Cts/min/ mg protein ($\times 10^{-3}$)
Control					
Cytoplasm	10·08	89	4398	55	436
Nuclei	1·26	11	—	45	—
(a) acid soluble	0·32	3	2180	27	6812
(b) acid insoluble	0·94	8	1416	18	1506
Actinomycin D					
Cytoplasm	2·41	82	1286	64	533
Nuclei	0·54	18	—	36	—
(a) acid soluble	0·08	3	215	11	2687
(b) acid insoluble	0·46	15	508	25	1106

Embryos at 64-cell stage were labeled with tritiated leucine for 30 min, washed and then incubated with 0·01 M unlabeled leucine. Cytoplasm, acid soluble and insoluble nuclear fractions were prepared at the mesenchyme blastula stage, or later, as described in Materials and Methods.

In examining the relative amount of radioactivity associated with nuclei, three major problems must be considered.

(i) Our method of nuclear isolation provides nuclei almost entirely free of visible cytoplasmic tags and outer nuclear membranes when examined by electron microscopy (Cognetti, Kedes & Gross, manuscript in preparation). Thus, even if the isolated nuclei were coated with cytoplasm, the observed lower specific activity of the contaminating protein (Table 1) precludes the possibility that it represents a large population of the radioactivity recovered with nuclei.

(ii) The second problem concerns the stability of newly synthesized proteins, and it is also, of course, a potential difficulty for the autoradiographic studies. If there is turnover of cytoplasmic protein relative to nuclear proteins, the fractional radioactivity could increase progressively in the nuclei. But no change is found in total embryo protein radioactivity even over very long intervals of the cold amino acid chase (Mangan, 1966; Fry & Gross, 1969), suggesting that most or all protein synthesized during cleavage is highly stable. Histones in the chromatin of mammalian cells also appear to be stable for many cell generations (Hancock, 1969).

(iii) Nuclear contamination of the cytoplasmic fraction would tend to decrease the estimated nuclear fraction of new proteins. The method of preparation described here is designed to yield the cleanest possible nuclei. Nuclear breakage and leakage of easily displaced nuclear proteins during subsequent preparative steps might contribute to contamination of cytoplasm with nuclear proteins at high specific radioactivity. For these reasons we believe that the nuclear percentages of total new proteins in Table 1 are somewhat less than the true value. Grain counts from the autoradiograms suggest that the nuclear protein fraction may be as much as 60% at the morula stage.

When embryos were grown in the presence of actinomycin, there was a decrease of about 20% in the fraction of labeled proteins chased into the nuclei (Table 1). The effect was primarily on the acid-soluble fraction. The pulses were executed when the animals reached a morphological stage comparable to the controls, usually one to two hours longer when labeling 64- to 128-cell embryos (Gross & Cousineau, 1964). Although a decrease in nuclear protein relative to total incorporation is detected in the actinomycin preparations (Table 1), we cannot exclude the possibility that the drug prevented migration of proteins into the nucleus or increased leakage of nuclear proteins back into the cytoplasmic fraction during or before preparation. On the other hand, grain counts of the autoradiograms certainly corroborate the evidence of decreased fractional protein accumulation in nuclei.

(c) *Histone-like proteins synthesized on a special class of polyribosomes*

As was shown in the pioneer studies of Robbins & Borun (1967), synchronously dividing HeLa cells synthesize nuclear proteins on a class of small polyribosomes which appear in the cytoplasm only during S phase. A previous report from this laboratory (Kedes & Gross, 1969*a*) also made note of the increase in prominence of one class of small polyribosomes (100 to 200 s) especially during the cleavage stage (i.e. rapid nuclear replication stage) of sea urchin embryogenesis. These light polyribosomes had earlier been observed by Nemer and coworkers (Nemer & Infante, 1967; Infante & Nemer, 1967; Nemer, 1967). They described them as inactive in protein synthesis (Spirin & Nemer, 1965), and suggested that the light polyribosomes are a cytoplasmic site for storage and accumulation of newly synthesized mRNA (Nemer & Infante, 1965). Recently, however, Nemer & Lindsay (1969) have reported results that agree with our earlier conclusion that light polyribosomes are active in translation (Kedes & Gross, 1969*a*), and they show that the light polyribosomes are engaged in production of peptides with a large arginine/lysine ratio.

During cleavage, most of the light polyribosomes are engaged in translation of new mRNA, i.e. templates other than those stored during oogenesis (Kedes & Gross, 1969*b*). The light polyribosomes function from the very beginning of development. An appreciable fraction of the proteins made on them depend upon information transcribed in the embryo's nuclei. Evidence to be presented elsewhere (Kedes & Gross, 1969*b*) shows that the most prominent class of the new mRNA molecules (the 9 s group described earlier; Kedes & Gross, 1969*a*) are templates for the synthesis of nuclear proteins as has also been found in HeLa cells (Borun, Scharff & Robbins, 1967).

We attempted to determine if histone-like proteins were synthesized specifically by these light polyribosomes by pulse labeling the embryos with radioactive lysine and tryptophan. Histones are uniquely rich in lysine and contain virtually no tryptophan (Murray, 1966). A significantly increased lysine/tryptophan ratio in any size class of polyribosome would suggest that it is the site of biosynthesis of histone-like proteins (i.e. the test of Robbins & Borun, 1967). The results of such an experiment

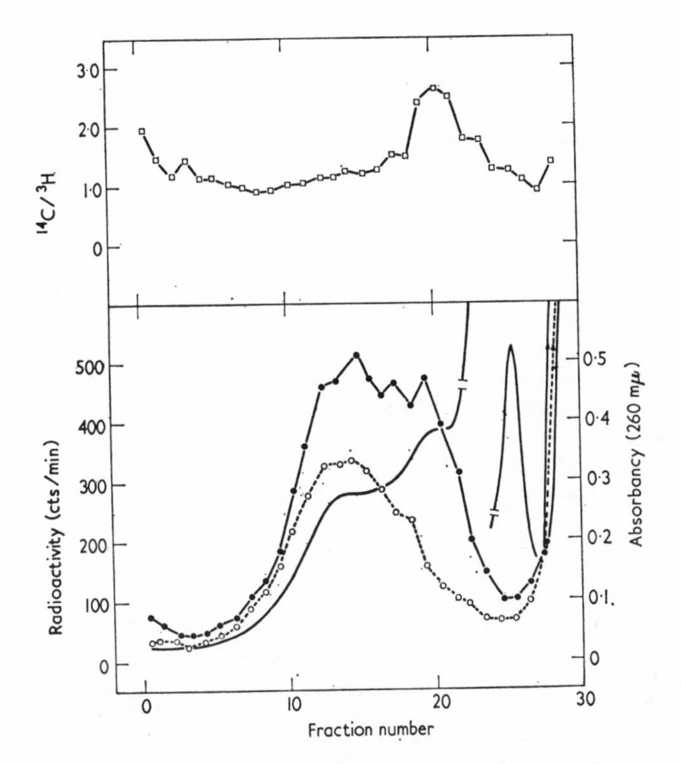

FIG. 1. Asymmetric distribution of lysine and tryptophan incorporation ratios.

Embryos at 32- to 64-cell stage were concentrated in 4 vol. sea water and labeled for 10 min with [^{14}C]lysine. (2 μc/ml.) and [^{3}H]tryptophan (10 μc/ml.) The embryos were washed twice with iced calcium–magnesium free sea water and polyribosomes prepared on sucrose density gradients as described previously (Kedes & Gross, 1969*a*). Final precipitation of proteins was with 10% trichloroacetic acid onto Millipore filters which were dried and counted in a Beckman liquid scintillation counter. Spill-over of ^{14}C into the ^{3}H channel was 24% and appropriate corrections were made. The ^{14}C/^{3}H ratio of each fraction was calculated and normalized to the average ratio of fractions from the heavy polyribosome region (fractions 6 to 15) as described in the text. The normalized ratios are plotted in (a); (b) represents the polyribosomes and radioactivity distribution in the sucrose density gradient: (———) O.D.$_{260}$; (—●—●—) [^{14}C]lysine cts/min; (--○--○--) [^{3}H]tryptophan cts/min.

are seen in Figure 1. Hot trichloroacetic acid-precipitable radioactivity in the polyribosome region increases with increased polymer number as described previously (Kedes & Gross, 1969*a*). The ratio of [^{14}C]lysine to [^{3}H]tryptophan counts was normalized by averaging the ratios of ten fractions from the heavy polysome region. The ratio for each fraction of the gradient was then divided by this number. The figure demonstrates that the light polyribosome region has a sharply increased lysine/tryptophan ratio. Thus, by the 16-cell stage lysine rich–tryptophan poor proteins are synthesized on light polyribosomes in sufficient amount to cause a significant local change in the incorporation ratio.

Identical results were obtained when [^{3}H]lysine and [^{14}C]tryptophan were used to label the embryos. Control experiments with [^{3}H]- and [^{14}C]leucine gave a constant ratio through the entire polysome region. The amino acid ratio deviation in the light polysome region probably reflects simultaneously increased lysine incorporation and decreased tryptophan incorporation, because examination of the specific activity–polymer number relationships (Kedes & Gross, 1969*a*) reveals a rise from the expected lysine specific activity and a decrease from the expected tryptophan specific activity in this region.

(d) *Effect of inhibition of DNA synthesis on polyribosome accumulation*

(i) *Selective inhibition of thymidine incorporation by hydroxyurea*

Effects of the DNA synthesis inhibitor, hydroxyurea (Young & Hodas, 1964; Yarbro, Kennedy & Barum, 1965) were tested systematically in sea urchin embryos. A concentration of 1 mM was found to be the lowest which inhibited cell division. After addition of the drug at this dose, embryos divided once or twice and then ceased to cleave. Lower concentrations (0·5 mM or less) allowed continued mitosis and development proceeded to the swimming blastula stage. The effect of the drug on incorporation of [^{14}C]thymidine or [^{3}H]uridine is demonstrated in Figure 2. Thymidine incorporation decreases upon exposure to hydroxyurea, and after one to two hours incorporation is almost stopped. Uridine incorporation, on the other hand, is unaffected, suggesting that over a two-hour exposure to the drug bulk RNA synthesis is not inhibited.

(ii) *Relationship of light polyribosomes to DNA synthesis and mitosis*

Morulae were exposed to hydroxyurea for 90 minutes and the polyribosomes compared to the polyribosomes of untreated controls. When equal numbers of

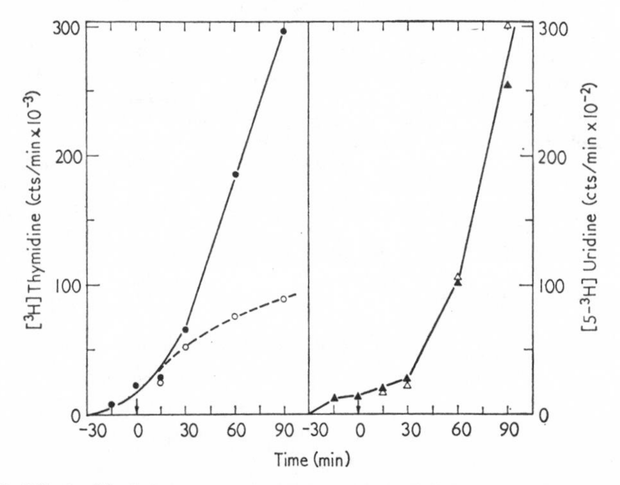

FIG. 2. Effect of hydroxyurea on incorporation of labeled thymidine and uridine.

Embryos at 16- to 32-cell stage were cultured continuously in the presence of [^{3}H]thymidine (5 μc/ml.) in the left panel or [5-^{3}H]uridine (2 μc/ml.) on the right. After 30 min half of each culture was made 1 mM in hydroxyurea. Samples were taken at frequent intervals. The embryos were concentrated by centrifugation and the sea water replaced by 1 ml. of 0·5% sodium dodecyl sulphate solution, which dissolved the material. The solution was precipitated by the addition of trichloroacetic acid to a final concentration of 6% and the precipitate collected on Millipore filters, washed with cold 6% trichloroacetic acid containing excess thymidine or uridine, and counted in a Beckman liquid scintillation counter. (●, ▲) Control points; (○, △) incorporation in the presence of hydroxyurea. The arrows point to the times of addition of hydroxyurea.

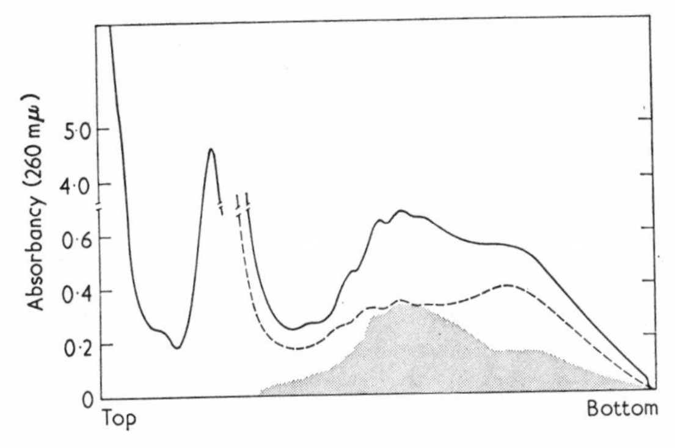

FIG. 3. Effect of hydroxyurea on polyribosome accumulation.

One-half of a culture of embryos at the 32-cell stage was made 1 mM in hydroxyurea. After 90 min polyribosomes were prepared from both cultures. (————) O.D.$_{260}$ of the control culture; (— — —) O.D.$_{260}$ from the culture exposed to the drug. The absorbancy profiles at the tops of the gradients were coincident. The shaded area represents the absolute O.D.$_{260}$ difference between the two profiles at each point.

embryos or equal amounts of cytoplasm are examined, a direct comparison of polyribosome patterns can be made (Kedes & Gross, 1969*a*). Figure 3 demonstrates that after hydroxyurea exposure there is a decrease of 45% in the O.D.$_{260}$ sedimenting faster than ribosomes (as measured by planimetry). The bulk of this decrease is from the light polyribosome region (shaded area). This result suggests that maintenance of an important fraction of the light polyribosomes is linked to continued DNA synthesis and/or to mitosis (Borun *et al.*, 1967).

(e) *Effect of inhibition of DNA or RNA synthesis on histone production*

(i) *Hydroxyurea inhibition of small polyribosome synthesis of nuclear proteins*

Hydroxyurea decreased markedly the asymmetry of incorporation of lysine rich–tryptophan poor nascent peptides over the light polyribosome region (Fig. 4). Thus, inhibition of DNA synthesis and/or mitosis, leads to a marked decrease in synthesis of histone-like proteins. It is important to note that the absolute count ratios of the heavier polyribosome region are similar in control and treated preparations. This suggests that the drug does not produce general and non-specific effects on protein synthesis, lysine or tryptophan incorporation, or precursor penetration. It would be difficult, furthermore, to argue that the drug acts at the level of translation, since it would then be required to prevent selectively the activity of a specific class of polyribosomes. While this is possible, it seems to us an unlikely, and therefore, for the moment, an unattractive explanation.

(ii) *Synthesis of histone-like proteins on new or "maternal" messages*

Sea urchin eggs are endowed with a set of "maternal" templates for protein synthesis, produced by transcriptions in the oocyte during oogenesis. These templates are sufficient to allow cleavage and nuclear multiplication (and presumably some nuclear protein production) to proceed when post-fertilization RNA synthesis is prevented by actinomycin (Gross & Cousineau, 1964). Such embryos are not normal, since they do not gastrulate and do not synthesize a normal complement of cytoplasmic proteins (Terman & Gross, 1965). Actinomycin-treated embryos therefore

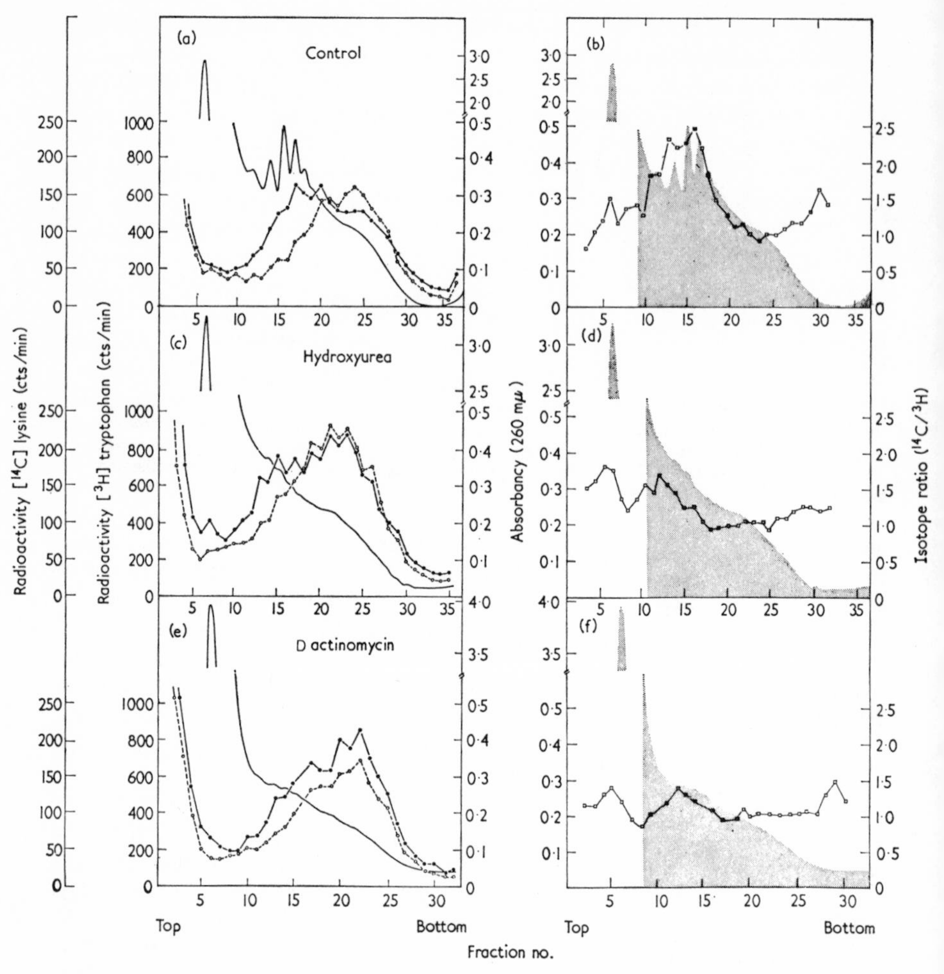

FIG. 4. Effect of hydroxyurea and actinomycin on the lysine/tryptophan ratios of nascent peptides on polyribosomes.

One-third of the eggs from a single female were fertilized after a 30-min exposure to actinomycin (25 μg/ml.) and cultured with the drug continuously. When the control culture reached 16- to 32-cell stage, half was made 1 mM in hydroxyurea. After 90 min the control and hydroxyurea treated cultures were concentrated and labeled with [^{14}C]lysine and [^{3}H]tryptophan as described in the text and legend of Fig. 1. The actinomycin culture was labeled 1 hr later when it had reached a morphologic stage comparable to the control. Polyribosomes were prepared by sucrose density gradient centrifugation and examined for optical density and radioactivity (left panels) as described in Fig. 1. The normalized isotope ratios are plotted in the right-hand panels superimposed on the optical density patterns of the polyribosomes (shaded areas) for reference.

lack new RNA, some of which is necessary for normal differentiation to proceed. New RNA templates are synthesized from fertilization and are utilized in protein synthesis (Rinaldi & Monroy, 1969; Kedes & Gross, 1969*a*). Most of these new templates accumulate in polyribosomes, and light polyribosomes especially are translating these newly synthesized messages (Kedes & Gross, 1969*a*).

The effect of actinomycin on the distribution of lysine and tryptophan radioactivity on polyribosomes was examined to determine if the histone-like proteins were synthesized on new or old templates. Eggs treated with 25 μg actinomycin/ml. for 30 minutes were fertilized and grown to the 32-cell stage. Labeling with lysine and tryptophan was performed as described above. Results are seen in the lower panel of Figure 4. Actinomycin decreases the lysine/tryptophan ratio peak in the light polysome region, *but some histone-like proteins continue to be synthesized.* The decreased production of histone-like proteins in the cytoplasm is apparently the cause of a decrease in accumulation of acid-soluble nuclear proteins (Table 1) in the presence of actinomycin.

Unfortunately, when actinomycin-treated embryos were exposed to hydroxyurea, cytolysis and almost complete polyribosome disaggregation occurred, preventing further analysis by this method of relative contributions to histone-like protein synthesis by new and "maternal" templates.

4. Discussion

Animal morphogenesis requires multicellularity. Hence, in addition to early chemodifferentiation, a self-evident necessity of early development is to make many cells from one. The experiments described here are designed to study macromolecular syntheses associated specifically with the rapid cell divisions of early embryos. These systems are in virtually continuous DNA synthesis during certain parts of preblastula development. The total mass of each embryo is constant while the nuclear mass increases exponentially. Accordingly, the proportion of the total biosynthetic activity that is devoted to production of new nuclei is far greater than in dividing and growing systems, such as cell cultures or regenerating organs.

Nuclear and cytoplasmic proteins of embryos, pulsed with radioactive amino acids at early cleavage stages and then chased with cold amino acids, were examined in two ways: by autoradiography and by direct cell fractionation. Autoradiography revealed that immediately after a short pulse new proteins are confined to the cytoplasm and are not found in the nuclei (nor with mitochondria or yolk). With slightly longer pulses, or after chasing, stable radioactive proteins accumulate in the nuclei. Some of these are located on or near metaphase chromosomes. Persistence of grains over the nuclei after all cells had undergone several division cycles suggests that the new proteins are a part of structures that maintain their integrity during mitosis and re-enter the daughter nuclei. These results are in agreement with the recent experiments of Hancock (1969) who demonstrated that histones are conserved in chromatin through at least eight generations in mammalian cells in culture. The rather uniform nuclear labeling suggests in addition that nuclear proteins of the kind detected are distributed to daughter cells in the divisions following a pulse.

During cleavage there is neither a normal nucleolus nor synthesis of ribosomal RNA (Gross, Kraemer & Malkin, 1965), ribosomal RNA precursors (Giudice & Mutolo, 1969), or ribosomal structural proteins (Terman, 1969). The absence of these activities shows that the proteins accumulated in nuclei are not ribosomal proteins.

The distribution of new proteins was examined directly by isolating nuclei from the pulsed–chased embryos (Table 1). The results of several experiments indicate that half the proteins produced in morula embryos are destined for nuclear structures. Of these, 40 to 60% are acid-soluble and presumably histones. By contrast, less than 10% of the chased radioactivity isolated from nuclei in non-synchronized HeLa cells is found in histones (Speer & Zimmerman, 1968).

Histones have an uncommonly high lysine content (8 to 27 moles per cent in various histone fractions) and no tryptophan (Murray, 1966). Taking advantage of the unique amino acid content of these proteins and of the fact that they account for such a large percentage of the total synthesis of nuclear proteins, we identified them on polyribosomes, using the approach taken earlier by Robbins & Borun (1967) in cultures of synchronously dividing HeLa cells. Cytoplasmic extracts from embryos labeled with lysine and tryptophan demonstrated that nascent peptides with very high lysine/tryptophan ratios localized on light polyribosomes only. Histones are of small molecular weight (8 to 57,000; Hnilica, 1967) and are probably synthesized on polyribosomes of polymer number 2 to 6. Light polyribosomes of *Xenopus laevis* embryos also exhibit a markedly increased lysine/tryptophan ratio (Gross & Crippa, unpublished observations), and behave in other respects just as do those of sea urchins.

When the number of nuclei per embryo is eight or less the amount of histone-like protein being synthesized is small compared to non-histone protein. In fact, when the radioactivity associated with nuclei is examined after labeling during the four- to eight-cell stage (and then chased to hatching), the fraction is only about 10% of the total (Cognetti, Kedes & Gross, manuscript in preparation). Only when the rate of synthesis of histone-like proteins per embryo increases beyond this level can a change in lysine/tryptophan ratio on polyribosomes be detected against the background of bulk protein production. An additional finding in accord with this interpretation is the fact that light polyribosomes begin to accumulate visibly only at the 16-cell stage and later (Kedes & Gross, 1969*a*; Infante & Nemer, 1967).

When DNA synthesis, nuclear replication, and cleavage are stopped with hydroxyurea, bulk RNA synthesis continues (Fig. 2), and some polyribosomes active in protein synthesis are maintained (Figs 3 and 4). Continued function of polyribosomes, other than the light classes examined here, and the normal production of nascent peptides (Figs 3 and 4) suggests that the drug has little or no general effect on peptide bond formation.

Although a direct relationship is unproven, it is probably more than coincidence that the decrease of polyribosome content (mainly light) with hydroxyurea (45%) is close to the fraction of the protein synthetic effort involved in nuclear protein production (Table 1). It is to be emphasized that inhibition of DNA synthesis does decrease the relative synthesis of those nascent proteins with high lysine/tryptophan ratios (Fig. 4). The fact that the light polyribosomes are affected selectively, and that nascent peptides do not accumulate in the monoribosome peak with hydroxyurea treatment, excludes the possibility that the drug's effect is caused by non-specific degradation of the polyribosomes.

When new template synthesis is prevented with actinomycin, fewer polyribosomes accumulate than in control embryos. At morula stage the smaller polyribosomes are suppressed selectively, and the over-all polysome loss is about 40% of the normal total (Kedes & Gross, 1969*a*). Autoradiograms and direct analyses both suggest that this drug decreases the fractional accumulation of protein radioactivity in the nuclei relative to control embryos. Histone accumulation especially is reduced and polyribosomal lysine/tryptophan ratios are also decreased in the presence of actinomycin. These results indicate that the synthesis of histone-like protein is in part linked to the production of new RNA templates. The present experiments do not allow a decision as to whether the new RNA's are messages for the lysine rich–tryptophan poor proteins, or are indirectly responsible for the activation of messages synthesized during oogenesis. Since nuclear proteins continue to be synthesized in the presence of actinomycin (specifically, those associated with chromosomes), it is possible that some of the new RNA serves as templates for a different class of proteins, perhaps regulating histone synthesis.

The authors are indebted to Miss Patricia Berkley and Dr Ester Rubino for expert technical assistance.

The work reported in this paper was supported by research grants from the National Institutes of Health (GM 13560), the American Cancer Society (no. 5050), the National Science Foundation (GB-6350), and the Medical Foundation, Inc.

During the performance of some of the experiments reported here, one of us (L. H. K.) was a Postdoctoral Fellow of the Medical Foundation, Inc. He is now a scholar of the Leukemia Society of America.

Note added in proof: Some of the proteins labeled during an amino acid pulse to post-cleavage embryos or to unfertilized eggs find their way into nuclei during a long chase, but unlike the situation during cleavage, only a small fraction of such "nuclear" proteins is histone-like. The upper limit appears to be about 10% of the total (Cognetti, Kedes & Gross, manuscript in preparation), recalling the findings of Speer & Zimmerman (1968) in asynchronous HeLa cell cultures.

REFERENCES

Borun, T. W., Scharff, M. D. & Robbins, E. (1967). *Proc. Nat. Acad. Sci., Wash.* **58**, 1977.
Fry, B. & Gross, P. R. (1969). *Devel. Biol.* in the press.
Giudice, G. & Mutolo, M. (1969). *Biochim. biophys. Acta*, in the press.
Gross, P. R. & Cousineau, G. H. (1964). *Expt. Cell. Res.* **33**, 368.
Gross, P. R., Kraemer, K. & Malkin, L. I. (1965). *Biochem. Biophys. Res. Comm.* **18**, 569.
Hancock, R. (1969). *J. Mol. Biol.* **40**, 457.
Harvey, E. B. (1956). In *The American* Arbacia *and Other Sea Urchins.* Princeton, N.J.: Princeton University Press.
Hinegardner, R. T., Rao, B. & Feldman, D. E. (1964). *Expt. Cell Res.* **36**, 53.
Hnilica, L. S. (1967). In *Progress in Nucleic Acid Research and Molecular Biology*, ed. by J. N. Davidson & W. E. Cohn, p. 25. New York: Academic Press.
Infante, A. A. & Nemer, M. (1967). *Proc. Nat. Acad. Sci., Wash.* **58**, 681.
Kedes, L. & Gross, P. R. (1969*a*). *J. Mol. Biol.* **42**, 559.
Kedes, L. & Gross, P. R. (1969*b*). *Nature*, in the press.
Lowry, O. H., Rosebrough, N. J., Farr, A. L. & Randall, R. J. (1951). *J. Biol. Chem.* **193**, 265.
Mangan, J. (1966). Doctoral Thesis, Brown University.
Millonig, G. (1961). *J. Biophys. Biochem. Cytol.* **11**, 736.
Murray, K. (1966). *J. Mol. Biol.* **15**, 409.
Nemer, M. (1967). In *Progress in Nucleic Acid Research and Molecular Biology*, ed. by J. N. Davidson & W. E. Cohn, p. 243. New York: Academic Press.
Nemer, M. & Infante, A. (1965). *Science*, **150**, 217.

Nemer, M. & Infante, A. (1967). In *The Control of Nuclear Activity*, ed. by L. Goldstein, p. 101. Englewood Cliffs, N.J.: Prentice-Hall.
Nemer, M. & Lindsay, D. T. (1969). *Biochem. Biophys. Res. Comm.* **35**, 156.
Penman, S. (1966). *J. Mol. Biol.* **17**, 117.
Rinaldi, A. M. & Monroy, A. (1969). *Devel. Biol.* **19**, 73.
Robbins, E. & Borun, T. W. (1967). *Proc. Nat. Acad. Sci., Wash.* **57**, 409.
Speer, H. L. & Zimmerman, E. F. (1968). *Biochem. Biophys. Res. Comm.* **32**, 60.
Spirin, A. S. & Nemer, M. (1965). *Science*, **150**, 214.
Terman, S. A. (1969). Doctoral Thesis, Massachusetts Institute of Technology.
Terman, S. A. & Gross, P. R. (1965). *Biochem. Biophys. Res. Comm.* **21**, 595.
Yarbro, J. W., Kennedy, B. J. & Barum, C. P. (1965). *Proc. Nat. Acad. Sci., Wash.* **53**, 1033.
Young, C. W. & Hodas, S. (1964). *Science*, **146**, 1172.

Old and New Protein in the Formation of the Mitotic Apparatus in Cleaving Sea Urchin Eggs

F. H. Wilt, H. Sakai and D. Mazia

1. Introduction

Studies during the past few years on the regulation of protein synthesis immediately following activation of the sea urchin egg have reopened interest in the relation of this early protein synthesis to the first cell division. New and improved methods of studying the structural proteins of the mitotic apparatus are now available, and we have taken advantage of them to re-study the contribution of early protein synthesis to its origin. The results show clearly that: (1) while some protein synthesis seems to be required for the first division of activated sea urchin eggs, the synthesis which is essential for cell division represents only a small portion of the newly synthesized protein; and (2) that the structural proteins of the mitotic apparatus are synthesized during this period at a rate not greatly exceeding the synthesis of bulk cell protein. We conclude that the essential role of protein synthesis for the first cell division of sea urchin eggs may be catalytic rather than structural, and that most of the mitotic apparatus for the first cell division is synthesized prior to egg activation.

The background leading to a re-investigation of the above questions is drawn from a number of sources. The arguments for preformation of mitotic apparatus protein in the sea urchin egg have been surveyed by Mazia (1961). Went (1959) showed by immunological methods that the proteins of the mitotic apparatus are present in unfertilized eggs. There is no net increase in total egg protein during cleavage (Monroy,

1965). The mitotic apparatus apparently represents something on the order of 5 to 10% of the total egg mass at the first cell division (Mazia & Roslansky, 1956).

Other experiments have been interpreted as showing that protein synthesis following egg activation is an essential and quantitatively important contribution to the formation of the mitotic apparatus for the first cell division. Hultin (1961) showed that the first cell division is inhibited by addition of puromycin to the sea water before insemination of the egg. Several investigators have recently claimed that the incorporation of isotope into the mitotic apparatus is much higher than into the remainder of the bulk proteins of the cell (Gross & Cousineau, 1963; Stafford & Iverson, 1964; Bibring & Cousineau, 1964). Autoradiographic evidence shows a clear localization of isotopically labeled amino acid in the region of the mitotic apparatus (Gross & Cousineau, 1963). In one study (Mangan, Miki-Noumura & Gross, 1965), the isolated mitotic apparatus was dissolved in urea and fractionated on Sephadex. The radioactive amino acids associated with mitotic apparatus did not accompany the bulk of the MA† protein; the authors suggest, however, that *de novo* synthesis of mitotic apparatus protein may be quantitatively important for cell division.

The technical difficulties in obtaining relatively pure and undenatured preparations of MA present obstacles in the interpretation of the results, as has been pointed out (Mangan *et al.*, 1965). The present experiments were carried out by allowing eggs to develop to metaphase in sea water to which radioactive leucine was added, and by preparing the MA with improved methods that allowed study of solubilized, undenatured MA structural protein by conventional methods.

2. Materials and Methods

Strongylocentrotus purpuratus eggs were used for all experiments. [^{3}H]Leucine (5 c/m-mole) and [^{3}H]thymidine (10 c/m-mole) were purchased from New England Nuclear and puromycin from Nutritional Biochemicals. [^{14}C]Valine (11·1 mc/m-mole) was obtained from Calbiochem. Eggs obtained from single females were washed, inseminated, and stripped of membranes by the usual procedure (Mazia, Mitchison, Medina & Harris, 1961). Development proceeded at 16°C, and fertilization was always 95% or better. Incorporation of isotope into the acid-insoluble fraction of whole eggs was determined by the methods of Hinegardner, Rao & Feldman (1964) for DNA and Mangan *et al.* (1965) for protein. The counting efficiency on a Packard Tri-Carb scintillation counter was about 7% for ^{3}H. Protein was determined in triplicate samples by the method of Lowry, Rosebrough, Farr & Randall (1951), using bovine pancreatic ribonuclease as a standard.

The mitotic apparatus was isolated from living cells by the dithiodipropanol method (Mazia *et al.*, 1961) as modified by Sakai (1966). In brief, washed, de-membranated eggs were lysed at metaphase with 0·15 M-dithiodipropanol, 0·1 mM-EDTA, 1·0 M-sucrose (pH 6·2) by gentle shaking at 15°C for a few minutes. The suspension was chilled on ice for 30 to 60 min, the density of the medium lowered by addition of 0·2 vol. of water, and the MA collected by centrifugation for 30 min at 2000 *g*. This results in isolation of an intact, undenatured MA which can be exhaustively washed free of contaminants, yet is easily soluble in salt solutions. Ribosomes were not detected in the solubilized MA by sucrose gradient centrifugation. The method was modified slightly by: (1) omitting centrifugation of the final, dissolved product at 100,000 *g* for 1 hr and (2) by sacrificing yield for purity by centrifuging the undissolved MA suspended in 5/6 concentrated isolation medium alternately at 2000 *g* for 30 min and 1800 *g* for 3 min to remove whole eggs, egg fragments, yolk and other contaminants. This was done 3 times. The MA obtained in experiments 1 and 2 was almost completely free of contaminants visible by phase microscopy, and in experiment 3 no contaminants of the preparation were observed.

† Abbreviation used: MA, mitotic apparatus.

3. Results

(a) *Effect of puromycin*

We have confirmed Hultin's (1961) finding that 10^{-4} M-puromycin added at the time of fertilization prevents the first division and substantially reduces amino acid incorporation (Fig. 1). Even in the presence of puromycin, however, the first replication of DNA occurring near the time of pronuclear fusion procedes on schedule (Hinegardner *et al.*, 1964) (Fig. 2). If one delays the addition of puromycin until 25 to 30 minutes following fertilization, the first cleavage occurs (although the second is blocked), even though protein synthesis is still inhibited by about 75%. We conclude that most of the early protein synthesis is unessential for the first cell division, and that perhaps some essential but minor component(s) is synthesized during the first 30 minutes following fertilization. While puromycin might also lower in some way production of ATP (Giudice, 1965) needed for cleavage, analysis of ATP levels in puromycin-blocked eggs by Dr David Epel did not support this suggestion.

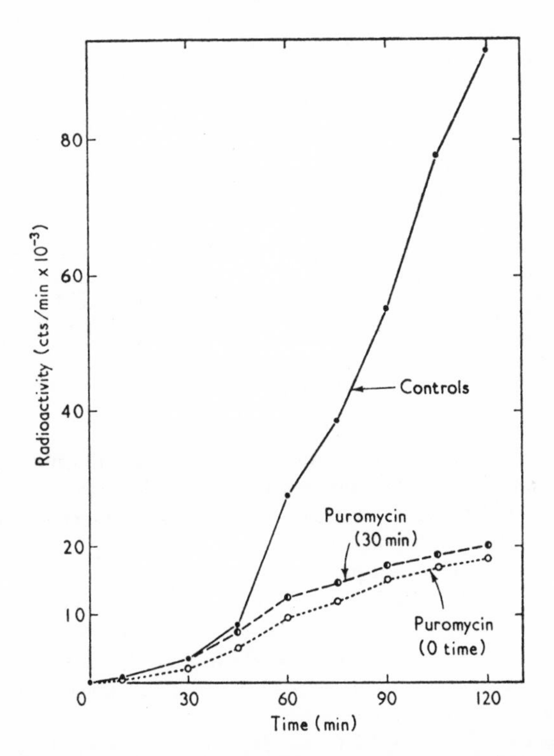

FIG. 1. Effect of the addition of puromycin (final concentration $= 10^{-4}$ M) on incorporation of [^{14}C]valine.

A 2% (v/v) suspension of eggs was inseminated in sea water containing 0·2 μc/ml. of L-[^{14}C] valine. The suspension was divided into 3 portions. Puromycin was added immediately to one portion (0 time, --○--○--) and to another portion 30 min after insemination (--◐--◐--). 5-ml. samples were withdrawn from the variously treated suspensions, pipetted into an equal volume of cold 10% trichloroacetic acid, and processed for amino acid incorporation. Control untreated (—●—●—).

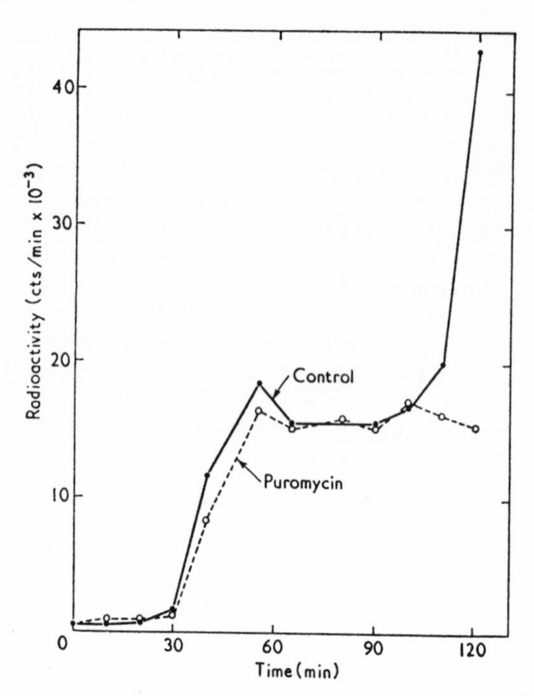

FIG. 2. Effect of puromycin on the incorporation of [^{3}H]thymidine into acid-insoluble product.

A 2% (v/v) suspension of eggs was fertilized and to half of it was immediately added puromycin (--○--○--) to a concentration of 10^{-4} M. One minute thereafter, [^{3}H]thymidine (final concentration 0·5 μc/ml.) was added to both batches and 4-ml. samples were removed at the times indicated. Metaphase was apparent in the control eggs (—●—●—) at 95 to 100 min following fertilization.

(b) *Synthesis of the mitotic apparatus*

The synthesis of the mitotic apparatus was detected by allowing cells to develop in the presence of radioactive leucine prior to the first cleavage, isolating the mitotic apparatus, and examining its constituent soluble proteins by sucrose density-gradient centrifugation. Sakai (1966) has shown that the MA isolated by this method seems free of the contamination of bulk cell protein and ribosomal nucleic acids, and is soluble in 0·53 M-KCl. The purified soluble MA protein is completely precipitable at pH 4·5. A small portion (12 to 15% in these studies) is insoluble in salt solution and is composed primarily of smooth-surfaced vesicles which invariably are found in association with the sea urchin MA (Harris, 1961; Harris & Mazia, 1962); their function is unknown, but their intimate and characteristic association with the MA region is regarded by us as indicating a possible role in cell division. The solubilized MA sediments in a sucrose density-gradient at 3·5 s, 13 s and 22 s (Fig. 3). Another component observed (Fig. 3) sediments at about 28 s and was probably removed in Sakai's (1966) earlier studies by a preliminary centrifugation. About 8% of the total protein collects at the bottom of the centrifuge tube; it may be aggregates of the above mentioned proteins (Sakai, 1966). The 3·5 s peak is the main component of the MA; it is a dimer of 2·5 s subunits bonded by disulfide bridges. Sakai's evidence suggests that the 13 s and 22 s components are higher states of aggregation of these subunits. The solubilized MA protein has recently been shown by Kiefer, Sakai, Solari & Mazia (1966) to be the protein of the microtubules of the MA.

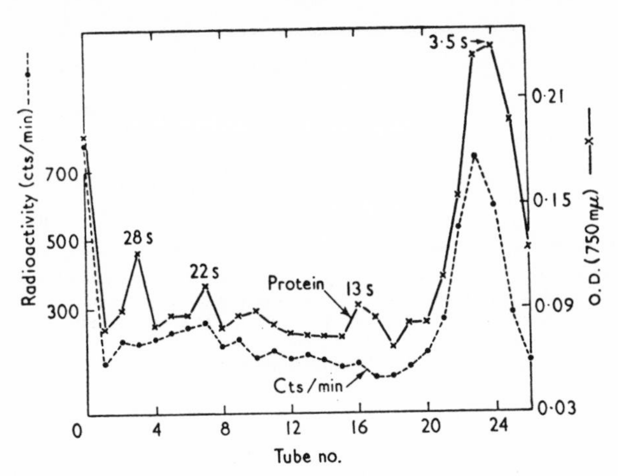

FIG. 3. Sucrose density-gradient centrifugation of the protein of the MA which is soluble in 0·53 M-KCl.

An 8% (v/v) suspension of eggs was fertilized. Forty-five seconds later, 2 vol. of calcium-free sea water containing 1 mg/ml. of mercaptoethylgluconamide, 0·1 M-EDTA (pH 8·0) was added. Four minutes later, 2000 μc of [^{3}H]leucine was added to the 350-ml. of suspension. After 30 min of development, membranes were removed by passing the eggs through bolting silk; after the eggs had settled, they were resuspended in calcium-free sea water 3 times. Development proceeded to metaphase as a 2% egg suspension in plain sea water. Isotope incorporation continued as if exogenous leucine was still present, even after washing of the eggs, because of the large radioactive amino acid pools.

The MA was prepared from metaphase eggs (95 min) by methods described in the text. After dissolution of the MA in 0·53 M-KCl, it was dialyzed against 0·003 M-Tris (pH 8·2) and centrifuged at 12,000 *g* for 20 min. The MA protein was precipitated by addition of acetic acid to pH 4·5, collected by centrifugation, and redissolved in 0·04 M-Tris (pH 8·4). This was layered over a 5 to 20% sucrose gradient containing 0·04 M-Tris (pH 8·4) and centrifuged for 3·5 hr at 39,000 *g* in the Spinco SW39 rotor. Fractions were collected after termination of the centrifuge run and 1·0 ml. of water was added to each fraction. 0·65 ml. of each diluted fraction was used for determination of protein, and 0·3 ml. for determination of isotope incorporation. The 3·5 s, 13 s, 22 s and 28 s peaks were reproducible. Radioactivity, cts/min (--●--●--); optical density (—×—×—).

If one labels the dividing cell, either throughout the period prior to first cleavage or between 30 minutes post-fertilization and cleavage, the isolated MA has a specific activity only slightly greater than that of the rest of the cell protein; but the KCl-insoluble portion of the MA has a much higher specific activity (Table 1). The amount of KCl-insoluble protein obtained from the MA in experiment 3 was too small to measure accurately and no specific activity could be determined.

The soluble MA protein was further examined by sucrose density-gradient centrifugation. The isotope is clearly incorporated into structural protein of the mitotic apparatus, and the specific activity of the several components does not differ dramatically (Fig. 3). Virtually identical results were obtained in all experiments regardless of whether the cells were labeled commencing at fertilization or 30 minutes thereafter.

Preparation number 3, which possessed an MA specific activity comparable to that of bulk cell protein, was also the one judged to be most free of contamination.

TABLE 1

Experiment	When labeled (min)	Spec. act. (cts/min/μg protein)		
		Soluble MA	KCl-insoluble	Bulk protein
1	30 to 90	86	196	47
2	0 to 90	358	612	204
3	0 to 90	136	—	120

The determination of protein and radioactivity was carried out as described in the Methods section. The bulk protein is the supernatant fraction of the dithiodipropanol cell lysate after the first centrifugation to sediment the total crude MA fraction. The soluble and insoluble MA represent portions of the final, purified MA fraction. The yield of purified MA was about 2% of the total cell protein in these experiments, and the KCl-insoluble portion constituted about 15% of the MA.

4. Discussion

Although these results are not altogether unlike those of previous investigations, they differ significantly in two ways which lead us to place a somewhat different interpretation on the events following fertilization. First, the results demonstrate that relatively uncontaminated and undenatured preparations of MA contain some newly synthesized structural proteins. Second, the specific activity of the solubilized MA is sufficiently close to that of whole cell protein that it is unlikely that early protein synthesis represents a heavy investment in the synthesis of structural MA protein. When considered in the light of the results with puromycin, it seems unlikely that synthesis of MA, while it may occur to a limited extent, is necessary or sufficient for the first cleavage to occur. The high specific activity of the portion of the MA insoluble in KCl is intriguing; this fraction contains the smooth-surfaced vesicles associated with the mitotic apparatus, and their role in its function deserves further exploration. Whether amino acid incorporation in cleaving sea urchin eggs represents *de novo* protein synthesis from yolk and amino acid pools, or turnover of pre-existent functional protein, or both, is not known. Furthermore, the absolute rate of polypeptide synthesis in cleaving sea urchin eggs is difficult to evaluate, although it probably corresponds to a small portion of the total egg protein during the period studied here. (Calculations made from unpublished data of Dr W. E. Berg indicate that the amount of protein synthesized between fertilization and the first cell division in *S. purpuratus* probably does not exceed 0·5% of the total egg protein.)

We wish to stress that from a functional point of view a relatively large investment in synthesis of structural proteins of the MA is not occurring and is not necessary. Most of the structural components of the MA may pre-exist in some form in the unfertilized egg; however, the synthesis of "catalytic components" functioning in cell division may be important.

This work was supported by grants from the National Institutes of Health (GM13882) and the National Science Foundation (GB2690). One of us (H.S.) is a United States Public Health Service International Postdoctoral Fellow.

REFERENCES

Bibring, T. & Cousineau, G. H. (1964). *Nature*, **204**, 805.

Giudice, G. (1965). *Develop. Biol.* **12**, 233.

Gross, P. R. & Cousineau, G. H. (1963). *J. Cell Biol.* **19**, 260.

Harris, P. (1961). *J. Biophys. Biochem. Cytol.* **11**, 419.

Harris, P. & Mazia, D. (1962). In *The Interpretation of Ultrastructure*, ed. by R. J. C. Harris, vol. 1, pp. 279–305. New York: Academic Press.

Hinegardner, R. T., Rao, B. & Feldman, D. E. (1964). *Exp. Cell Res.* **36**, 53.

Hultin, T. (1961). *Experientia*, **17**, 410.

Kiefer, B., Sakai, H., Solari, A. J. & Mazia, D. (1966). *J. Mol. Biol.* **20**, 75.

Lowry, O. H., Rosebrough, N. J., Farr, A. L. & Randall, R. J. (1951). *J. Biol. Chem.* **193**, 265.

Mangan, J., Miki-Noumura, T. & Gross, P. R. (1965). *Science*, **147**, 1575.

Mazia, D. (1961). In *The Cell*, ed. by J. Brachet & A. E. Mirsky, vol. 3, pp. 77–412. New York: Academic Press.

Mazia, D. & Roslansky, J. D. (1956). *Protoplasma*, **46**, 528.

Mazia, D., Mitchison, J. M., Medina, H. & Harris, P. (1961). *J. Biophys. Biochem. Cytol.* **10**, 467.

Monroy, A. (1965). *Chemistry and Physiology of Fertilization*. New York: Holt, Rinehart & Winston.

Sakai, H. (1966). *Biochim. biophys. Acta*, **112**, 132.

Stafford, D. W. & Iverson, R. M. (1964). *Science*, **143**, 580.

Went, H. A. (1959). *J. Biophys. Biochem. Cytol.* **6**, 447.

Gene Transcription in the Early Embryo

*PROPERTIES OF NUCLEAR RNA IN SEA URCHIN EMBRYOS**

BY ARTHUR I. ARONSON† AND FRED H. WILT

There is now considerable evidence in a variety of eukaryotic cell systems for a class of RNA molecules (neither ribosomal nor tRNA) that are confined to the nucleus.[1, 2] Both kinetic and actinomycin chase experiments provide evidence that a large fraction of this RNA turns over rapidly relative to the time required for cell division. There has been some speculation on the function of this RNA, with most suggestions falling within the realm of regulation.[1, 3, 4]

It was evident from earlier work[5] that this type of RNA existed in pregastrula sea urchin embryos and that at these early stages this RNA may be the only general class synthesized. We have been able to extend these studies by using early stages of sea urchin embryos and the excellent cell-fractionation techniques available for this system. We have found that a large majority of the nuclear RNA turns over rapidly relative to the cell division time and that only 6 per cent of this RNA is transported to the cytoplasm primarily as polysomes.[15] The large majority of this nuclear RNA exists as membrane-bound polysomes.

Materials and Methods.—(1) *Embryo cultures and labeling:* Eggs from *Strongylocentrotus purpuratus* were washed and inseminated, and their fertilization membranes removed when necessary by previously described techniques.[6] Suspensions of 1% (v/v) were cultured at 15°C in Millipore-filtered sea water containing 50 μg/ml streptomycin. Only cultures showing greater than 98% normal fertilization and development were used.

Radioactive precursors were diluted in sea water and added to cultures at the following final concentrations: 5-H^3-uridine (0.5–1.0 μc/ml); 2-C^{14}-uridine (0.1 μc/ml); L-ul-C^{14}-leucine (0.1 μc/ml). Labeling was terminated by pouring the embryo culture into chilled tubes or onto crushed frozen sea water for labeling periods of 1 min or less. Nuclei and cytoplasm were prepared by the method of Hinegardner.[7] In some instances his method was modified by omitting a 2.38 *M* sucrose layer from the bottom of the step gradient so that the nuclei were centrifuged to the bottom of the tube.

(2) *Chemical analysis:* RNA, DNA, and protein fractions were prepared by a modified Schmidt-Thannhauser procedure.[8, 9] Acid-insoluble radioactivity was determined directly in cells, cell fractions, and sucrose-gradient fractions by addition of bovine serum albumin (variable to 200 μg) and cold trichloroacetic acid to a final concentration of 7%. Precipitates were collected and washed on glass-fiber filters (Reeve Angel 934AH). Filters were heated at 70°C for 60 min before addition of fluor.

Nucleotides were separated by chromatography.[10, 11]

Sucrose-gradient zonal centrifugation was carried out by conventional techniques. Sedimentation coefficients were determined by employing the tables published by McEwen.[12]

(3) *Nuclear fractionation:* The nuclear pellets were suspended in either TKM (0.05 *M* Tris, 0.1 *M* KCl, 0.007 *M* MgAc, pH 7.8) at 0°C or T-KCl (TKM with MgAc omitted) buffer containing bentonite (200 μg/ml). The latter buffer was used for detergent treatment with 1% deoxycholate (DOC) plus 0.5% Lubrol W. After lysis, 1.0 *M* MgAc was added to 0.007 *M* and the suspension was briefly sonicated (1 sec two or three times) in a Branson model S-75 sonifier at setting no. 4. The suspension was centrifuged at 18,000 × *g* for 15 min. The supernatant from this centrifugation was then analyzed in a 15–30% linear sucrose gradient in TKM buffer underlaid with 0.2 ml of 40% sucrose.

(4) *RNA preparation:* The material was suspended in TKM buffer, bentonite (400 μg/ml), and sodium dodecyl sulfate (SDS) to 1%. An equal volume of redistilled phenol saturated with buffer and containing 0.1% 8-hydroxyquinoline was added and the suspension shaken at room temperature. After centrifugation, the aqueous phase was incubated with pronase (100 μg/ml) for 3 hr at room temperature. The phenol extraction was repeated twice and the aqueous phase was precipitated with 2 vol of 95% ethanol. The precipitate was dissolved in 0.1 *M* NaAc–0.001 *M* Na_3 EDTA, pH 5.5, and dialyzed at 4°C for 16 hr against the same buffer.

(5) *Counting:* Aqueous samples were counted by addition of 0.1–1.0 ml to 10 ml of Bray's solution. The extent of quenching was determined by either the channels ratio method or external standard method on a Nuclear-Chicago scintillation counter. Filters with radioactive precipitates were counted in 5 ml of toluene fluor.

(6) *Materials:* 5-H^3-uridine (26.6 c/mM), 2-C^{14}-uridine (55 mc/mM), and ul-C^{14}-leucine (273 mc/mM) were purchased from New England Nuclear Corp.

Results.—(1) *Kinetics and distribution:* Upon addition of H^3-uridine to a suspension of embryos, there is a very rapid uptake into the acid-soluble pool (Fig. 1). Fractionation of the acid extract by thin-layer chromatography[11] has provided evidence that >95 per cent of the radioactivity is in phosphorylated compounds (UMP, UDP, UTP). There is also rapid incorporation of the precursor into RNA as revealed by the increased specific activity of the 2′(3′) UMP released from RNA.

Since there is an excess of precursor[5] and the quantity of RNA per embryo remains essentially constant,[13a] these kinetics indicate rapid turnover of the newly synthesized RNA. The same kinetics of RNA synthesis are found at stages when cell division is rapid (64–200 cells) and when cell division has slowed down considerably (200–500 cells).[6]

It appeared from earlier radioautographic studies[5] that most of the RNA was localized in the nucleus and that all the nuclei were labeled. A more quantitative determination was made by directly isolating nuclei. It was first necessary to establish that the procedures for nuclear isolation resulted in no leakage of macromolecules or extensive contamination by cytoplasm. The former criterion was met by pulse-labeling embryos with H^3-uridine and isolating nuclei (Table 1). Less than 1 per cent of the radioactive RNA and nuclear DNA was found in the cytoplasm. Cytoplasmic contamination of nuclei appeared to be 1 per cent or less on the basis of a reconstruction experiment with purified pulse-labeled RNA (or a crude double-labeled extract of embryos[18]) added to washed embryos just prior to homogenization.[15] Recycling the nuclei through a second sucrose gradient did not alter their specific activity.

Table 1 shows that the specific activity of the nuclei is about 3.5 times that of the total extract. If all the pulse-labeled RNA belongs with the nucleus, this difference means that about 28 per cent of the total cellular RNA is associated

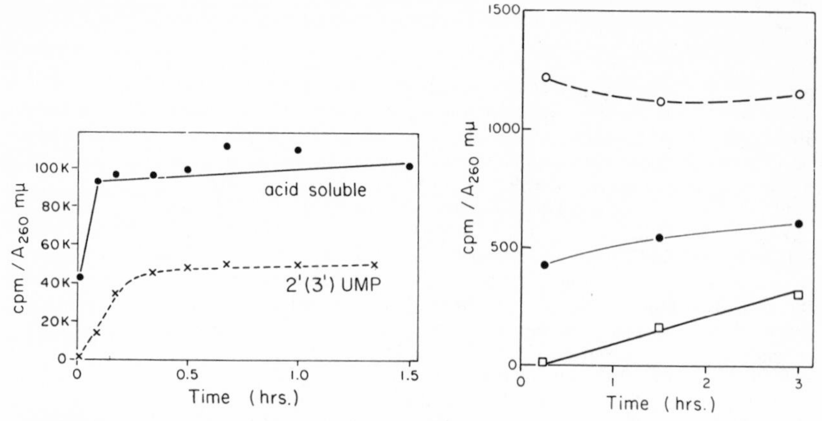

(*Left*) FIG. 1.—Kinetics of incorporation of 5-H^3-uridine (1 μc/ml) into hatched blastulae (23.5 hr).

(●—●), cpm/A_{260} mμ in acid-soluble fraction,[10] (×—×), cpm/A_{260} mμ of 2′(3′) UMP from hydrolyzed RNA.

(*Right*) FIG. 2.—Changes in specific activity with time of the RNA of various cell fractions; 24-hr embryos (hatched blastulae) incubated with 0.8 μc/ml 5-H^3-uridine.

(○—○), nuclei; (●—●), total extract; (□—□), cytoplasm.

with the nucleus. Measurements of the RNA content of nuclei from hatched blastulae (22 hr) gave values of 23 and 28 per cent.

When embryos are incubated continuously with H^3-uridine, the specific activities of the cytoplasm and nuclei change as shown in Figure 2. The nuclei rapidly reach a constant specific activity. After an initial lag of about 15 minutes, the specific activity of the cytoplasmic RNA appears to increase linearly over a 3-hour period. Many other experiments of varying time periods up to 3 hours have been done with embryos at the 64-cell stage to late mesenchyme blastulae with similar results. If all RNA is made in the nucleus, then the rate of increase of specific activity of the cytoplasm is consistent with a transfer of 6 per cent of the RNA made every 15–20 minutes. Radioautographic studies show no extensive heterogeneity in terms of cytoplasmic labeling; it does not appear likely, therefore, that a few cells are releasing all their nuclear RNA to account for the average value. Nor do the specific activities change when unbroken cells and embryos are rehomogenized and fractionated.

TABLE 1. *Localization of pulse-labeled RNA in subcellular fractions of hatched sea urchin blastulae.*

Fraction*	Total cpm† (× 10^{-3})	Cpm/A_{260} mμ (× 10^{-3})
Total extract	540	12.3
Cytoplasm	4	0.35
Nuclei	490	40.5

* Hatched embryos incubated with 4 μc/ml 5-H^3-uridine for 10 min. Fractionation as described in *Materials and Methods*.

† Corrected for 80% recovery of cytoplasm (losses in centrifugation and sampling) and 60% recovery of nuclei (based on DNA determinations).

(2) *Nature of nuclear pulse-labeled RNA:* Nuclei were isolated from hatched embryos that had been incubated for ten minutes with H^3-uridine. Various treatments were then used in an attempt to solubilize the radioactive RNA (Table 2). While sonication completely destroyed the integrity of the nucleus, as judged by phase microscopy or by release of DNA into the soluble fraction, very little of the radioactive RNA was released (Table 2, *a–f*). Recoveries of DNA and RNA were not good, presumably owing to release (or activation) of nucleases by sonication. About 75 per cent of the labeled RNA is released after treating nuclei with DOC-Lubrol. The RNA in the Sorvall pellet cannot be released by further detergent treatment nor with DNase. RNase or pronase can solubilize this radioactivity. Examination of this pellet in the electron microscope showed what seemed to be extensive aggregates of polyribosomes with membrane fragments.[15]

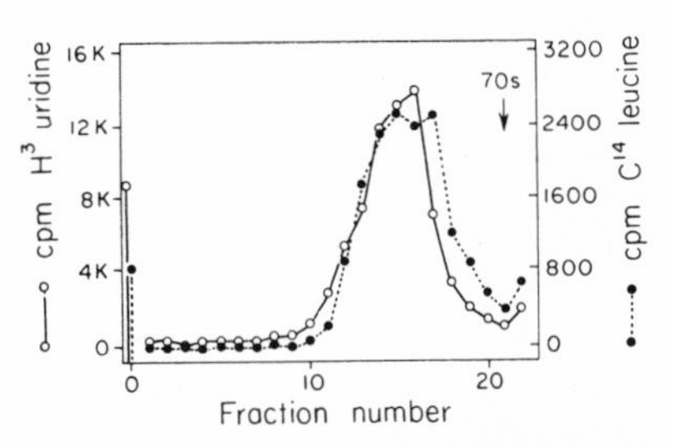

FIG. 3.—Sucrose gradient fractionation of nuclear extract solubilized by treatment with DOC-Lubrol (as in Table 2*g*); 21-hr embryos incubated for 10 min with 5-H^3-uridine and for 45 sec with ul-C^{14}-leucine. Centrifugation in a 10–30% sucrose gradient (TKM buffer), SW39 rotor, 20,000 rpm/20 min. Pellet removed by washing with TKM; these values are plotted along the ordinate. 98% of input radioactivity recovered.

The nuclear supernatant (Table 2*g*) was fractionated on a sucrose gradient (Fig. 3). In this particular experiment the embryos were also pulse-labeled (45 sec) with C^{14}-leucine. Both precursors are found in components that cosediment in a broad peak at greater than 70*S* (approximately 150–400*S*). Even after this brief, low-speed sedimentation, a large fraction of the radioactivity is in the pellet. More than 80 per cent of the labeled RNA placed on the gradient is found in these rapidly sedimenting fractions. After a 45-second pulse with amino acid, about 10 per cent of the labeled polypeptide is found associated with the nucleus. About one half of this (5–6% of the total) is solubilized by detergent treatment, and 80 per cent of that solubilized cosediments with the labeled RNA. Examination of both the pellet and the peak tubes (14–16) in the electron microscope showed a predominance of polyribosomes with a few membrane fragments.[15]

By a number of other criteria, the radioactive RNA appeared to be associated with polysomes. (1) The rapidly sedimenting radioactive components were destroyed by RNase and pronase. (2) Most of the labeled RNA remained near the top of the gradient as a result of extraction of nuclei in T-KCl buffer in the presence of 0.01 *M* EDTA. (3) The radioactive profile was not altered by treatment with DNase except for removal of some counts near the top of the gradient (see Fig. 5). (4) Extraction of the peak tubes (14–16 of Fig. 3) with phenol and sedimentation of the RNA showed (Fig. 4) that the 28*S* component is unlabeled and that there is a partial overlap of the radioactive RNA with the 18*S* peak. (5) A very short pulse-labeling with radioactive amino acids results in

TABLE 2. *Fractionation of pulse-labeled nuclei.*

Fraction*	Per cent total nuclear cpm†
a. Total nuclear extract, 15-sec sonication (in TKM)	100
b. Supernatant‡ of *a*	4 (1–5)
c. Pellet‡ of *a*	70 (70–85)
d. Total nuclear extract (in TKM), 45-sec sonication	100
e. Supernatant‡ of *d*	3
f. Pellet‡ of *d*	58
g. Supernatant of nuclei treated with DOC-Lubrol‡	72 (60–80)
h. As in *g*-pellet‡	22 (20–35)

* Hatched blastulae (21 hr) incubated for 10 min with 0.8 μc/ml 5-H^3-uridine. Nuclei isolated as described in *Materials and Methods.*

† 8 × 10^5 cpm (TCA precipitate) in purified nuclei. Figures in parentheses refer to range of % recovery in several experiments.

‡ Centrifuged at 12,500 rpm/15 min in a Sorvall SS-34 rotor.

some labeled polypeptide cosedimenting with the newly made RNA (see Fig. 3). This same distribution of radioactive RNA in the nucleus was also found for pulse-labeled 12-hour embryos and 90-minute labeled 21-hour embryos (hatched blastulae).

As a control, pulse-labeled purified RNA (a broadly sedimenting fraction with a peak at about 30*S*) was added to nuclei that were fractionated by detergents and sucrose gradients. None of the added RNA behaved as does that described above; i.e., its sedimentation properties were unaltered.

An *in vivo* control was provided by incubating embryos for a prolonged period with 2-C^{14}-uridine (Table 3). Under appropriate conditions, the C^{14}-uridine (or UTP) pool can eventually be depleted via synthesis of stable cytoplasmic

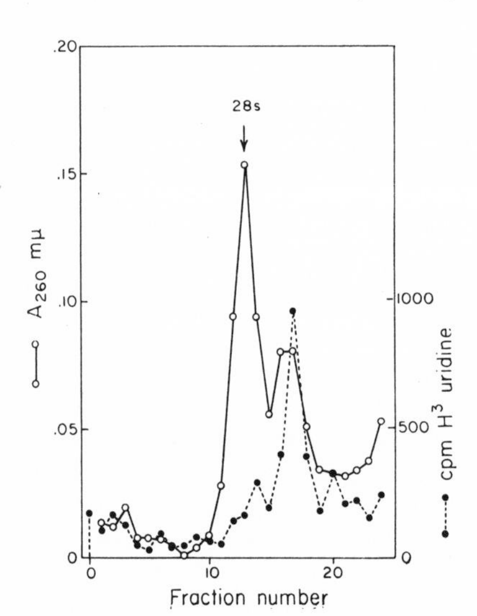

FIG. 4.—Sucrose gradient fractionation of RNA prepared from the polyribosome region of a nuclear extract (as in Fig. 3 except embryos incubated solely with 5-H^3-uridine for 10 min). RNA centrifuged in a 10–30% sucrose gradient in 0.1 *M* NaAc–0.001 *M* Na_3EDTA, pH 5.5, plus 0.1% SDS for 12 hr at 20,000 rpm in an SW39 rotor at 23°C.

TABLE 3. *Specific activities of embryo fractions after a long-term incubation and chase.*

Fraction*	Cpm/A_{260} mμ (× 10^{-3}) C^{14}†	H^3‡
Total extract	1.45	7.0
Cytoplasm	1.80	0.6
Nuclei§	0.41	23.0

* 23-hr embryos (hatched) fractionated as described in *Materials and Methods*. Specific activities determined on alkali-labile fraction.

† 2-C^{14}-uridine (0.1 μc/ml) added to 7-hr embryos for 12 hr. Embryos then washed and incubated with 10^{-5} *M* uridine for 2.5 hr. They were again washed and resuspended for 2.5 additional hr.

‡ 5-H^3-uridine added to 0.5 μc/ml for final 10 min of incubation.

§ Specific activity measured after centrifuging nuclei through two sucrose gradients.

RNA, but mainly due to extensive DNA synthesis. Under these conditions, the specific activity of the nuclei (C^{14}-RNA) is considerably less than that of the cytoplasm and the nuclei contain only 5–10 per cent of the C^{14}-RNA found in the cytoplasm. The nuclear radioactivity may be due to contamination or (more likely) to the presence of a small amount of C^{14}-UTP in the precursor pool. After the prolonged incubation, the embryos were also pulse-labeled for ten minutes with H^3-uridine. The nuclei were lysed with DOC-Lubrol, and the soluble fraction was separated on a sucrose gradient (Fig. 5). In this case, the majority of the C^{14} in the pellet and at the top of the gradient is found in a DNase-sensitive form. Only a marginal quantity of C^{14} is found to sediment faster than 70*S*.

We calculate that only about 2 per cent of the total cytoplasmic RNA is radioactive.[13b] If we assume that this labeled cytoplasmic RNA is representative of that which may contribute to nuclear contamination, then the results of Figure

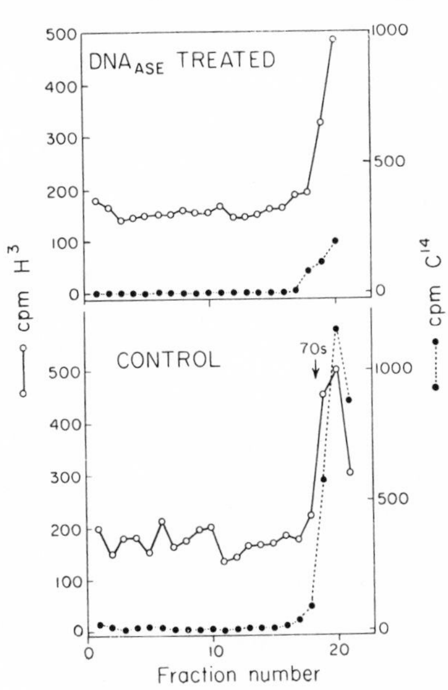

FIG. 5.—Sucrose gradient fractionation of a nuclear extract (prepared as in Fig. 3) from 23-hr embryos (hatched blastulae) incubated for 12 hr with 2-C^{14}-uridine and for 10 min with 5-H^3-uridine, as described in footnote to Table 3. Sedimentation in a 15–30% sucrose gradient (in TKM buffer), 45 min at 35,000 rpm in an SW39 rotor. For DNase treatment, 0.2 ml of nuclear extract incubated for 10 min at room temperature with 10 μg DNase. The radioactivity in the pellets was: control: H^3, 11,000 cpm and C^{14}, 2,000 cpm; DNase-treated: H^3, 9,000 cpm and C^{14}, 150 cpm.

5 (correcting for yield) indicate that there is less than 1 per cent contamination by cytoplasmic RNA components of this rapidly sedimenting nuclear fraction.

Discussion.—The results of kinetic, autoradiographic, and cell fractionation experiments with pregastrula sea urchin embryos directly demonstrate that over 90 per cent of the newly synthesized RNA rapidly turns over in association with nuclei. This confirms the original suggestion by Harris and the subsequent experiments by several groups.[1] Since these pregastrula embryos do not synthesize ribosomal or transfer RNA, there is no confusion with synthesis and modification of the precursors of ribosomal and transfer RNA. Labile nuclear RNA apparently is a prominent feature of eucaryotic cells—we would suggest that it is a distinctive feature of virtually all eucaryotes throughout their life history.

It is generally agreed that this RNA is in a very large particulate form[2] and has a very short lifetime.[1] While it is difficult to eliminate artifactual aggregation completely, our reconstruction experiments rule out important participation of cytoplasmic contaminants in this process. Since the labile nuclear RNA is probably associated with membranes, aggregation of attached polysomes with each other or with membrane fragments is possible. We feel that our evidence is best explained at present by proposing that the nuclear RNA is present and functioning in membrane-bound "nuclear" polysomes (cf. ref. 14). The aggregates are large, contain labile RNA and pulse-labeled polypeptide, and are sensitive to EDTA, RNase, and pronase (but not DNase). They have the morphology of polysomes under the electron microscope, and 28*S* and 18*S* ribosomal RNA can be extracted from them. The latter fact may have gone undetected in other systems because the RNA content of this nuclear subfraction represents $\leq$ 1 per cent of the total embryo RNA.

The nuclei of hatched blastulae of sea urchins are bounded by a double membrane. Electron micrographs of thin sections of isolated nuclei show large clusters of ribosomes on the outer membrane (Wartiovaara, unpublished results). Since nuclei interact structurally and functionally with the rest of the cell, it is difficult to define "pure" nuclei. The present results do show, however, that the nuclei as isolated comprise a distinct functional compartment, since virtually all the pulse-labeled RNA is found associated with them.

Our results also indicate that nuclear RNA is a precursor to a cytoplasmic RNA fraction which may be messenger; it is nonribosomal, nontransfer RNA, which is associated with polysomes.[15] The rate of accumulation of radioactive cytoplasmic RNA is slower than that of nuclear RNA, and it does not seem to turn over as rapidly. Most importantly, it appears only after a lag of 10–15 minutes, and thereafter increases at a constant rate of 6 per cent of the initial rate of nuclear RNA synthesis.

The present experiments do not directly bear upon the similarity of cytoplasmic and nuclear RNA. The simplest hypothesis is that labile nuclear RNA is a family of messengers, some of which pass to the cytoplasm and are stabilized there. Selection of RNA species (or more specifically polysomes) in the nuclear family may occur, and the cytoplasm may play a role in this selection.[3] In contrast to bacteria, transcription may represent a scanning process (primarily of very long repetitive DNA sequences[16]), and posttranslational selection may then dictate the flow of information to the cytoplasm. Perhaps small molecules in the cytoplasm can stabilize nuclear polysomes through interaction with the nascent polypeptide.[17] We suggest that in eucaryotes control of gene activity (qualitative and quantitative) may occur without invoking repressor proteins and operator loci.

* Supported by research grants from the National Institutes of Health (F. H. W.) and National Science Foundation (A. I. A.). We thank Lois Atkinson and Margaret Kay for assistance.

† Guggenheim fellow (1967–1968).

[1] Harris, H., in *Progress in Nucleic Acid Research*, ed. R. N. Davidson and W. E. Cohn (New York: Academic Press, 1963), vol. 2, p. 19; Warner, J., R. Soeiro, C. Birnboim, and J. Darnell, *J. Mol. Biol.*, **19**, 349 (1966); Attardi, G., H. Parnas, M. Hwang, and B. Attardi, *J. Mol. Biol.*, **20**, 145 (1966); Scherrer, K., L. Marcaud, F. Zaidela, I. London, and F. Gros, these PROCEEDINGS, **56**, 1571 (1966).

[2] Penman, S., C. Vesco, and M. Penman, *J. Mol. Biol.*, **34**, 49 (1968).

[3] Church, R., and B. J. McCarthy, these PROCEEDINGS, **58**, 1548 (1967).

[4] Bonner, J., and J. Widholm, these PROCEEDINGS, **57**, 1379 (1967).

[5] Kijima, S., and F. H. Wilt, *J. Mol. Biol.*, in press.

[6] Hinegardner, R. T., in *Methods in Developmental Biology*, ed. F. H. Wilt and N. K. Wessells (New York: Thomas Crowell Co., 1967), p. 139.

[7] Hinegardner, R. T., *J. Cell Biol.*, **15**, 503 (1962).

[8] Munro, H., and J. Fleck, *Methods Biochem. Anal.*, **14**, 113 (1962).

[9] Ceriotti, G., *J. Biol. Chem.*, **198**, 297 (1952).

[10] Katz, S., and D. G. Comb, *J. Biol. Chem.*, **238**, 3065 (1963).

[11] Neuhard, J., E. Randerath, and K. Randerath, *Anal. Biochem.*, **13**, 211 (1965).

[12] McEwen, C. R., *Anal. Biochem.*, **20**, 114 (1967).

[13] (*a*) Tocco, G., A. Orengo, and E. Scarano, *Exptl. Cell Res.*, **31**, 52 (1963); (*b*) This calculation is based on the assumption that about 0.5% of the RNA in the embryo is synthesized every 15–20 min,[5] and that 6% of this material is transferred to the cytoplasm as stable RNA.

[14] Sadowski, P. D., and J. A. Howden, *J. Cell Biol.*, **37**, 163 (1968); Bach, M. K., and H. G. Johnson, *Nature*, **209**, 893 (1966).

[15] Aronson, A. I., and F. H. Wilt, manuscript in preparation.

[16] Britten, R., and D. Kohne, *Science*, **161**, 529 (1968).

[17] Schimke, R. T., E. W. Sweeney, and C. M. Berlin, *Biochem. Biophys. Res. Commun.*, **15**, 214 (1964); Schimke, R. T., *J. Biol. Chem.*, **239**, 3808 (1964).

[18] Embryos were labeled with 2-C^{14}-thymidine for 18 hr and 5-H^3-uridine for 15 min prior to collection at 22 hr.

Rate of Nuclear Ribonucleic Acid Turnover in Sea Urchin Embryos

S. Kijima and Fred H. Wilt

1. Introduction

Students of development have been concerned for a long time with the causal relationships of differential genetic activity and the observable phenomena of growth and development. The use of genetic techniques to probe these questions remains limited. The model stating that the sole product of a gene is RNA allows the investigation of some aspects of gene action and development by strictly chemical methods. The purpose of the present paper is to report a quantitative characterization of RNA synthesis in developing sea urchin embryos. The initiation of this research arose from the belief that eventually quantitative information on RNA synthesis would have to be supplied for some developing systems if a chemical approach to gene action in development was going to be useful. Furthermore, such information might help re-evaluate some ideas about the molecular basis of development.

Our approach is direct; an isotopically labeled precursor of RNA is administered to developing embryos, and measurements of the specific activity of the precursor pool and the RNA are made as a function of time. Guanosine was used as a precursor in all the experiments because it simplifies the technical procedures and interpretation: guanosine and its phosphorylated derivatives are not easily converted to other RNA precursors in most cells, and this is also true for the sea urchin. Hence, incorporation may be equated to synthesis and cannot be confused with non-synthetic turnover of the CMP–CMP–AMP† portion of transfer RNA. Second, an immediate precursor of the RNA, GTP, is relatively easy to isolate from the acid-soluble pool. The assumption underlying this type of analysis is that the precursor pool is not heterogeneous, and the data obtained are consistent with that assumption.

Some measurements have been made of the rate and amount of RNA synthesized during embryonic development. Pannbacker (1966) has shown that rate of RNA synthesis in developing slime molds does not vary much from stage to stage. Brown & Littna (1964,1966) and Brown & Gurdon (1966) have used indirect methods to measure the synthesis of different types of RNA during development of the toad embryo. They found that the rate of accumulation of various types of RNA changed markedly with different developmental stages, and they postulated very precise regulation of synthesis of the different types of RNA. The developing sea urchin embryo has been the subject of a number of recent investigations on the patterns of RNA synthesis (see Siekevitz, Maggio & Catalano, 1966, for references). There is fairly general agreement on the qualitative description of the various types of RNA being synthesized at various times. Synthesis of heterogeneous, non-ribosomal, non-transfer RNA seems to predominate prior to gastrulation, and subsequently labeling of transfer RNA and ribosomal RNA can be detected. Really all that is known, however, is the ease of labeling of a component, and almost nothing is known about the amount synthesized, rate of synthesis and stability of the RNA. The pool sizes of the nucleoside triphosphates apparently decrease during development, and it is possible that RNA synthesis early in development is not negligible (Yanagisawa & Isono, 1966; Hultin, 1957; Nilsson, 1959,1961). Furthermore, some autoradiographic studies of cells labeled with RNA precursors have been carried out, and they show that virtually all the RNA synthesis early in development is restricted to the nucleus, although by the time of gastrulation some radioactivity can be detected in the cytoplasm (Ficq, Aiello & Scarano, 1963).

The cells of the sea urchin embryo differ in a fundamental respect from many other procaryotic and eucaryotic cells which have been examined—they are a closed system. There is no growth, and no net accumulation of ribonucleic acid or change in dry weight. DNA content does increase markedly, of course. Formally, all synthesis in these cells is turnover. It must be remembered, however, that a good deal of this turnover is probably a special kind of synthesis—degradation of storage materials, i.e. yolk, and synthesis of functional macromolecules.

The results we have obtained show that RNA synthesis is very active as early as mid-cleavage, and on a per cell basis actually becomes less active as development proceeds. The RNA synthesized is primarily restricted to the nucleus and turns over rapidly, although later in development some RNA accumulates and can be detected in the cytoplasm.

2. Materials and Methods

(a) *Biological materials*

Embryos of *Strongylocentrotus purpuratus* derived from a single female were used for each experiment. Fertilization, removal of fertilization membranes and embryo culture were carried out by the usual techniques (Hinegardner, 1967). Embryos were cultured as a 1 to 2% (v/v) suspension in Millipore-filtered sea water containing 50 μg each of penicillin and streptomycin sulfate/ml. A 1% suspension of these eggs contains about $1{\cdot}8 \times 10^4$

† Abbreviations used: GMP, UMP, AMP and CMP, the monophosphates of guanosine, uridine, adenosine and cytidine, respectively; GTP, UTP, ATP, the triphosphates of the same nucleosides.

eggs/ml. In cultures that were carried through gastrulation, the embryo concentration was reduced to 0·3%. Development proceeded with stirring at 15°C.

Bacterial contamination was examined with labeling techniques several times during the course of the experiments. If shed eggs are washed with filtered sea water and inseminated with inactive sperm (fertilizing capacity inactivated by heating at 45°C for 20 min), there is no incorporation of RNA precursor into acid-insoluble material during an 8-hr period; these eggs do, nonetheless, take up precursor slowly and convert it to the triphosphate (unpublished results). Furthermore, in the presence of streptomycin, less than 1% of the acid-insoluble radioactive material of a sample taken during a labeling experiment was present in the sea water from which embryos were removed by centrifugation. Incorporation of label by bacteria can therefore be completely eliminated from these experiments.

(b) *Labeling of embryos*

All experiments were initiated by addition of [^{3}H]guanosine to a final concentration of 0·5 to 1·5 μc/ml. ([8-^{3}H]guanosine, 2·7 c/m-mole, International Chemical and Nuclear) to a stirred culture of embryos (about a 1% concentration of embryos). Samples of the embryo suspension of known volume were removed periodically and fixed in 3:1 alcohol–glacial acetic acid. The fixed embryos were used for autoradiography and determination of embryo concentration. Incorporation was monitored on samples of 5 to 30 ml. that were removed and quickly centrifuged in a hand centrifuge (this took 15 to 20 sec) and washed 3 times with ice-cold sea water containing 10^{-5}M unlabeled guanosine; the embryos were then frozen on dry ice and maintained at −20°C until analysis. This washing procedure removed more than 99% of the extracellular guanosine.

(c) *Acid-soluble fraction*

The frozen embryos were treated to obtain the acid-soluble pool by 3 extractions lasting 10 to 20 min with 2 to 4 ml. of ice-cold 0·5 N-perchloric acid. During the first extraction, the embryos were briefly sonicated. No further radioactivity could be liberated following the third extraction. The acid extracts were clarified by centrifugation, adjusted to pH 8 with KOH, the $KClO_4$ removed by centrifugation, and samples taken for determination of optical density at 260 mμ and radioactivity. The remaining portions were analyzed for the specific activity of the GTP.

(d) *Acid-insoluble fraction*

The acid-insoluble residue was washed once with cold 5% trichloroacetic acid, twice with methanol, and incubated in 0·3 N-KOH at 37°C for 16 to 20 hr after addition of 2 mg of calf thymus DNA. After hydrolysis, 0·5 N-perchloric acid was added until pH 2 was attained, and the samples were allowed to stand in the cold for at least 6 hr. The resultant precipitate was removed by centrifugation, and a portion of the supernatant fraction was used for determination of absorbance at 260 mμ and radioactivity. The remainder was analyzed for the specific activity of the 2′-3′ GMP derived from the RNA by hydrolysis.

(e) *Isolation of guanosine monophosphate*

Guanosine monophosphate was isolated from the alkaline hydrolysate on Dowex-50 by the method of Katz & Comb (1963); occasionally it was isolated by paper chromatography (Lane, 1963; Brown & Littna, 1966). Agreement between the two methods was better than 5%. There was enough material in each experiment for re-analysis of GMP specific activity, and the reproducibility is as good as the method (±3%).

Figure 1 shows a typical separation of the acid-insoluble, alkali-labile material on Dowex-50. Not all the radioactive material is located in the GMP fraction. Some of it (about 30%) is eluted by 0·05 N-HCl behind the UMP peak. This material has subsequently been identified as ^{3}HHO arising from a base-catalyzed exchange reaction (Jardetzky, 1964) between ^{3}H on the 8-position of guanine and H_2O (Wilt, 1969). Therefore, all the radioactivity in the alkaline hydrolysate is derived from labeled GMP.

(f) *Isolation of guanosine triphosphate*

Guanosine triphosphate was isolated from the acid-soluble extracts on small columns of Dowex-1-formate based on the experience of Hurlbert, Schmitz, Brumm & Potter

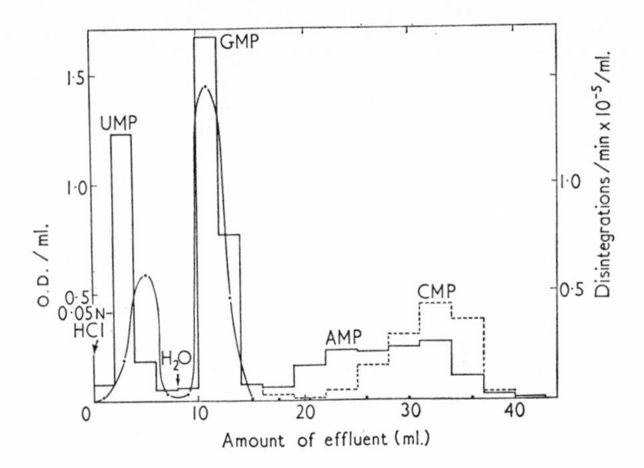

Fig. 1. Fractionation of alkali-labile, acid-insoluble radioactive material on a Dowex-50 column according to the directions of Katz & Comb (1963).

The wavelengths used for determination of absorbance of the different nucleotides were UMP, 260 mμ; GMP, 257 mμ; AMP, 259 mμ; CMP, 279 mμ. Recovery of total counts was 98%. The acid-insoluble fraction was prepared from hatching blastulae labeled for 30 min with [^{3}H]guanosine (1 μc/ml.). The block pattern represents absorbance, and the smooth curve represents radioactivity. Amounts of CMP and AMP present in the eluate were determined by spectrophotometric means (Katz & Comb, 1963).

(1954). A simplified method was devised using stepwise gradients of formate. Column size, flow rate, formate concentration and column capacity were systematically examined using pure standards of various nucleotides. Nucleotides and nucleosides were purchased from Calbiochem.

A 0·5 cm × 4·5 cm column of Dowex-1-formate was poured into a disposable Pasteur pipette plugged with glass wool. After washing the column with water, the sample (in 0·05 N-NH_4OH) was applied to the column, and the column was washed with 5 ml. of water. Nucleotides were eluted by gravity flow with 5 ml. of 1 N-formic acid, 0·5 N-ammonium formate (solution A), followed by 22 ml. of 1 N-formic acid, 0·75 N-ammonium formate (solution B). Finally, the GTP was eluted with 1 N-formic acid, 1·8 N-ammonium formate (solution C); three 2-ml. fractions of the latter were collected. Other triphosphates and diphosphates were completely eluted prior to solution C. A column which was not charged with acid-soluble fraction was run in parallel in every experiment to provide spectrophotometric blanks. Six columns could be run simultaneously and the analysis required about 6 hr. Authentic [^{14}C]GTP (New England Nuclear, 300 μc/m-mole) was included with some experimental samples and was recovered (> 90%) in the putative GTP fractions. Desalting was accomplished by diluting the GTP fraction fivefold in water, passing this over a 0·5 cm × 2·0 cm column of Dowex-1-formate, and eluting the GTP with 6 ml. of 0·5 N-HCl. The eluant was concentrated by flash evaporation. Thin-layer chromatography (Randerath & Randerath, 1964) of the GTP fractions showed only one ultraviolet-absorbing spot, which corresponded to the position of GTP.

Figure 2 shows results obtained using the above procedure. Authentic GTP is clearly separated from other likely contaminants (Fig. 2(a)) and the acid-soluble fraction from embryos labeled with [^{3}H]guanosine contains material which elutes from the column as if it were GTP (Fig. 2(b)). We did occasionally encounter anomalies in the behavior of the columns; e.g., some of the GTP emerged in the later part of solution B rather than with solution C. This was probably due to poor column packing and irregular flow rates. Using this elution schedule, no other common nucleoside polyphosphates were found to contaminate the GTP fraction. In the experiments reported in this paper, the spectra at acid,

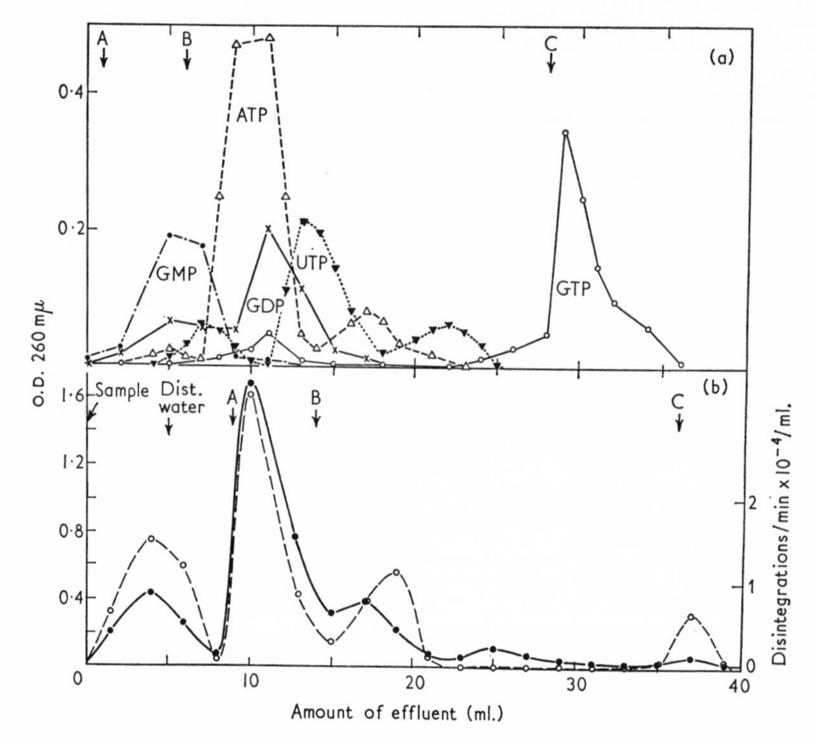

Fig. 2. Fractionation of nucleoside phosphates on Dowex-1-formate.

(a) Elution pattern of some authentic nucleotides from Dowex-1-formate as described in Materials and Methods. GMP, GDP, GTP, ATP and UTP were run on parallel columns with identical elution schedules and all the data are presented as one graph. A, B and C represent the addition of different eluting solutions. After absorption of nucleotides and prior to elution with buffer A, the columns were washed with distilled water. The elution peaks of the various nucleotides are indicated in the diagram.

(b) The same procedure when applied to fractionation of the acid-soluble fraction from labeled embryos. The acid-soluble fraction was prepared from hatching blastula (18 hr post-fertilization) exposed to 1 μc of [^{3}H]guanosine/ml. for 5 min. The solid line indicates the optical density at 260 mμ, and the dashed line the radioactivity.

neutral and alkaline pH of all GTP fractions were determined after desalting, and the spectral ratios were close to pure GTP. Specific activity of the GTP in the desalted sample was close to that determined on column effluents.

(g) *Counting*

All samples were counted by mixing 0·1 ml. of sample with 10 ml. of Bray's solution (1960). Quenching and absolute efficiency were determined by the channels ratio of a ^{133}Ba external standard on a Nuclear Chicago scintillation counter. Efficiencies in a group of samples worked up together were always within 3% of one another, and typical efficiencies were about 23%.

(h) *Autoradiography*

Autoradiography was carried out by dipping 5-μ sections of labeled embryos into Ilford K5 emulsion. After exposure, the autoradiograms were developed with D19 developer, and lightly stained through the emulsion with acid hematoxylin. Prior to dipping, all sections were treated with DNase as outlined by Wessells (1964). Controls were also carried out in which sections were digested with RNase after DNase treatment. Grain counts overlying nuclei, cytoplasm and unlabeled cells were made in at least 75 cells selected at random and counted under oil immersion. Grain counts of another 75 cells on the same slide did not differ from the first counts by more than 10%.

3. Results

(a) *Rate of RNA synthesis*

Four stages of development were selected for detailed study; several other intermediate stages were examined less intensively. Figure 3 shows the kinetics of incorporation of guanosine into the GTP pool and into the 2′-3′ GMP of RNA at these

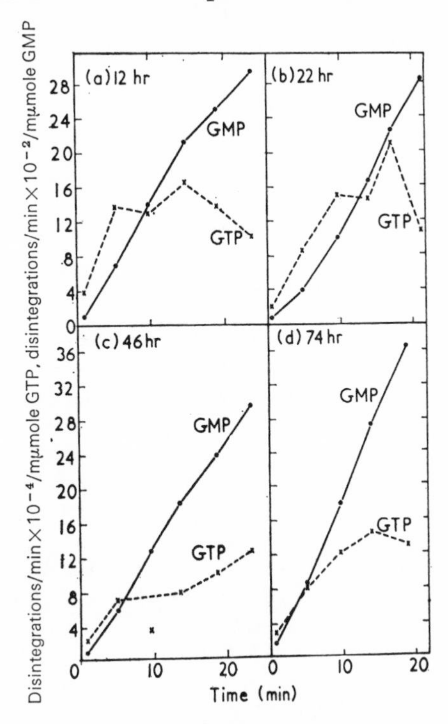

FIG. 3. Kinetics of incorporation of [^{3}H]guanosine into RNA and precursor. The results are expressed as disintegrations per min per mμM of nucleotide. Notice the difference in specific activity scale of GTP and GMP, the specific activity of GTP being 100 times greater.

four different stages of development. At all post-fertilization stages the cells accumulate guanosine from the sea water (cf. Nemer, 1962). In the stages studied here there is a rapid phosphorylation of guanosine, and the GTP pools reach seemingly rather steady specific activities within five to ten minutes ($\pm 30\%$). The total acid-soluble pool behaves differently (cf. Fig. 4) and will be discussed later. The incorporation into RNA is approximately linear for these short time-periods. Since the specific activity of GTP is fairly constant in the period of 10 to 20 minutes after addition of label, and the incorporation of radioactive material into the 2′-3′ GMP of RNA is increasing linearly, the apparent rate of RNA synthesis may be calculated for these time-periods. The maximum observed specific activity of GTP in the 10 to 20 minute time-period was used for these calculations; this would, of course, give the minimum apparent rate of RNA synthesis. The results of duplicate determinations of apparent rate of RNA synthesis were reproducible within 20%.

TABLE 1

Comparison of rate of RNA synthesis

Stage	Time of development (hr)	Cells/embryo†	RNA synthesis (% turnover in 10 min)
Mid-cleavage	12	140	0·71
Mesenchyme blastula	22	420	0·80
Late gastrula	46	700	1·23
Pluteus	74	1000	1·34

The rate of RNA synthesis at different stages is compared by calculation of the turnover of total embryo RNA. The molar amount of radioactive 2′-3′ GMP which accumulates between 10 and 20 min after the addition of label is divided by the amount of total RNA and multiplied by 100 (cf. Pannbacker, 1966). The total RNA/embryo is constant.

† The number of cells per embryo was calculated from the data of Hinegardner (1967).

Table 1 shows the results obtained from this calculation. Since the total RNA of the embryo remains constant (9×10^{-12} mole of nucleotide) (Whitely, 1950), one may compare the rates at different stages by expressing the amount of synthesis as the percentage of the total GMP in RNA which turns over during the ten-minute time-period. It is obvious that from mid-cleavage to larval stages, RNA synthesis is active; there seems to be an increase in the apparent rate (about twofold) during this period of $2\frac{1}{3}$ day. Any undetected turnover of newly made RNA occurring during these labeling periods would reduce the apparent rate of RNA synthesis. These values of the apparent rate of RNA synthesis will also be affected by the base composition of the new RNA, which might vary from 19% in DNA-like RNA to 33% guanosine in ribosomal RNA. Hence, a 1·7-fold increase in rates might be obtained solely due to changes in base composition. The number of cells per embryo is increasing during this period, and at a much more rapid rate than the increase in the rate of RNA synthesis. This means that the rate of RNA synthesis per cell genome is actually decreasing somewhat during development.

(b) *Turnover of newly made RNA*

Similar incorporation experiments were carried out for longer labeling periods; as found by Siekevitz *et al.* (1966), at all stages examined the initial rate of incorporation was not maintained (Fig. 4).

Figure 4 shows the rate of incorporation of radioactive material into the 2′-3′ GMP of the RNA decreases as the labeling time is extended. A steady-state rate does not seem to have been attained by one hour after addition of label. The kinetics of labeling of the *total* acid-soluble pool are also shown in Figure 4, and they differ from the labeling of GTP (cf. Fig. 3). After a rapid increase to a maximum, the specific activity of the total pool remains level (at 12 hr) or declines rapidly. Only after 20

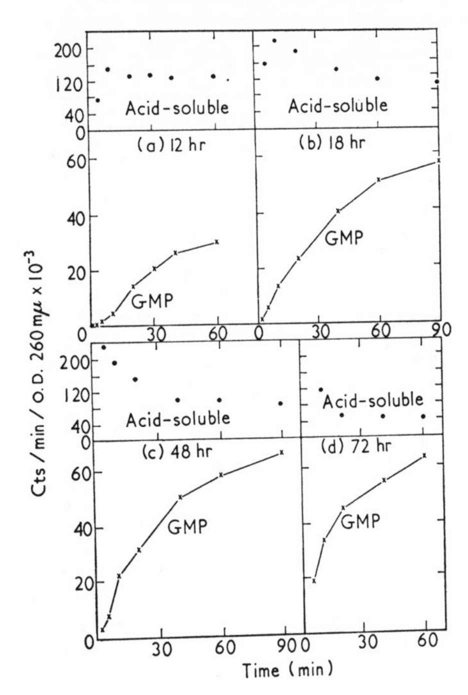

FIG. 4. Kinetics of incorporation of [^{3}H]guanosine into the acid-soluble pool and into the 2′-3′ GMP of hydrolyzed RNA is shown for four different developmental stages. The counting efficiency for all counts shown was 22 ± 1·5%.

minutes (at 18, 48 and 72 hr) does the decline in total pool specific activity begin to level off, and after 30 minutes changes in total pool are not great. (GTP-specific activity was determined in these same experiments and it mimicked the behavior in Fig. 3.) We presume that early changes in total pool radioactivity during a time when GTP specific activity is rather steady may be a reflection of equilibration of cellular compartments with isotope in the sea water. Our interpretation of these kinetics is that some or all of the newly made RNA is unstable, breaks down, and contributes to the acid-soluble pool. The decreasing rate of incorporation is not an artifact. It has been noted by others in these cells (Comb, Katz, Branda & Pinzino, 1965; Siekevitz *et al.*, 1966), occurs reproducibly and at all times during development. The decrease in rate is not due to a pathological effect of isotope; labeled embryos develop normally. The data do not allow precise calculation of how many distinct lifetimes of RNA molecules there are, and it is not improbable that they range widely. However, an indication of the quantitative importance of short-lived species of RNA can be made by comparing the initial rates of accumulation of radioactive RNA (10 to 20 min) to the rates of accumulation of label between 50 and 60 minutes after beginning the experiment; the comparison is shown in Table 2. The specific activities of the precursor, GTP, at these two time-periods was the same (within the reproducibility of the method, ±20%), and corrections for changes in precursor specific activity were unnecessary. The apparent rate of synthesis is much lower during the 50 to 60 minute

TABLE 2

Stability of newly synthesized RNA

Time of development (hr)	Accumulated label† / extrapolated accumulation	Rate of labeling‡ 50 to 60 min/10 to 20 min
12	0·53	0·15
18	0·79	0·53
48	0·83	0·42
72	0·68	0·35

† The amount of radioactivity which accumulated in 2′-3′ GMP of RNA in 1 hr in the experiments of Fig. 4 is divided by the amount which would have accumulated in 1 hr if the initial rate (Δ10 to 20 min) had been maintained. Corrections for changes in precursor specific activity were not necessary (see text).

‡ From the experiments shown in Fig. 4. The slope of accumulation of 2′-3′ GMP into RNA between 50 and 60 min after addition of label is divided by the slope of accumulation between 10 and 20 min.

time-period, presumably because breakdown of labeled RNA is taking place. Decrease in GTP specific activity would have to be twofold or greater to account for this alternative explanation of the decreasing rate of incorporation and this does not occur. The rate of incorporation for the 10- to 20-minute time-period can be extrapolated, and the amount of label accumulated at a given time period can be compared to the amount which would have accumulated without turnover. This comparison is also shown in Table 2. In one hour, from 0·2 to 0·5 of the newly made RNA seems to have broken down.

TABLE 3

Grains per cellular element

Hr of development	DNase		DNase + RNase		Difference (=RNA)	
	Nuc.	Cyt.	Nuc.	Cyt.	Nuc.	Cyt.
1. 12 (cleavage)						
1 hr label	14·8	4·4	—	—	—	—
2 hr label	16·8	4·2	4·9	3·5	11·9	0·7
2. 18 (hatching)						
1 hr label	7·9	3·1	—	—	—	—
2 hr label	7·9	4·4	2·3	3·6	5·6	0·8
3. 24 (mesenchyme blast.)						
1 hr label	4·5	1·5	—	—	—	—
2 hr label	5·2	1·1	2·4	0·5	2·8	0·6
4. 36 (gastrula)						
1 hr label	5·7	1·7	—	—	—	—
2 hr label	3·1	2·3	0·9	0·5	2·2	1·8
5. 72 (pluteus: gut)						
1 hr label	3·3	2·0	—	—	—	—
2 hr label	6·6	3·6	1·4	2·1	5·2	1·5

Embryos were exposed to [^{3}H]guanosine in sea water (10 μc/ml.) at the stage indicated for 1 or 2 hr. Numbers are averages from counts of at least 75 cells. The columns indicate the number of grains remaining after DNase treatment, or after both DNase and RNase treatment. The latter is taken to represent background. The difference between the two columns should represent incorporation into RNA, and is recorded in the last column. Grain counts at the pluteus stage were made only on cells of the invaginated archenteron.

(c) *Cellular localization of RNA synthesis*

Autoradiographic determination of isotope distribution was made on sectioned material. Counts of the distribution of grains over nuclear and cytoplasmic regions after different labeling times are given in Table 3. The counts were made on sections previously digested with DNase. Sections digested with both DNase and RNase had grain levels only slightly above background, and are shown in Table 3. The predominance of nuclear labeling prior to gastrulation is striking, and has been reported previously (Ficq *et al.*, 1963). The decrease in total cell grain-density from cleavage to mesenchyme blastula is consistent with the argument (Table 1) that the apparent rate of RNA synthesis per cell decreases during development. Furthermore, there is little change in total grains/cell between one and two hours of labeling at pregastrular stages, supporting the evidence for equilibrium between synthesis and breakdown at a time when the specific activity of the total pool is not changing. An increase in total cellular grain density with labeling time is only apparent after the gastrula stages. The grain counts for only one post-gastrular stage are shown, because after 36 hours of development the labeling of different cells becomes very heterogeneous, i.e. some cells are heavily labeled and some lightly labeled; this is probably a reflection of their specialized status. The amount of radioactive material in the cytoplasm is rather low prior to gastrulation, but becomes more obvious later in development.

4. Discussion

There are three important aspects to these findings. First, the rate of RNA synthesis can be measured at all stages examined, and it seems substantial. Although shifts upward in the mole-percentage of guanosine in new RNA made during development would produce an apparent rise in the RNA synthesis, our values are based on rates between 10 and 20 minutes after the addition of label; it is unlikely the new RNA would attain the base composition of ribosomal RNA after such a short period, even at the pluteus stage. We cannot state with assurance that there is a gradual increase in rate per embryo. It is very likely the rate per embryo is higher after gastrulation than before. It is certain that the apparent rate of RNA synthesis per cell does decrease considerably during development.

An approximation of the amount of RNA synthesis during mid-cleavage may be illustrated by the following calculation. At 12 hours post-fertilization, there are 140 cells per embryo; they would contain about 7×10^{-13} mole of nuclear DNA nucleotide. The embryo contains 9×10^{-12} mole of RNA nucleotide at all stages of development. A turnover of 0·7% of the total RNA every ten minutes (Table 1) represents synthesis corresponding to an equivalent of about 9·3% of the weight of nuclear DNA, or the equivalent of about 50% of the weight of nuclear DNA in one hour. Synthesis of RNA equivalent to 50% of the weight of the genome per hour at 15°C is certainly not inconsequential; it may represent copying of extensive portions of the genome, or very intense transcription of small regions of the genome.

Second, a portion of the newly transcribed RNA fails to persist for very long. As shown in Table 2, this unstable RNA may represent *at least* 0·2 to 0·5 of the RNA made during the preceding one-hour period. Complete equilibrium between breakdown and synthesis of RNA is not attained within 90 minutes; some newly made RNA may accumulate during the time-course of the experiment. Recent experiments (Aronson & Wilt, 1969) have shown by direct application of cell fractionation methods that more than 90% of the RNA made at pregastrular stages turns over rapidly; about 5% of the new RNA is exported to the cytoplasm when it seems stable. Indications of turnover were found by Siekevitz *et al.* (1963) in another species, so the phenomenon may be general. Brown & Littna (1966) and Denis (1966) have also found that much of the RNA made early in development of *Xenopus* is unstable.

Third, most of the RNA made during a two-hour period prior to the gastrula stage is restricted to the nucleus, and the greater part of the RNA does not seem to accumulate in the cytoplasm. The data do show small amounts of RNA in the cytoplasm at two hours, and quantitative methods have confirmed this (Aronson & Wilt, 1969). The experiments are consistent with a model for the metabolic behavior of RNA in which most of it is synthesized in the nucleus and breaks down there without passing to the cytoplasm.

An alternative model is one in which the newly made RNA is restricted to the nucleus during most of its lifetime and then passes to the cytoplasm and quickly breaks down. The autoradiographic data are not sensitive enough to eliminate this possibility, and the site of demise of the RNA molecules cannot be determined. We favor the first alternative for two reasons. The almost complete absence of grains located over the cytoplasm of early-stage embryos shows that the major portion of the lifetime of the RNA is nuclear. Nor have short-term labeling experiments (after 10 and 30 min) on cleavage-stage embryos revealed cytoplasmic labeling. Second, rapid destruction of RNA which might be transferred to the cytoplasm cannot be correlated with any function of this RNA as a postulated unstable template for cytoplasmic protein synthesis (Gross, Malkin & Moyer, 1964). Absence of RNA synthesis in cleavage does not depress the rate of amino acid incorporation into protein.

The present findings bear on some general aspects of cell function. Cleavage stages of sea urchin embryos have been used to provide evidence for particles which function in the delivery of messenger RNA to the cytoplasm, the informosomes (Spirin & Nemer, 1965). The fact that newly made RNA is primarily found in the nucleus at these early stages prompts the suggestion that informosome-type particles may have a solely intra-nuclear function. Second, the proposal for unstable intranuclear RNA has been made from other experiments using very different kinds of cells (Harris, Fisher, Rodgers, Spencer & Watts, 1963; Shearer & McCarthy, 1967; Attardi, Parnas, Hwang & Attardi, 1966; Warner, Soeiro, Birnboim, Girard & Darnell, 1966). The results obtained with early sea urchin embryos complement this evidence, because of the absence of ribosomal and transfer RNA synthesis (Nemer, 1963; Comb *et al.*, 1965; Giudice & Mutolo, 1967). The nuclear RNA represents a substantial portion of the RNA made in these cells; perhaps it has a general significance in the economy of many cells. Critical experiments to evaluate its function are necessary, and the sea urchin embryo offers favorable material for this purpose.

The data place restrictions on different views of the role of RNA synthesis in developing systems. Models which postulate a *massive* activation of genes at the time of gastrulation are unlikely. DNA transcription is quantitatively important at all stages examined. We feel some emphasis should be placed on the role of the passage of RNA from nucleus to cytoplasm, and the possible stabilization of transcribed gene products in the cytoplasm in general models of the molecular basis of development.

This work was supported by grant GM13882 from the National Institutes of Health, U.S.A.

REFERENCES

Aronson, A. I. & Wilt, F. H. (1969). *Proc. Nat. Acad. Sci., Wash.* in the press.
Attardi, G., Parnas, H., Hwang, M. & Attardi, B. (1966). *J. Mol. Biol.* **20**, 145.
Bray, G. A. (1960). *Analyt. Biochem.* **1**, 279.
Brown, D. D. & Gurdon, J. B. (1966). *J. Mol. Biol.* **19**, 399.
Brown, D. D. & Littna, E. (1964). *J. Mol. Biol.* **8**, 669.
Brown, D. D. & Littna, E. (1966). *J. Mol. Biol.* **20**, 81.
Comb, D. G., Katz, S., Branda, R. & Pinzino, C. J. (1965). *J. Mol. Biol.* **14**, 195.
Denis, H. (1966). *J. Mol. Biol.* **22**, 285.
Ficq, A., Aiello, F. & Scarano, E. (1963). *Exp. Cell Res.* **29**, 128.
Giudice, G. & Mutolo, V. (1967). *Biochim. biophys. Acta*, **138**, 278.
Gross, P. R., Malkin, L. I. & Moyer, W. A. (1964). *Proc. Nat. Acad. Sci., Wash.* **51**, 407.
Harris, H., Fisher, H. W., Rodgers, A., Spencer, T. & Watts, J. W. (1963). *Proc. Roy. Soc.* B, **157**, 177.
Hinegardner, R. (1967). In *Methods in Developmental Biology*, ed. by F. H. Wilt & N. K. Wessells, p. 139. New York: T. Y. Crowell.
Hultin, T. (1957). *Exp. Cell Res.* **12**, 413.
Hurlbert, B., Schmitz, H., Brumm, A. F. & Potter, V. R. (1954). *J. Biol. Chem.* **209**, 23.
Jardetzky, O. (1964). *J. Org. Chem.* **29**, 1988.
Katz, S. & Comb, D. G. (1963). *J. Biol. Chem.* **238**, 3065.
Lane, B. G. (1963). *Biochim. biophys. Acta*, **72**, 110.
Nemer, M. (1962). *J. Biol. Chem.* **237**, 143.
Nemer, M. (1963). *Proc. Nat. Acad. Sci., Wash.* **50**, 230.
Nilsson, R. (1959). *Acta Chem. Scand.* **13**, 395.
Nilsson, R. (1961). *Acta Chem. Scand.* **15**, 583.
Pannbacker, R. G. (1966). *Biochem. Biophys. Res. Comm.* **24**, 340.
Randerath, K. & Randerath, E. (1964). *J. Chromatog.* **16**, 111.
Shearer, R. W. & McCarthy, B. J. (1967). *Biochemistry*, **6**, 283.
Siekevitz, R., Maggio, R. & Catalano, C. (1966). *Biochim. biophys. Acta*, **129**, 145.
Spirin, A. & Nemer, M. (1965). *Science*, **150**, 215.
Warner, J., Soeiro, R., Birnboim, H. C., Girard, M. & Darnell, J. E. (1966). *J. Mol. Biol.* **19**, 349.
Wessells, N. K. (1964). *J. Cell Biol.* **20**, 415.
Whitely, A. H. (1950). *Amer. Naturalist*, **83**, 249.
Wilt, F. H. (1969). *Analyt. Biochem.* in the press.
Yanagisawa, T. & Isono, N. (1966). *Embryologia*, **9**, 170.

Regulation of DNA-Like RNA and the Apparent Activation of Ribosomal RNA Synthesis in Sea Urchin Embryos: Quantitative Measurements of Newly Synthesized RNA[1]

CHARLES P. EMERSON, JR., AND TOM HUMPHREYS

INTRODUCTION

Significant changes in RNA metabolism occur during development of the sea urchin embryo. RNA having a low G + C base composition and heterogeneous sedimentation properties (DNA-like RNA) is the only detectable class of RNA synthesized during cleavage and blastula stages (Glisin and Glisín, 1964; Wilt, 1964; Gross *et al.*, 1964a), whereas 28 S and 18 S ribosomal RNA (rRNA) are a large fraction of newly synthesized RNA at late gastrula and pluteus stages (Nemer, 1963; Nemer and Infante, 1966). These qualitative differences in RNA metabolism have suggested that the transcription of ribosomal genes is repressed during the early stages of sea urchin development and activated at about gastrulation (Comb *et al.*, 1965; Nemer and Infante, 1966; Giudice and Mutolo, 1967, 1969). Although this interpretation is consistent with the available data, other explanations, such as differences in the synthesis or accumulation of DNA-like RNA, could also account for the observations.

This paper presents experiments designed to reduce such ambiguities by defining more precisely the changes which occur in RNA metabolism during development of sea urchin embryos. They include a more extensive analysis of the contributions which synthesis and preferential accumulation of both rRNA and DNA-like RNA make to these changes. The sedimentation and base composition of RNA pulse-labeled with radioactive precursors were compared with those of the more stable RNA molecules which preferentially accumulated

[1] Supported by Grant No. HD 03480 from the National Institutes of Health. The senior author was supported by a predoctoral fellowship from the National Institutes of Health, Training Grant No. HD 00211.

during longer labeling periods. After a pulse label most radioactivity was in DNA-like RNA and none was detectable in RNA having the properties of 28 S and 18 S rRNA or precursor rRNA. Thus, rRNA was not a significant proportion of the RNA synthesized at any stage of development. However, after longer labeling periods rRNA was accumulated preferentially relative to newly synthesized DNA-like RNA. This preferential accumulation occurred to a much greater extent at later stages than at earlier stages, but the accumulation of newly synthesized rRNA could be detected in embryos as early as the first half of blastula stage.

The absolute quantities of newly synthesized DNA-like RNA and rRNA must be known in order to interpret these differences in the relative accumulation of these two classes of RNA. Therefore, the accumulation of RNA was measured quantitatively at cleavage, blastula, and pluteus stages when the relative levels of accumulation of DNA-like and ribosomal RNA were very different. These measurements demonstrated that there was a large quantitative decrease in the accumulation of newly synthesized DNA-like RNA during development, and that this decrease was of sufficient magnitude to account for the change in relative accumulation of DNA-like RNA and rRNA without requiring that the synthesis of rRNA be regulated during development.

MATERIALS AND METHODS

Culture conditions. Gametes were collected from the sea urchin, *Strongylocentrotus purpuratus*, by injection of 0.5 *M* KCl into the coelomic cavity. Eggs were washed twice by settling in 500 volumes of Woods Hole formula artificial seawater (NaCl, 24.7 gm; $CaCl_2$, 1.0 gm; $MgCl_2 \cdot 6H_2O$, 9.9 gm; Na_2SO_4, 3.7 gm; $NaHCO_3$, 0.2 gm per liter) containing 50 μg/ml of potassium penicillin G and 30 μg/ml of streptomycin sulfate, fertilized by addition of semen at a final dilution of 1/50,000, and then washed twice to remove excess sperm. Fertilized eggs were cultured suspended in 20 volumes of antibiotic-containing seawater at 18°C in Erlenmeyer flasks on a gyratory shaker. Embryos cultured beyond gastrula stage were washed twice at 30 hours postfertilization by centrifugation at 100 *g* for 15 seconds and reincubated in shaker flasks in 100 volumes of antibiotic seawater. Under these conditions, embryos developed according to the schedule in Fig. 1.

Isotopic labeling. Radioactive RNA precursors, adenosine-8-^{3}H

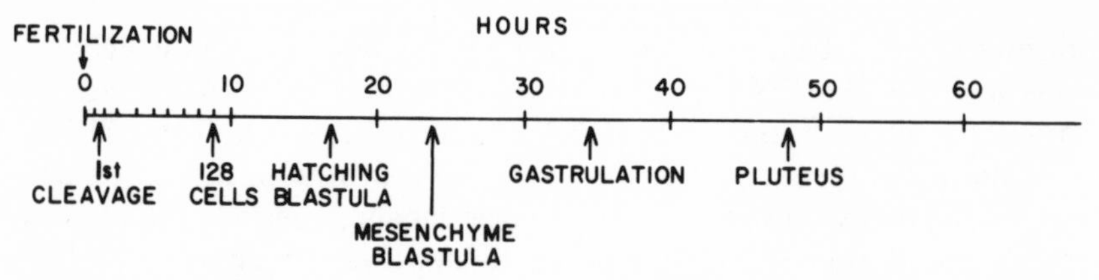

FIG. 1. Schedule of development of *Strongylocentrotus purpuratus* embryos at 18°C.

(9 Ci/mmole, Schwarz Bio-Research, Inc.) ^{32}P, inorganic phosphate (carrier free, New England Nuclear), or uridine-5-^{3}H (24 Ci/mmole, Schwarz Bio-Research, Inc.) were introduced into shaker flask cultures for specified labeling periods, after which embryos were washed twice at 4°C in acid seawater (artificial seawater made up in 0.01 *M* sodium acetate buffer, pH 4.5) prior to extraction of RNA or acid-soluble nucleotide pools.

Purification of RNA. RNA was purified from embryos by phenol extractions. The number of embryos to be extracted was determined within ±7% by drawing a sample of suspension into a 10 μl capillary pipette and counting the embryos under a dissecting microscope (Hinegardner, 1967). A centrifuged pellet of a known number of embryos was then homogenized with 20 strokes of a cold glass Dounce homogenizer in 20 volumes of cold acetate extraction buffer (0.01 *M* sodium acetate buffer, pH 5.1, 0.1 *M* NaCl, 10^{-3} *M* $MgCl_2$) which was made 0.5% in sodium dodecyl sulfate (SDS) immediately before use. Two volumes of phenol saturated with acetate extraction buffer were immediately added. The mixture was incubated for 20 minutes at 4°C with intermittent mixing. Phases were separated by centrifugation at 10,000 *g* for 10 minutes, the phenol phase was removed and replaced with fresh phenol, and a second cold extraction was performed. After centrifugation, the phenol phase was discarded and the aqueous phase was collected and saved. The interface was reextracted with 1 volume of fresh acetate extraction buffer containing 0.5% SDS and 2 volumes of phenol at 60°C for 3 minutes, then rapidly cooled on ice, centrifuged, and the aqueous phase collected. This hot extraction procedure was repeated once.

The residual interface material remaining after the hot phenol extractions was then incubated for 10 minutes at 60°C in 1 volume acetate extraction buffer made 0.5% in SDS and 1 *M* in $NaClO_4$, and 2 volumes of phenol–chloroform (2:1). After cooling on ice, the aqueous phase was separated by centrifugation. The aqueous phases from the cold, hot, and $NaClO_4$ extractions were pooled and the RNA was precipitated with 2 volumes of 95% ethanol for at least 1 hour at −15°C.

The RNA was collected by centrifugation at 27,000 *g* for 15 minutes at 4°C, washed for 5 minutes with cold 95% ethanol to remove residual phenol, and then air dried. The RNA pellet was resuspended in acetate extraction buffer and incubated for 30 minutes at room temperature with 100 μg/ml of pancreatic DNase I (Worthington Biochemical, purified electrophoretically). This enzymatic treatment hydrolyzed DNA to low molecular weight fragments which sedimented in sucrose gradients at less than 10 S. The reaction was terminated by the addition of SDS, to a concentration of 0.5%, and of two volumes of acetate buffer-saturated phenol. The RNA was reextracted twice for 10 minutes each, at room temperature, and precipitated with 2 volumes of 95% ethanol at −15°C. RNA was collected by centrifugation, and the pellet was washed in 95% ethanol before resuspension in Tris-SDS buffer (0.01 *M* Tris-HCl, pH 7.6, 0.1 *M* NaCl, 10^{-3} *M* Na_2EDTA, 0.5% SDS) for sedimentation in sucrose gradients. The RNA recovered using this extraction procedure had 260/280 mμ ratios ranging from 1.95 to 2.03.

RNA yields. Recoveries of labeled RNA from the purification procedures were greater than 90% as assayed by measuring the amount of alkali-hydrolyzable radioactivity isolated from embryos which had been incubated in uridine-5-^3H. The amount of labeled RNA was calculated as the difference between the total trichloroacetic acid (TCA)-precipitable radioactivity and TCA-precipitable radioactivity remaining after treatment with alkali. Alkali sensitivity was determined by incubation of samples in 0.3 *N* NaOH at 37°C for 18 hours. RNA was precipitated in 15% TCA at 4°C, collected by filtration onto Millipore filters (0.45-μ pore size), and washed with cold 5% TCA. The amount of radioactivity bound to dried filters was determined by counting the filters directly in toluene scintillation mix (5 gm PPO, 0.05 gm POPOP, in 1 liter of toluene) in a Beckman scintillation counter, or by eluting the sample from the filters with 0.1 *N* NH_4OH and then counting in Beckman scintillation mix (100 gm naphthalene, 4 gm PPO, to 1 liter with dioxane). Appropriate standard samples or external standards were used to measure the relative counting efficiencies of the samples.

Sedimentation analysis. Purified embryo RNA was centrifuged at 25°C in 15–30% (w/w) linear sucrose gradients buffered with Tris-SDS buffer in the Spinco SW 25.1 or SW 27 rotors. Fractions containing equal numbers of drops were collected with a continuous-flow Gilford spectrophotometer automatically recording the optical density at 260 mμ. Yeast RNA (50 μg) was added to samples having less than 1 optical density unit absorption at 260 mμ and the RNA was precipitated for 15 minutes at 4°C by the addition of an equal volume of cold 30% TCA. Precipitates were collected on membrane filters, washed with cold 5% TCA, and dried. Filters carrying precipitates of ^{32}P-labeled RNA to be subsequently analyzed for base composition were counted in toluene scintillation mix. All other samples were dissolved in 0.1 *N* NH_4OH and counted in Beckman mix. Sedimentation coefficients were determined by assuming that embryo ribosomal RNA sediments at 28 S and 18 S and that there is a linear relationship between distance sedimented and the sedimentation coefficient.

Specific activities of the acid-soluble ATP and GTP pools. Acid-soluble extracts of washed embryos were prepared by homogenizing a centrifuged pellet of (2 to 4×10^6) embryos in two volumes of cold 0.5 *N* $HClO_4$ with 25 strokes of a cold glass Dounce homogenizer. This extract was centrifuged at 10,000 *g* for 10 minutes, the supernatant was removed, and the insoluble residue was reextracted for 10 minutes with cold 0.5 *N* $HClO_4$ and then recentrifuged.

The nucleotides in the combined acid-soluble supernatant fractions were adsorbed to 200 mg of acid-washed Norit (Tsuboi and Price, 1959). The Norit was washed twice with cold distilled water by centrifugation at 10,000 *g* for 10 minutes. The nucleotides were eluted by incubation in 0.1 *N* NH_4OH, 50% ethanol for 1 hour at 37°C, and then Norit was removed by centrifugation at 10,000 *g* for 20 minutes at room temperature. Exchange of tritium from the adenosine-8-^3H during the elution from Norit was undetectable. Nucleotide samples were dried under a stream of air, resuspended in 0.01 *N* NH_4OH, and fractionated either by chromatography on 0.3×23 cm columns of Dowex-formate 1-X8 or on Whatman No. 1 paper. Dowex columns were washed with water, and the nucleotides were eluted according to the ammonium formate buffer system described by Hulbert *et al.* (1954). The eluted components were identified by their spectral properties and elution characteristics. Nucleotides to be resolved on Whatman No. 1 paper were chromatographed with isobutyric acid, H_2O, conc. NH_4OH (66:33:1) (Tsuboi and Price, 1959).

The major contaminants of the ATP and GTP preparations recovered from the Dowex columns were the deoxy derivatives of ATP and GTP. In order to achieve further purification, the fractions from Dowex chromatography corresponding to ATP and GTP were each adsorbed to 50 mg of acid-washed Norit, washed with water to remove salts, and hydrolyzed to monophosphates by incubation of charcoal-bound triphosphates in 1 *N* HCl for 30 minutes in a boiling water bath (Brown and Littna, 1966). After rewashing with water, the nucleotides were recovered by incubation of Norit in 0.1 *N* NH_4OH in 50% ethanol for 1 hour at 37°C, followed by removal of the Norit by centrifugation. The exchange of tritium in the adenosine-8-^3H during acid hydrolysis was less than 10%, and the ribonucleotide triphosphates were recovered from this hydrolysis procedure as monophosphates without detectable destruction of the purine ring. Deoxynucleotides, however, were largely degraded to an unphosphorylated derivative.

The nucleotides recovered after acid hydrolysis of ATP were rechromatographed on Dowex-formate 1-X8 columns (0.3 × 20 cm), using the formic acid system (Hurlbert *et al.*, 1954), which separates nucleotides in a different elution sequence than the ammonium formate system used in the initial purification step. The AMP-^3H fractions recovered from this column were dried under a stream of air.

GMP-^3H recovered from acid hydrolysis and the AMP-^3H from chromatography were resuspended in 0.1 *N* NH_4OH and electrophoresed at 4°C in 0.05 *M* borate buffer, pH 9.2 (Smith, 1955) on Whatman No. 1 chromatography paper. Under these conditions AMP-^3H and GMP-^3H were separated from their corresponding deoxy monophosphates and the degradation products of the acid hydrolysis. The ribose monophosphates were eluted from the electrophoresis paper in 0.1 *N* NH_4OH. The spectral characteristics of these purified compounds were identical to those of known standards, and the concentrations of nucleotides were calculated from their molar extinction coefficients at 260 mμ: AMP = 15.4×10^3; GMP = 11.8×10^3.

Samples of ATP purified by paper chromatography were resolved from other contaminating radioactive nucleotides by this single-step purification procedure. ATP was eluted from chromatograms, and the amounts of ATP were determined by the luciferinase reaction using a Beckman scintillation counter (Cole *et al.*, 1967; Emerson, in preparation).

The specific activities of samples of AMP-^{3}H, GMP-^{3}H and ATP-^{3}H were determined by counting known concentrations of these nucleotides in vials containing 15 ml of Beckman scintillation mix and a blank Millipore filter. The counting efficiencies of the sucrose gradient samples prepared from solubilized TCA-precipitated fractions and of the AMP-^{3}H and GMP-^{3}H pool samples were equivalent, as measured using external standards. Calculation of the rates of RNA accumulation could, therefore, be made directly from these radioactivity measurements without correction.

Number of nuclei per embryo. Embryos were fixed at various developmental stages with acetic acid–ethanol (3:1) and nuclei were stained with Azure A (Swift, 1955). The embryos were squashed on a slide, and the number of nuclei per embryo were counted using a grid ocular for reference. The time course of increase of nuclei during development is shown in Fig. 2.

Methylated albumin kieselguhr [MAK] chromatography. Methylated albumin, kieselguhr, and methylated albumin-coated kiesel-

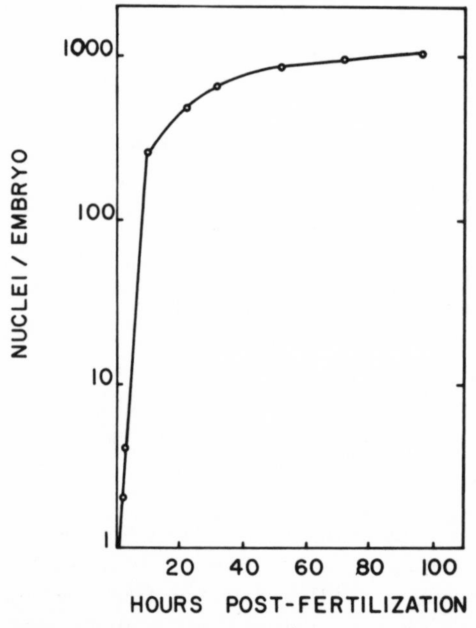

FIG. 2. The developmental time course of the increase in the number of nuclei per embryo, as determined by counting the number of Azure A-staining nuclei in embryo squash preparations.

guhr were prepared as described by Mandell and Hershey (1961). Columns, 3 cm in diameter, were constructed of four layers; the bottom layer consisted of 0.5 g paper powder, the second layer was formed with 4 g of kieselguhr in 0.1 *M* NaCl, 0.05 *M* phosphate buffer, pH 6.5, which was coated with methylated albumin immediately before use, the third layer was formed in 0.4 *M* NaCl, 0.05 *M* phosphate buffer, pH 6.5 from 3 gm of kieselguhr mixed with 1 gm of methylated albumin-coated kieselguhr which had been prewashed to remove unbound protein, and the fourth layer was 0.5 gm kieselguhr, which served as a protective layer on the top of the column, RNA samples were loaded onto columns in 0.4 *M* NaCl, 0.05 *M* phosphate buffer, pH 6.5, at concentrations of 50 μg/ml. Loaded columns were washed first with 0.7 *M* NaCl, 0.05 *M* phosphate buffer, pH 6.5, and then the RNA was eluted with linear NaCl gradients at room temperature. Five-milliliter fractions were collected and the optical density was measured. Aliquots were precipitated with TCA in the presence of 100 μg of carrier yeast RNA and collected on Millipore filters. Filter-bound radioactivity was measured in a scintillation counter using toluene-based mix. NaCl concentration of fractions was determined with a refractometer.

Marker 28 S ribosomal RNA could be recovered with yields in excess of 90% from these MAK columns; however, only 40% to 70% of the radioactive RNA in the 28 S fraction from embryos at gastrula and blastula stages could be eluted with NaCl.

RNA base composition. After scintillation counting in toluene-based mix, appropriate filters from sucrose gradient or MAK column chromatography fractions were washed in toluene and chloroform, dried, and incubated for 2 hours in 0.1 *N* NH_4OH to dissolve filter-bound RNA (Brown and Littna, 1964). The ^{32}P-RNA was hydrolyzed in 0.3 *N* KOH for 18 hours at 37°C in the presence of 10 mg of yeast RNA. The pH of the hydrolyzate was adjusted to 8.5–9.0 with 2 *N* HCOOH, and the nucleotides were separated on 0.9 × 2.6 cm Dowex-formate 1-X8 columns by stepwise elution (Hayashi and Spiegelman, 1961), using solutions of (1) 0.025 *N* HCOOH; (2) 0.15 *N* HCOOH; (3) 0.05 *N* $HCOONH_4$, 0.05 *N* HCOOH; and (4) 0.1 *N* HCOOH, 0.2 *N* $HCOONH_4$ to elute the 2′–3′ nucleotides in the order of C, A, U, G. The optical density of each of the fractions at 260 mμ was measured, and aliquots were dried in scintillation vials with 0.01% SDS and counted in toluene-based fluor. The relative counting efficiencies for the ^{32}P nucleotides were equal, as determined using internal standards. The base compositions of the ^{32}P-RNA were calculated from the fraction of the total ^{32}P radioactivity in each of the nucleotides as determined by measurements of the specific activity (radioactivity/UV absorption) of peak fractions and the sum UV absorption of the four nucleotides, or by measuring the total ^{32}P in each of the nucleotides. The results were identical by either method and were accurate to ± 0.5 mole percent.

The base composition of ribosomal RNA was measured by Dowex chromatography of the 2′–3′ nucleotides prepared from unlabeled 28 S and 18 S ribosomal RNA which had been purified from unfertilized eggs by phenol extraction and sucrose gradient sedimentation (Table 1). The molar extinction coefficients used to calculate the composition were: CMP, $\epsilon_{260m\mu} = 6.5 \times 10^3$; AMP, $\epsilon_{260m\mu} = 14.2 \times 10^3$; UMP, $\epsilon_{260m\mu} = 10.0 \times 10^3$; GMP, $\epsilon_{260m\mu} = 11.4 \times 10^3$.

DNA base composition. DNA was isolated from seawater-washed sperm by incubation with 50 μg/ml of self-digested Pronase (CalBiochem) for 8 hours at 37°C, followed by extraction with phenol–chloroform–isoamyl alcohol (2:1:0.03), and ethanol precipitation. DNA was hydrolzyed in 5×10^{-2} *M* Tris-HCl, pH 7.3, and 5×10^{-3} *M* $MgCl_2$ with DNase I (Worthington), followed by further hydrolysis at pH 9.0 with snake venom phosphodiesterase (Worthington). The 5′ nucleotides were separated on Dowex-formate 1-X8 (17 × 0.9 cm) columns using a linear gradient of 0.1 to 0.7 *M* $HCOONH_4$, pH 4.35 (Canellakis and Mansavinos, 1958). The base composition was calculated from the molar extinction coefficients of the fractionated nucleotides at pH 4.35: dCMP, $\epsilon_{280m\mu} = 10.6 \times 10^3$; dTMP, $\epsilon_{260m\mu} = 8.8 \times 10^3$; dAMP, $\epsilon_{260m\mu} = 15.3 \times 10^3$; dGMP, $\epsilon_{260m\mu} = 11.4 \times 10^3$.

RESULTS

RNA Synthesis

RNA molecules which incorporate radioactivity during a short duration with isotope are a representative sample of the various species of RNA molecules being synthesized at that time if the length of the radioactive pulse is short relative to the turnover rate of any unstable molecules that are being synthesized (Soeiro *et al.*, 1968; also see Nierlich, 1967). Since the half-life of newly synthesized RNA in sea urchin appears to be about 30 minutes (Kimjima and Wilt, 1969), a 10 minute incubation with isotopic precursor was chosen. At all stages the radioactivity was incorporated into a heterogeneous population of RNA molecules when either ^{32}P (Fig. 3) or uri-

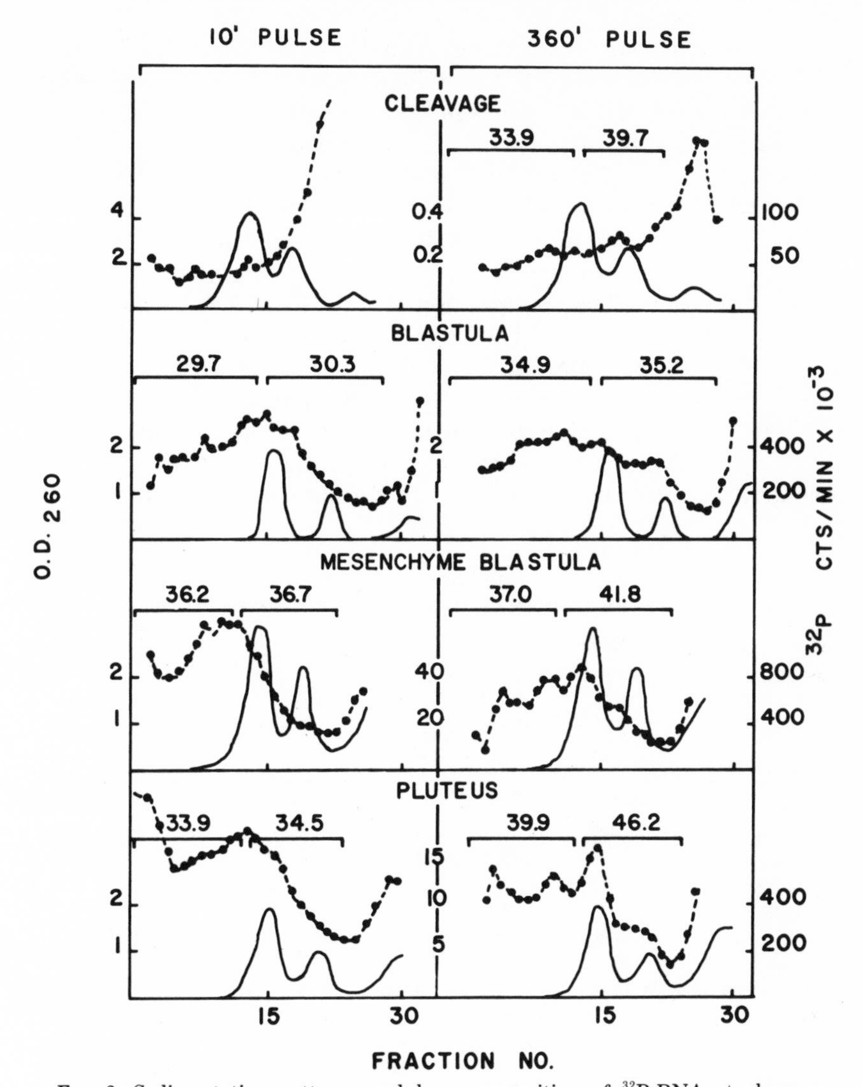

FIG. 3. Sedimentation patterns and base composition of ^{32}P-RNA at cleavage, blastula, mesenchyme blastula, and pluteus stages after 10-minute and 360-minute incubations with ^{32}P. Cleavage stage embryos (3 hours) were labeled for 10 minutes with 100 μCi/ml ^{32}P and for 360 minutes with 40 μCi/ml ^{32}P. Blastula embryos (14 hours) were labeled for 10 minutes and 360 minutes with 10 μCi/ml of ^{32}P and mesenchyme blastula (27 hours) and pluteus (49 hours) stage embryos were labeled with 40 μCi/ml of ^{32}P. Purified RNA from cleavage, mesenchyme blastula and pluteus embryos were sedimented for 12 hours at 24,000 rpm in the SW 25.1 rotor. Blastula stage RNA was sedimented for 12 hours at 26,000 rpm in a SW 27 rotor. Numbers indicate the moles percent G + C base composition of the pooled fractions enclosed within brackets. RNA pelleted in the bottom of each gradient tube was included in the analysis of the heavy (>30 S) RNA fraction. Complete base composition analyses are described in Table I. Radioactivity, cpm ^{32}P (●——●); optical density at 260 mμ (——).

dine-^{3}H (not shown) was used as a precursor. During early cleavage, molecules smaller than 28 S predominated in the heterogeneous RNA. Beginning at late cleavage and continuing into later stages, increasingly larger proportions of the radioactive molecules sedimented as a broad peak centered at 36 S. Predominance of this broad 36 S peak was a characteristic feature of the sedimentation patterns of RNA synthesized in embryos from early blastula until pluteus stage.

At all stages the base composition of the ^{32}P-RNA were characterized by high U (36%) and low G + C (30–37%) contents (Table 1). Although the base compositions of this RNA were asymmetric and the G + C contents somewhat variable compared to those of DNA (Table 1), this class of RNA will be referred to as DNA-like RNA (Brown and Littna, 1966). There was no indication of radioactive ribosomal RNA precursors or 28 S and 18 S ribosomal RNA after any of

TABLE 1
BASE COMPOSITION OF ^{32}P RNA LABELED AT DIFFERENT DEVELOPMENTAL STAGES

Stage (hours after fertilization)	Duration of ^{32}P labeling	Size class of ^{32}P RNA	Moles % 2′–3′ nucleotides				
			C	A	U(T)	G	%GC
	min						
Cleavage (3 hours)	360	10–30 S	18.8	27.6	32.5	20.9	39.7
		> 30 S	15.6	30.0	36.2	18.3	33.9
Early blastula (14 hours)	10	10–30 S	9.8	33.6	36.2	20.5	30.3
		> 30 S	10.1	30.1	40.2	19.6	29.7
	360	10–30 S	11.5	30.8	34.1	23.7	35.2
		> 30 S	12.4	31.6	33.5	22.5	34.9
Mesenchyme blastula (27 hours)	10	10–30 S	15.6	28.6	35.0	20.9	36.5
		> 30 S	15.2	27.5	36.4	21.0	36.2
	360	10–30 S	19.3	28.7	29.8	23.0	42.3
		> 30 S	17.6	31.0	32.0	19.4	37.0
Pluteus (48 hours)	10	10–30 S	12.9	28.0	36.4	21.6	34.5
		> 30 S	13.5	30.6	35.5	20.4	33.9
	360	10–30 S	20.7	26.2	27.8	25.5	46.2
		> 30 S	18.2	29.4	31.4	21.0	39.9
Sperm		DNA	18.8	30.7	31.1	19.5	38.3
Unfertilized egg		28 S rRNA	26.1	20.2	18.7	34.9	61.0
		18 S rRNA	22.6	23.8	20.8	31.8	54.4

these short incubations with isotopes. No distinct peaks were observed in sucrose gradients and the ^{32}P base compositions were always much lower than the expected 55–60% G + C content of ribosomal RNA (Table 1; Willems *et al.*, 1968).

Accumulation of Stable RNA Molecules

In order to determine the characteristics of the relatively more stable species of RNA molecules synthesized at the various stages of development, radioactive RNA was isolated from embryos which were incubated with ^{32}P for periods longer than 30 minutes. The sedimentation properties of the radioactive RNA from pregastrula embryos incubated for 360 minutes with isotope were not found to be strikingly different from the radioactive RNA after a 10-minute incubation, indicating that no size class of molecules constituting a detectable portion of the newly synthesized RNA is more stable and preferentially accumulating (Fig. 3). At later stages, changes in the sedimentation distribution of radioactive RNA were more marked after the longer incubations and these changes were characterized by the appearance of peaks of 28 S and 18 S rRNA. Base compositions of the radioactive RNA accumulated after the longer incubation with ^{32}P at both early and late stages were consistently higher in G + C content than after a 10-minute incubation (Table 1). These increases in the G + C contents were larger at later stages than at earlier stages, but occurred even at early blastula. At all stages most of the increase occurred in RNA sedimenting between 10 and 30 S, and very little in the RNA sedimenting more rapidly than 30 S, suggesting that the increase in G + C content was due largely to synthesis of 28 S and 18 S rRNA.

These results indicate that stable rRNA preferentially accumulates relative to the less stable DNA-like RNA much more extensively at later stages than at early stages. Quantitative information on the actual amounts of newly synthesized RNA accumulated at the various stages must be available before such changes in relative amounts of DNA-like RNA and rRNA can be interpreted in terms of absolute increases or decreases in synthesis of either class of RNA. Therefore, quantitative measurements of the amount of radioactive RNA accumulated per nucleus were made during cleavage, early blastula, and pluteus stages by determining the incorporation of adenosine-8-^{3}H into RNA and the specific activity of the nucleotide triphosphate precursor pools.

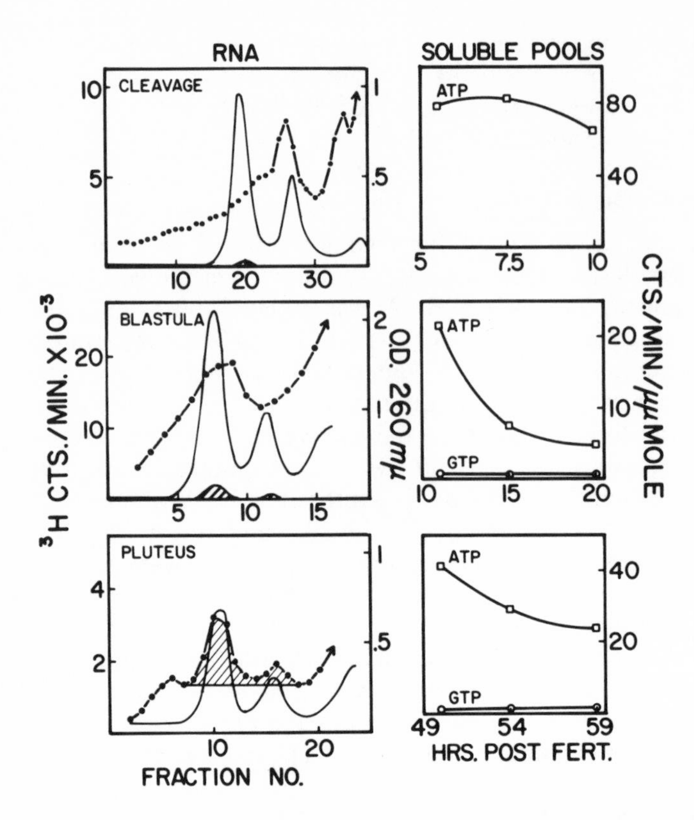

FIG. 4. The incorporation of adenosine-8-^{3}H into RNA and acid-soluble nucleotide triphosphates at cleavage, early blastula, and pluteus stages. Embryos were incubated with adenosine-8-^{3}H at cleavage stage, 5–10 hours after fertilization (18 μCi/ml), at early blastula stage, 10–20 hours after fertilization (2 μCi/ml), and at pluteus stage, 49–59 hours after fertilization (0.3 μCi/ml). Acid-soluble pools were purified at intervals during the period of incubation, and the specific activities of the nucleotide triphosphate precursors were determined (right). RNA was extracted and purified from 6 $\times$ 10^4 cleavage stage embryos, 3 $\times$ 10^5 early blastula stage embryos, and 8 $\times$ 10^4 pluteus stage embryos and fractionated by sucrose gradient centrifugation (left). In the gradient of RNA extracted from pluteus stage embryos, the shaded areas indicate graphical estimates of the amounts of radioactive 28 S and 18 S rRNA. The shaded areas in gradients of RNA isolated from cleavage and blastula stage embryos indicate the calculated amounts of radioactive 28 S and 18 S rRNA expected if the rate of accumulation of newly synthesized rRNA per nucleus is assumed to be equal to that measured during pluteus stage. RNA gradients, radioactivity (cpm) (●——●); optical density at 260 mμ, (——). Nucleotide soluble pools (cpm/pmole): ATP (□——□), GTP (○——○).

Quantitative Measurement of Newly Synthesized RNA

The incorporation of adenosine-8-^{3}H into RNA and the specific activities of the acid-soluble ATP and GTP pools were measured at early pluteus stage of development, 49–59 hours after fertilization, when ribosomal RNA synthesis was easily detectable. At this stage a large fraction of the adenosine-8-^{3}H in RNA after the 10-hour incubation sedimented as peaks at 28 S and 18 S (Fig. 4). This radioactive RNA appeared to be rRNA since its base composition was similar to rRNA when ^{32}P is used as a precursor at this same stage (Table 1; Giudice and Mutolo, 1967). The amount of radioactivity in ribosomal RNA isolated from a known number of embryos was estimated graphically from these sedimentation distributions, as illustrated by the shaded areas (Girard *et al.*, 1965). The radioactivity not in these peaks and sedimenting more rapidly than 10 S was designated DNA-like RNA.

The specific activities of purified ATP and GTP from the total acid-soluble pools during the 10-hour incubation were determined from aliquots of embryos withdrawn at intervals of 1, 5, and 10 hours after addition of adenosine-8-^{3}H to the cultures (Fig. 4). The ATP pools reached a maximum specific activity within the first hour and declined slowly throughout the incubation. Some interconversion of adenosine to guanosine nucleotides was also measurable, but compared to the ATP pools, the specific activities of GTP were almost negligible.

The gram quantities of newly synthesized 28 S and 18 S ribosomal and DNA-like RNA accumulated per nucleus were calculated from the radioactivity on the sucrose gradients, from the average specific activity of the precursor pools, and from the known AMP and GMP base compositions of these RNA classes. All nuclei were assumed to be equally active in RNA synthesis. With respect to rRNA synthesis, this assumption is consistent with the observation that a large fraction of nuclei at late stages of development contain morphologically typical nucleoli which actively incorporate precursors into RNA (Karasaki, 1968). The accumulation of 28 S and 18 S rRNA during this 10-hour period was calculated to be 1.2×10^{-14} gm per nucleus, which is equivalent to about 3000 molecules per nucleus per 10 hours. The accumulation of DNA-like RNA was calculated to be 3.4×10^{-14} gm per nucleus (Table 2). Estimation of the errors inherent in the measurements suggest that these values should be accurate to within $\pm 20\%$ (specific activity of pools, $\pm 5\%$; radioac-

TABLE 2
QUANTITATIVE ACCUMULATION OF RIBOSOMAL AND DNA-LIKE RNA[a]

Labeling period (stage, hours after fertilization)	Average number of nuclei/ embryo	Average specific activity of acid-soluble pools[b] (cpm/pmole)		28 S ribosomal RNA (cpm/10^2 embryos)	Accumulation of 28 S + 18 S ribosomal RNA[c] (gm/nucleus)	DNA-like RNA[e] (cpm/10^2 embryos)	Accumulation of DNA-like RNA (gm/ nucleus)
		ATP	GTP				
Pluteus, 49–59 hours	820	28.7	1.2	12	1.2×10^{-14}	75	3.4×10^{-14}
Blastula, 10–20 hours	350	9.1	0.8	[1.7][d]	[1.2×10^{-14}][d]	46	14×10^{-14}
Cleavage, 5–10 hours	50	74.1	—	[0.8][d]	[0.6×10^{-14}][d]	177	59×10^{-14}

[a] Calculations based on the data of Fig. 4.

[b] Calculated assuming a linear increase in the specific activity of the ATP and GTP pools between the time of addition of adenosine-8-^{3}H and the first measurement.

[c] Ribosomal accumulation at pluteus stage was estimated graphically from the incorporation of adenosine-8-^{3}H into 28 S ribosomal RNA (shaded area, Fig. 3c), assuming that 28 S ribosomal RNA is composed of 20.2 moles percent of AMP and 34.9 moles percent of GMP. It was assumed that 18 S ribosomal RNA has half the mass of the 28 S ribosomal RNA molecule and that these two species are accumulated coordinately.

[d] The numbers in brackets indicate the expected accumulation of ribosomal RNA at cleavage and blastula stages if this accumulation occurs at the same rate per nucleus as measured at pluteus stage.

[e] The amount of DNA-like RNA was calculated assuming that this class of molecules has a DNA-like base composition, 30.7 moles percent of AMP and 19.5 moles percent of GMP. All RNA-^{3}H sedimenting greater than 12 S including the RNA pelleted in the gradient tubes, was defined to be DNA-like RNA, except for the 28 S and 18 S ^{3}H-ribosomal RNA (shaded areas, Fig. 4). The gram quantity for cleavage and blastula stages is a minimal estimate, based on the assumption that these molecules are stable and that incorporation reflects the average specific activities of the precursor pools.

tivity in ribosomal and DNA-like RNA, ±7%; number of nuclei, ±2%, and the number of embryos, ±7%). Independent measurements of RNA accumulation using both Dowex and luciferinase methods for determination of the specific activity of ATP precursor pools agree within 15%.

The accumulation of newly-synthesized RNA at pluteus stage was compared quantitatively to RNA accumulation at cleavage and blastula stages, when newly synthesized, high molecular weight RNA is predominantly DNA-like and newly synthesized rRNA is not detectable by sucrose gradient sedimentation. Cleavage stage and early blastula stage embryos were incubated with adenosine-8-^3H for 5 and 10 hours, respectively, the specific activities of the total acid-soluble ATP and GTP pools were assayed at intervals, and the amounts of radioactivity accumulated into high molecular weight RNA during the incubation period were determined after sucrose gradient centrifugation (Fig. 4). As expected at these stages, labeled RNA sedimented as a heterogeneous population of molecules with no detectable peaks of rRNA. From the average specific activities of the pools during the period of labeling, the minimum amount of DNA-like RNA accumulated by cleavage stage embryos was calculated to be 59×10^{-14} gm per nucleus, and by early blastula stage embryos, 14×10^{-14} gm per nucleus (Table 2).

The net accumulation per nucleus of DNA-like RNA at cleavage and blastula stages were, respectively, at least 18 times and 4 times greater than at pluteus stages. Since DNA-like RNA has a short half-life in sea urchin embryos (Kimjima and Wilt, 1969) and the specific activity of the precursor pools decreased during the incubation periods, the calculated amounts of DNA-like RNA are minimal estimates and could be somewhat larger.

Although no 28 S and 18 S peaks of rRNA were detected at these earlier stages, it was possible to calculate the amount of radioactive rRNA which would be expected if rRNA was accumulating per nucleus at the rate measured in pluteus stage embryos (i.e., at 1.2×10^{-14} gm per nucleus per 10 hours). Radioactivity would accumulate in newly synthesized rRNA to the extent shown by the shaded areas under the 28 S and 18 S rRNA optical density peaks in the sucrose gradients of RNA extracted for these stages (Fig. 4). These amounts are equivalent to approximately 2% of the radioactivity sedimenting at 28 S at cleavage stage and 8% at blastula stage (Table 2) and would not be expected to form peaks in the sucrose gradients or increase the base ratios extensively.

Detection of rRNA Synthesis during Early Development

A fractionation procedure, involving a combination of sucrose gradient sedimentation and MAK chromatography was devised to separate the DNA-like RNA from the rRNA so that rRNA synthesis could be measured at early stages. Sucrose gradient sedimentation in high ionic strength buffers fractionates RNA mainly on the basis of molecular weight. MAK chromatography appears to separate RNA on the basis of molecular weight, G + C content, and degree of hydrogen bonding (Sueoka and Chen, 1962). If RNA molecules of

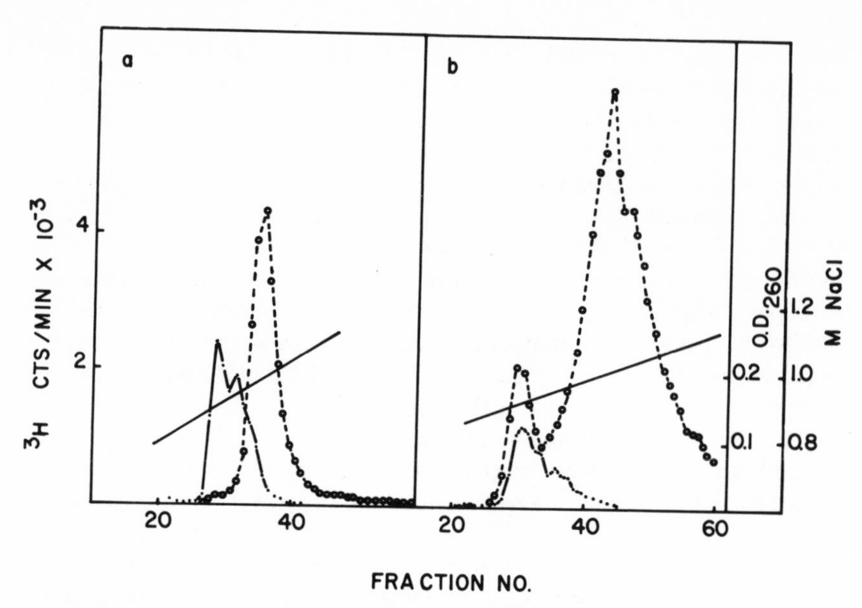

FIG. 5. MAK chromatography of 28 S RNA isolated from mesenchyme blastula embryos pulse-labeled for 10 and 180 minutes. RNA, purified from mesenchyme blastula embryos (27 hours) labeled for 10 and 180 minutes with 30 μCi/ml ^{32}P, was fractioned by sucrose gradient centrifugation. RNA sedimenting at 28 S was isolated, a carrier 28 S ribosomal RNA from unfertilized eggs was added as an optical density marker, and the RNA was chromatographed on MAK. 28 S RNA isolated from embryos pulse-labeled for 10 minutes (a) was eluted with a linear gradient of 0.7 to 1.5 *M* NaCl, and the 28 S RNA isolated from embryos labeled for 180 minutes (b) was eluted with a linear gradient of 0.8 to 1.3 *M* NaCl. Radioactivity, counts per minute ^{32}P (O——O); optical density at 260 mμ (●——●); molarity NaCl (——).

approximately one size which sediment at the same rate on a sucrose gradient are chromatographed on a MAK column, molecules with base compositions as different as rRNA and DNA-like RNA should be readily separated.

The technique was tested by fractionating the RNA from 27-hour mesenchyme blastula embryos which had been incubated for 180 minutes with ^{32}P. This RNA was fractionated on a sucrose gradient and the RNA sedimenting at 28 S was chromatographed on MAK columns. The MAK column separated the RNA into two distinct peaks (Fig. 5b). A small fraction, which eluted at the lower salt concentrations, chromatographed with 28 S ribosomal RNA marker. Most of the other radioactive molecules eluted as a broad peak at higher salt concentrations. Although ribosomal RNA markers were recovered from these columns with yields greater than 90%, only 56% of the labeled RNA was recovered during salt elution up to concentrations of 4 *M* NaCl.

The base composition of radioactive RNA cochromatographing with the ribosomal marker had a 59% G + C content, similar to that of 28 S ribosomal RNA (Fig. 6 and Table 3). The ^{32}P-RNA eluted from these columns at higher salt concentrations than the ribosomal RNA marker had base compositions ranging from 42% to 35% G + C. The molecules having lower G + C base contents eluted at higher salt concentrations, as would be expected from the relationship between G + C content and salt elution previously estab-

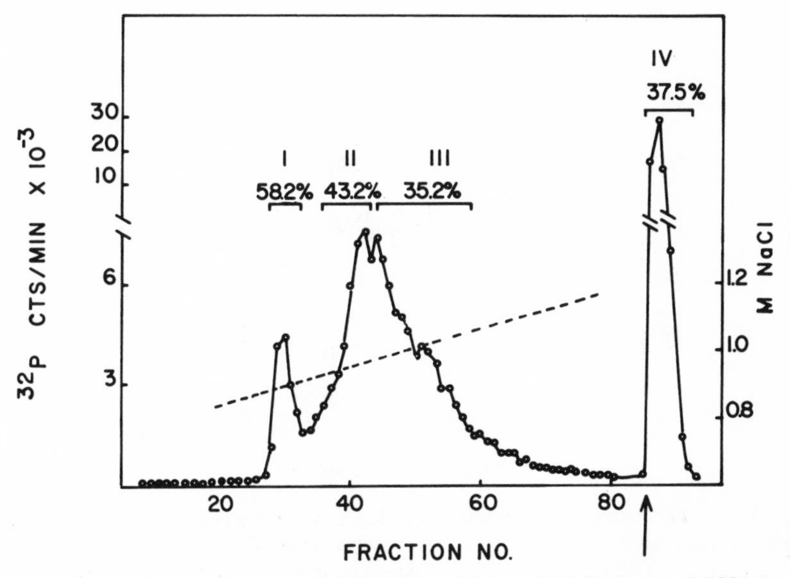

FIG. 6. MAK chromatography and base composition of 28 S ribosomal RNA isolated from mesenchyme blastula embryos labeled 180 minutes with ^{32}P. 28 S, ^{32}P-labeled RNA isolated from mesenchyme blastula embryos and purified by sucrose gradient centrifugation was chromatographed on MAK using a 0.8 to 1.3 *M* NaCl gradient, followed by washing these columns with 8 *M* urea in 0.05 *M* phosphate buffer, pH 6.5. The ^{32}P RNA recovered by urea treatment was designated fraction IV. Fractions contained in regions I, II, III, and IV were then pooled, and hydrolyzed in alkali; the base compositions of the nucleotides were measured. Numbers represent the moles percent of G + C, and the complete base compositions are described in Table 2. Radioactivity, counts per minute ^{32}P (O——O); molarity NaCl (——).

lished for the separation of DNA (Sueoka and Chen, 1962). These elution patterns provide an independent indication that the variations in ^{32}P base compositions reflect variations in the true base composition of the RNA molecules and are not an artifact of changes in specific activity of the α-phosphates in the nucleotide triphosphates (Y'cas and Vincent, 1960). The RNA molecules, which on the average may have a "DNA-like" (37% G + C) base composition, are thus apparently somewhat heterogeneous with respect to base composition.

Radioactive RNA not removed from the MAK columns by NaCl was recovered by washing with 8 *M* urea. This fraction was similar to the DNA-like RNA eluted by NaCl (37% G + C), and any unique properties it may have are unknown. The high recoveries of

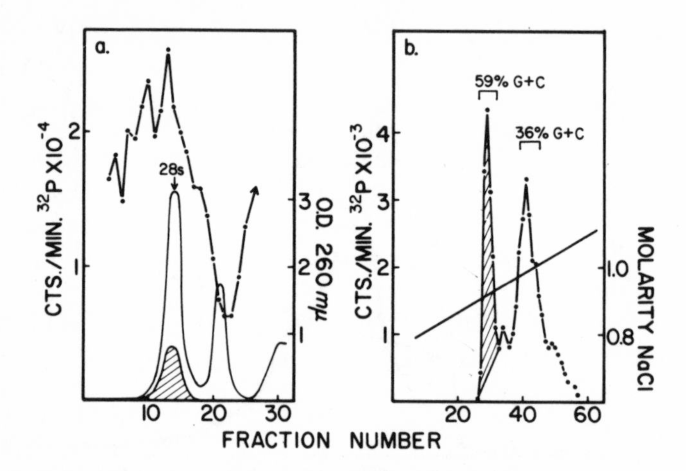

FIG. 7. Sedimentation distribution and MAK chromatography of ^{32}P-RNA accumulated during early blastula stages. (a) Embryos were labeled for 10 hours with ^{32}P, 5 μCi/ml, beginning at early blastula stage (10 hours). The RNA from these embryos was sedimented for 12 hours at 26,000 rpm in a Spinco SW 27 rotor; 1.2-ml fractions were collected, and 0.2 ml of each fraction was precipitated with TCA and radioactivity was determined. The shaded area at 28 S represents the amount of ^{32}P labeled, 28 S ribosomal RNA, calculated as described in the text from results of the MAK chromatography (7b). (b) 28 S RNA fractionated on sucrose gradients (0.8 ml of fraction 14, Fig. 7a) was chromatographed on MAK columns with linear 0.8 to 1.3 *M* NaCl gradient. The base compositions of fractions pooled from regions I and II were measured and are described in Table 3. The numbers indicate the moles percent of G + C in the fractions and the arrow indicates the approximate NaCl concentrations at which known 28 S ribosomal RNA elutes from these columns. Radioactivity, counts per minute ^{32}P (● —— ●); molarity of NaCl (——).

TABLE 3
BASE COMPOSITION OF 28 S RNA FRACTIONATED BY MAK CHROMATOGRAPHY

Stage (time labeled after fertilization)	Region (fraction number)	Moles % 2′-3′ nucleotides				
		C	A	U	G	G + C
Mesenchyme blastula, (27–30 hours)	I (28–32)	25.5	21.8	20.0	32.7	58.2
	II (36–43)	19.6	30.7	27.0	22.7	42.3
	III (43–58)	12.0	31.8	33.0	23.2	35.2
	IV (urea wash)	18.5	31.2	31.4	19.0	37.5
Early blastula, (10–20 hours)	I (27–32)	21.6	21.6	20.0	36.9	58.5
	II (39–46)	13.4	30.2	33.5	22.9	36.3
Egg 28 S rRNA		26.1	20.2	18.7	34.9	61.0

ribosomal RNA optical density with NaCl would suggest that all the radioactive ribosomal RNA is also eluted from the columns with the marker. A fraction of 28 S RNA isolated from mesenchyme blastula embryos which had been pulse-labeled for only 10 minutes with ^{32}P eluted on MAK columns at salt concentrations higher than marker 28 S ribosomal RNA (Fig. 5a). There was no indication that any ^{32}P-RNA chromatographed with 28 S ribosomal RNA. This indicates that the high G + C content fraction of 28 S RNA observed after 180 minutes labeling is preferentially accumulated, as would be expected of ribosomal RNA.

When this procedure was proved effective, it was applied to analysis of radioactive RNA synthesized during the first half of blastula stage when only heterogeneous RNA could be detected by sedimentation analysis of total RNA (Figs. 3 and 4). A 28 S fraction of radioactive RNA from embryos incubated with ^{32}P from 10 to 20 hours was fractionated on a MAK column (Fig. 7a). It separated into two distinct populations (Fig. 7b). One class eluted at the NaCl concentration of 28 S ribosomal RNA and had a 58% G + C base composition typical of ribosomal RNA. The other eluted at higher salt concentrations and had a low G + C base composition (Table 3). As a control, RNA from areas other than the 28 and 18 S peaks were fractionated on MAK columns. Only one broad peak not coincident with ribosomal RNA was obtained with these fractions indicating that the high G + C RNA was predominantly in the peaks of rRNA. About 12% of the radioactive RNA in the 28 S fraction isolated from embryos incubated with ^{32}P during the first half of blastula stage chromatographed on MAK coincident with rRNA. The amount of ^{32}P-

labeled 28 S ribosomal RNA in the sucrose gradient fractions was estimated by assuming that this high G + C RNA is rRNA and is distributed in the sucrose gradient as the marker of optical density 28 S rRNA. This amount, shown by the shaded area in Fig. 7, represented only a small portion of the total labeled RNA sedimenting at 28 S; however, this amount is similar to the amount (8%) predicted on the basis of the quantitative measurements of RNA accumulation (compare Figs. 3 and 7). This result, therefore, indicates that the accumulation of newly synthesized rRNA occurs at much the same rate during the first half of blastula stage and during pluteus stage. The rRNA which is synthesized during early blastula stage also appeared to be associated with ribosomes since uridine-^{3}H was incorporated into RNA which sedimented as a distinct peak at 28 S when RNA was isolated from purified ribosomes (unpublished observation).

The synthesis of rRNA during cleavage stages has also been analyzed using these techniques, but radioactive rRNA could not be definitively detected in RNA isolated from embryos incubated with either ^{32}P or adenosine-8-^{3}H as precursors. Although the MAK column was unable to resolve the predicted amounts of rRNA over the large background of DNA-like RNA at this early stage, the amount and base composition of the radioactive RNA cochromatographing with rRNA optical density were compatible with rates of rRNA synthesis occurring at later stages. Since actual amounts of radioactivity per embryo (as opposed to per nucleus) were much lower at cleavage than at blastula stage, the inability to detect ribosomal RNA synthesis during cleavage does eliminate the possibility that the rRNA detected at early blastula stages was synthesized by immature oocytes contaminating the preparation of embryos.

DISCUSSION

Synthesis of RNA during development. Base composition and sedimentation analyses of pulse-labeled RNA have demonstrated that DNA-like RNA is the predominant class of high molecular weight RNA being synthesized during development from cleavage to pluteus stages of *S. purpuratus* sea urchin embryos. The proportion of newly-synthesized DNA-like RNA which sediments more rapidly than 28 S increased markedly during late cleavage stages and accounted for a large fraction of synthesis at all later stages. Most of this rapidly sedimenting RNA was not a high molecular weight rRNA precursor (Nemer and Infante, 1966), but had heterogeneous sedimentation properties, and a high U and low G + C base composition. The pulse-labeled RNA at these stages is very unstable (Kimjima and Wilt, 1969). These particular sedimentation and base composition properties indicate that this class of RNA is the unstable "nuclear" RNA described in other animal cells (Attardi *et al.*, 1966; Houssais and Attardi, 1966; Soeiro *et al.*, 1966; Shearer and McCarthy, 1967).

The observed increase in the proportion of large "nuclear" RNA molecules during the transition from cleavage to blastula stages may be related to the observation that the majority of RNA molecules synthesized during early cleavage stages are produced in the cytoplasm, not in the nucleus (Chamberlain, 1967, 1968). The increasing predominance of "nuclear" RNA synthesis during late cleavage and early blastula stages, therefore, may simply reflect the rapid increase in the numbers of nuclei which occurs during this period of development. Further analysis of this transition at late cleavage is required to verify this suggestion.

Accumulation of RNA during development. RNA molecules with G + C base compositions higher than those of pulse-labeled RNA were shown to be preferentially accumulated at all stages. At later stages of development a large fraction of this high G + C RNA was rRNA, and such may be true for earlier stages as well. However, because considerable heterogeneity in the G + C content of DNA-like RNA molecules was demonstrated by the fractionation on MAK columns, a portion of such preferentially accumulating RNA may be DNA-like RNA molecules with higher G + C compositions.

The variation of G + C content from 30% to 42% in DNA-like RNA was unexpected since analyses of total heterogeneous RNA usually give base ratios rather similar to DNA. These data show that "DNA-like" RNA molecules do not necessarily have similar average base compositions but, in sea urchins at least, may vary considerably from the average base composition of the DNA. This indicates that many genes with base compositions considerably different from the average exist in the DNA and are active during sea urchin development.

It was recognized that the preferential accumulation of rRNA at late developmental stages could represent stage-specific differences in the quantitative accumulation of either rRNA or DNA-like RNA. The quantitative measurements of RNA accumulation presented in this paper have demonstrated that the rate of accumulation of DNA-like RNA per nucleus is much greater at cleavage and blastula stages than at pluteus stage. If the rate of accumulation of rRNA per nucleus were assumed to be identical at all stages, these quantitative differences in the accumulation of DNA-like RNA were sufficiently large to obscure the accumulated rRNA in sucrose gradients of RNA extracted from cleavage and early blastula stages. By use of more sensitive purification procedures, newly synthesized rRNA was found in early blastula in the approximate proportions predicted on the basis of quantitative measurements of rRNA synthesis at pluteus stage. These large quantitative differences in the accumulation of DNA-like RNA, therefore, must account completely for the changes in the relative accumulation of DNA-like RNA and rRNA which have been repeatedly observed to occur at early gastrula stages (Nemer, 1963; Comb *et al.*, 1965; Giudice and Mutolo, 1967, 1969).

During cleavage stages, the accumulation per nucleus of newly synthesized DNA-like RNA was found to be even greater than at blastula stage, and consequently the calculated synthesis of rRNA was an even smaller fraction of the total RNA accumulated. The fact that the synthesis of only DNA-like RNA has been detectable during cleavage also must be the result of the greater accumulation of DNA-like RNA at this stage, not a result of the repressed synthesis of rRNA.

Although these results do explain the biochemical data which originally gave rise to the hypothesis, they do not, in themselves, resolve the question of whether rRNA synthesis is regulated during the cleavage stage. Other evidence has suggested that rRNA synthesis may be abnormal during early stages of sea urchin development. Nucleoli appear to be atypical during early developmental stages (Millonig, 1966; Karasaki, 1968; Longo and Anderson, 1968). However, this atypical form is most likely due to the rapidity of cell division at these early stages since nucleoli have been observed to undergo rapid growth throughout the short interphase of cleavage stage nuclei and typical nucleoli can be induced by inhibition of cell division (Humphreys, T., in preparation). Also, the incorporation of $^{14}CH_3$-methionine into rRNA has not been detectable during early stages (Comb *et al.*, 1965). The interpretation of this observation, however, is not clear since these results are not subject to any quantitative restrictions from which it would be possible to calculate the amounts of precursor expected to be incorporated into rRNA. Furthermore, studies with high specific activity $^{3}H_{3}C$-methionine have indicated that this compound is not a specific precursor for rRNA in these embryos, but is incorporated into all RNA classes, presumably by the pathway for purine biosynthesis (unpublished observations; also see Scarano *et al.*, 1965). These additional considerations, therefore, show that no conclusions regarding regulation of rRNA can be drawn from these observations.

Changes in RNA metabolism during development of *Xenopus laevis* have been well characterized, and both the qualitative and quantitative changes in patterns of RNA accumulation are similar to those described for the sea urchin embryo (Brown and Littna, 1964, 1966). These observations suggested that the ribosomal genes were repressed during cleavage and blastula stages and were activated at gastrulation of *Xenopus*. Reconsideration of these data in the light of this analysis of sea urchin RNA synthesis leads to the conclusion that changes in DNA-like RNA synthesis also account for the changes in RNA metabolism during development of *Xenopus* embryos. Recalculation of the data (Brown and Littna, 1966) indicates that accumulation of newly synthesized rRNA, expressed on a per nucleus basis, is similar at all stages at which ribosomal synthesis was actually detectable, and that accumulation of newly synthesized DNA-like RNA per nucleus were large enough during cleavage and blastula stages to obscure the predicted amount of newly synthesized rRNA if synthesis was occurring at the same rate in these early stages.

The metabolic or developmental significance of the striking changes in the accumulation of DNA-like RNA observed during development of these organisms is presently unknown, but is of considerable interest. The greater accumulation of DNA-like RNA molecules during early stages may reflect either a greater stability of the DNA-like RNA molecules or a more rapid synthesis of RNA resulting from a more rapid transcription of genes or from transcription of a larger percentage of the genome. Measurements of the kinetics of labeling of RNA during sea urchin development have indicated that the rate of RNA synthesis per cell is greater at early than at late stages (Kimjima and Wilt, 1969), which would lend support to the latter interpretation. This high rate of DNA-like RNA metabolism could be related to the rapid cell division occurring at the early stages or may have considerable significance for the important steps in the determination of the various cell types during these stages of development. Further analysis of these and other possibilities should prove to be very interesting.

SUMMARY

RNA synthesis was analyzed in sea urchin embryos in order to determine the contribution which changes in synthesis of heterogeneous DNA-like RNA and of ribosomal RNA make to the changes in TNA metabolism which occur during embryogenesis. The base composition and sedimentation behavior of RNA molecules incorporating radioactivity during a short incubation of embryos with radioactive RNA precursors indicate that heterogeneous, DNA-like RNA was the predominant class of RNA synthesized in embryos at all stages of development. During longer incubations with radioactive precursors, radioactivity preferentially accumulated in ribosomal RNA. This preferential accumulation of newly synthesized rRNA was much more extensive at later than at early stages of development. However, accumulation of newly synthesized 28 S ribosomal RNA could be demonstrated as early as the first half of blastula stage when rRNA was extensively purified from DNA-like RNA by sucrose gradient sedimentation and MAK chromatography.

In order to interpret these developmental changes in the relative accumulation of ribosomal and DNA-like RNA, the amounts of newly synthesized RNA accumulated during cleavage, blastula, and pluteus stages were quantitatively estimated by measurement of the incorporation of radioactive precursors into RNA and simultaneous determination of the specific activity of the nucleotide triphosphate precursor pools. The results indicated that the accumulation of newly synthesized DNA-like RNA per nucleus was quantitatively much greater at cleavage and blastula stages than at pluteus stage, whereas the accumulation of newly synthesized rRNA per nucleus at blastula and pluteus stages were similar. Therefore, decreases in the quantitative accumulation of DNA-like RNA account entirely for the change in the relative accumulation of ribosomal RNA and DNA-like RNA which occurs during development of sea urchin embryos.

REFERENCES

Attardi, G., Parnas, H., Hwang, M.-I., and Attardi, B. (1966). Giant-size rapidly-labeled nuclear ribonucleic acid and cytoplasmic messenger ribonucleic acid in immature duck erythrocytes. *J. Mol. Biol.* **20,** 145–182.

Brown, D. D., and Littna, E. (1964). RNA synthesis during the development of *Xenopus laevis*, the South African clawed toad. *J. Mol. Biol.* **8,** 669–687.

Brown, D. D., and Littna, E. (1966). Synthesis and accumulation of DNA-like RNA during embryogenesis of *Xenopus laevis*. *J. Mol. Biol.* **20,** 81–94.

Canellakis, E. S., and Mansavinos, R. (1958). The conversion of ^{14}C-deoxynucleoside-5′-monophosphates to the corresponding di- and triphosphates by soluble mammalian enzymes. *Biochim. Biophys. Acta* **27,** 643–645.

Chamberlain, J. (1967). RNA synthesis in non-nucleate fragments of *Arbacia* eggs. *Biol. Bull.* **133,** 461.

Chamberlain, J. (1968). Extranuclear RNA synthesis in sea urchin embryos. *J. Cell Biol.* **39,** 23a.

Cole, H. A., Wimpenny, J. W. T., and Hughes, D. E. (1967). The ATP in *E. coli*. I. Measurement of the pool using a modified luciferinase assay. *Biochim. Biophys. Acta* **143,** 445–453.

Comb, D. G., Katz, S., Branda, R., and Pinzino, C. (1965). Characterization of RNA species synthesized during early development of sea urchins. *J. Mol. Biol.* **14,** 195–213.

Girard, M., Latham, H., Penman, S., and Darnell, J. E. (1965). Entrance of newly formed messenger RNA and ribosomes into HeLa cell cytoplasm. *J. Mol. Biol.* **11,** 187–201.

Giudice, G., and Mutolo, V. (1967). Synthesis of ribosomal RNA during sea urchin development. *Biochim. Biophys. Acta* **138,** 276–285.

Giudice, G., and Mutolo, V. (1969). Synthesis of ribosomal RNA during sea urchin development. II. Electrophoretic analysis of nuclear and cytoplasmic RNA's. *Biochim. Biophys. Acta* **179,** 341–347.

Glisin, V. R., and Glisin, M. V. (1964). Ribonucleic acid metabolism following fertilization in sea urchin eggs. *Proc. Nat. Acad. Sci. U. S.* **52,** 1548–1553.

Gross, P. R., Kraemer, K., and Malkin, L. I. (1964). Base composition of RNA synthesized during cleavage of the sea urchin embryo. *Biochem. Biophys. Res. Commun.* **18,** 569–575.

Gross, P. R., Malkin, L. I., and Moyer, W. A. (1964). Templates for the first proteins of embryonic development. *Proc. Nat. Acad. Sci. U.S.* **51,** 407–414.

Gurdon, J. B. (1967). Control of gene activity during the early development of *Xenopus laevis*. *In* "Heritage from Mendel" (R. A. Brink, ed.), pp. 203–214. Univ. of Wisconsin Press, Madison, Wisconsin.

Hayashi, M., and Spiegelman, S. (1961). The selective synthesis of informational RNA in bacteria. *Proc. Nat. Acad. Sci. U. S.* **47,** 1565–1580.

Hinegardner, R. T. (1967). Echinoderms. *In* "Methods in Developmental Biology" (F. H. Wilt and N. K. Wessels, eds.), pp. 139–155. Crowell, New York.

Houssais, J. F., and Attardi, G. (1966). High molecular weight nonribosomal-type nuclear RNA and cytoplasmic messenger RNA in HeLa cells. *Proc. Nat. Acad. Sci. U. S.* **56,** 616–623.

Hulbert, R. B., Schmitz, H., Brumm, A. F., and Potter, V. R. (1954). Nucleotide metabolism. II. Chromatographic separation of acid-soluble nucleotides. *J. Biol. Chem.* **209,** 23–29.

Karasaki, S. (1968). The ultrastructure and RNA metabolism of nucleoli in early sea urchin embryos. *Exp. Cell Res.* **52,** 12–26.

Kimjima, S., and Wilt, F. H. (1969). Rate of nuclear RNA turnover in sea urchin embryos. *J. Mol. Biol.* **40,** 235–246.

LONGO, F. J., and ANDERSON, E. (1968). The fine structure of pronuclear development and fusion in the sea urchin, *Arbacia punctulata*. *J. Cell Biol.* **39,** 339–368.

MALKIN, L. I., GROSS, P. R., and ROMANOFF, P. (1964). Polysomal protein synthesis in fertilized sea urchin eggs: The effect of actinomycin treatment. *Develop. Biol.* **10,** 378–394.

MANDELL, J. D., and HERSHEY, A. D. (1960). A fractionating column for analysis of nucleic acids. *Anal. Biochem.* **1,** 66–77.

MILLONIG, G. (1966). The morphological changes of the nucleolus during oogenesis and embryogenesis of Echinoderms. *Proc. Int. Congr. Electron Microsc. 6th* Vol. 2, pp. 345–346.

NEMER, M. (1963). Old and new RNA in the embryogenesis of the purple sea urchin. *Proc. Nat. Acad. Sci. U. S.* **50,** 217–221.

NEMER, M., and INFANTE, A. A. (1966). Early control of gene expression. *In* "The Control of Nuclear Activity" (L. Goldstein, ed.), pp. 101–127. Prentice-Hall, Englewood Cliffs, New Jersey.

NIERLICH, D. (1967). Radioisotope uptake as a measure of synthesis of messenger RNA. *Science* **158,** 1186–1188.

PENMAN, S., SMITH, I., and HOLTZMAN, E. (1966). Ribosomal RNA synthesis and processing in a particulate site in the HeLa cell nucleus. *Science* **154,** 786–789.

SCARANO, E., IACCARINO, M., GRIPPO, P., and WINCKELMANS, D. (1965). On methylation of DNA during development of the sea urchin embryos. *J. Mol. Biol.* **14,** 603–607.

SHEARER, R. W., and MCCARTHY, B. J. (1967). Evidence for ribonucleic acid molecules restricted to the cell nucleus. *Biochemistry* **6,** 283–289.

SMITH, J. D. (1955). The electrophoretic separation of nucleic acid components. *In* "The Nucleic Acids" (E. Chargaff and J. N. Davidson, eds.), Vol. 1, pp. 267–284. Academic Press, New York.

SOEIRO, R., BIRNBOIM, H. C., and DARNELL, J. E. (1966). Rapidly labelled HeLa cell nuclear RNA. II. Base composition and cellular localization of heterogeneous RNA fractions. *J. Mol. Biol.* **19,** 362–372.

SOEIRO, R., VAUGHN, M. H., WARNER, J. R., and DARNELL, J. E. (1968). The turnover of nuclear DNA-like RNA in HeLa cells. *J. Cell Biol.* **39,** 112–118.

SUEOKA, N., and CHEN, T. (1962). Fractionation of nucleic acids with the methylated albumin column. *J. Mol. Biol.* **4,** 161–172.

SWIFT, H. (1955). Cytochemical techniques for nucleic acids. *In* "The Nucleic Acids" (E. Chargaff and J. M. Davidson, eds.), Vol. 2, pp. 51–92. Academic Press, New York.

TSUBOI, K. K., and PRICE, T. D. (1959). Isolation, detection and measure of microgram quantities of labeled tissue nucleotides. *Arch. Biochem. Biophys.* **81,** 223–237.

WILLEMS, M., WAGNER, E., LAING, R., and PENMAN, S. (1968). Base composition of ribosomal RNA precursors in the HeLa cell nucleolus: further evidence of non-conservative processing. *J. Mol. Biol.* **32,** 211–220.

WILT, F. (1964). Ribonucleic acid synthesis during sea urchin embryogenesis. *Develop. Biol.* **9,** 299–313.

Y'CAS, M., and VINCENT, W. S. (1960). A ribonucleic acid fraction from yeast related in composition to deoxyribonucleic acid. *Proc. Nat. Acad. Sci. U.S.* **46,** 804–811.

Mitochondrial Nucleic Acids in Sea Urchin Eggs And Embryos

Cytoplasmic Synthesis of RNA in the Sea Urchin Embryo

SUSAN E. SELVIG, PAUL R. GROSS, AND ANNE L. HUNTER

INTRODUCTION

Early autoradiographic studies on RNA synthesis in cleaving sea urchin embryos suggest that a fraction of labeled precursor may be incorporated at cytoplasmic sites of transcription. It appears, in particular, that the actinomycin-resistant fraction of incorporation during early cleavage is due largely to turnover of the pCpCpA terminal of transfer RNA, and autoradiograms of actinomycin-treated embryos labeled with RNA precursors show cytoplasmic, rather than the normally predominant nuclear, radioactivity (Gross, 1964).

End-labeling of transfer RNA does occur during the period of cleavage, and this process is actinomycin-resistant, whereas true synthesis of RNA cán be suppressed almost completely with an adequate dose of the drug (Gross *et al.*, 1965; Malkin *et al.*, 1964; Giudice *et al.*, 1968). Nevertheless, it remains important to examine directly the possibility that some transcription, as opposed to end-labeling, may take place in the cytoplasm. The RNA synthesis of late cleavage and of later stages is overwhelmingly nuclear. This has been demonstrated by autoradiographic and biochemical means (Aronson and Wilt, 1969; Kijima and Wilt, 1969; Hunter and Gross, unpublished observations). But the situation in early cleavage is not so clear. Recent molecular hybridization experiments suggest that cytoplasmic as well as nuclear genes may be transcribed during pregastrular development (Hartmann and Comb, 1969), and there have appeared other indications of a cytoplasmic locale for some transcription processes of early cleavage (Chamberlain, 1968).

This research has been supported by grants from the National Institutes of Health GM 13560-04, the American Cancer Society 1356-C-1, and the National Science Foundation GB6350.

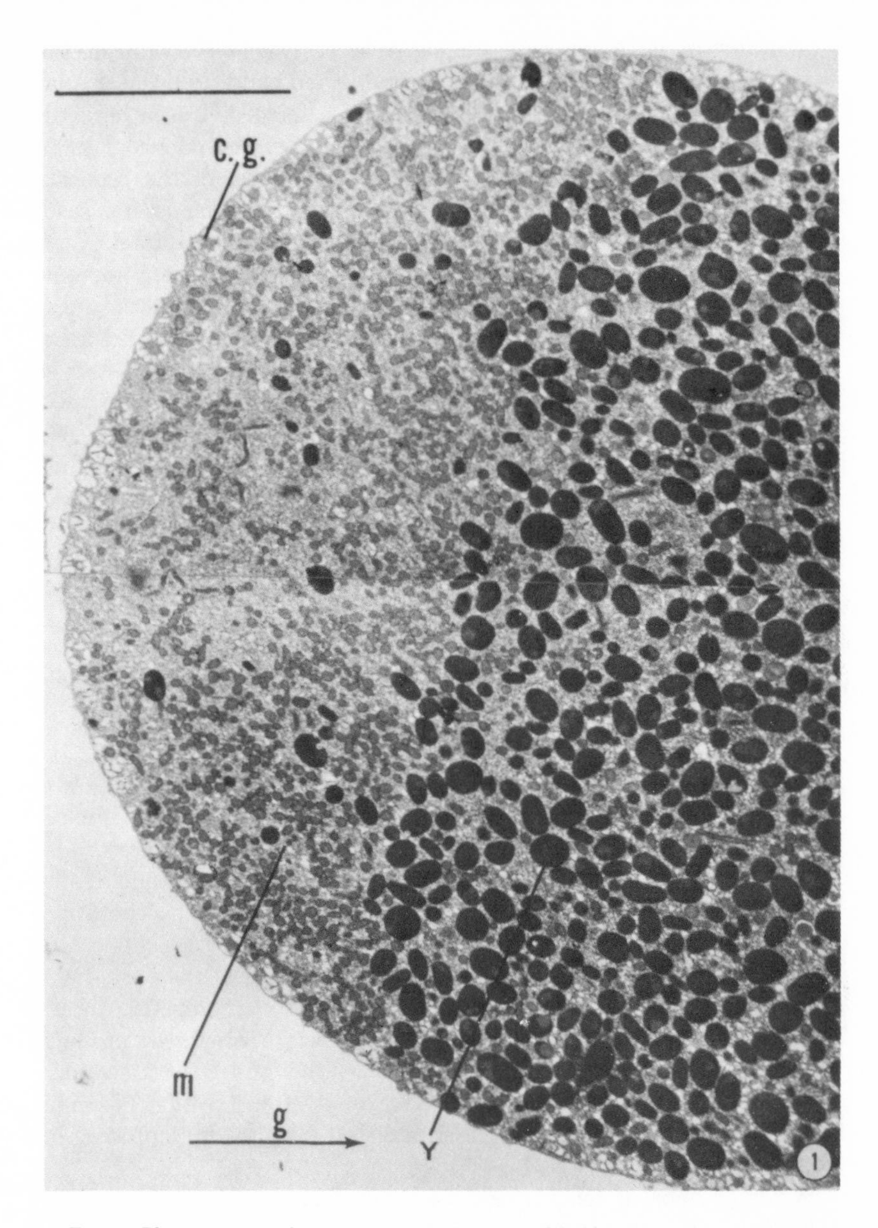

FIG. 1. Photomontage electron micrograph of a red half fixed soon after collection from the centrifuge tube. *g* shows the direction of the centrifugal field; *y* is a yolk particle; *c.g.* directs attention to the row of cortical granules, with their characteristically complex ultrastructure, just beneath the plasma membrane. The mitochondrial layer

The fact of RNA synthesis during early cleavage—even during the first mitotic cycle—has now been established, and it is clear that some of the new RNA serves as a template for protein synthesis almost immediately after its transcription (Rinaldi and Monroy, 1969; Kedes and Gross, 1969). This makes urgent a test of the proposal that some of the transcription is cytoplasmic. If such is the case, for example, earlier interpretations of experiments on protein synthesis in enucleate egg fragments will have to be reexamined, although the conclusion stands that "maternal" or untranslated messenger RNA exists in egg cytoplasm, since it does not depend upon the enucleate egg studies (Gross, 1964, 1968; Spirin, 1966; Gross *et al.*, 1965). There is, in any case, a large cytoplasmic DNA component in eggs, most or all of which resides in the mitochondria (see, for example, Pikó *et al.*, 1968; Dawid, 1965, 1966; but see also Balthus *et al.*, 1968). It seems, on the basis of our preliminary studies, that this DNA fails to be replicated during sea urchin development (Yanover, Hunter, and Gross, unpublished observations), but there is no reason why, in principle, it might not be transcribed. There may, of course, be other kinds of cytoplasmic DNA that serve as templates for RNA synthesis.

A first step in the examination of these problems must be the unequivocal demonstration of cytoplasmic RNA synthesis. This demonstration must show:

(a) whether or not the incorporation is into molecules of moderate and/or high molecular weight, higher than that of the end-labeled transfer RNA; (b) whether the high molecular weight products are in fact RNA (i.e., rather than DNA labeled by deoxynucleoside metabolites of the labeled precursor); (c) whether the incorporation is truly intracellular and truly cytoplasmic, and whether or not it might be attributed, in the half-egg preparations necessary for these experiments, to contaminant oocytes, nucleate fragments, interstitial cells from the ovary, etc.; (d) the extent to which the product is localized in particular organelles; (e) whether the RNA has properties (e.g., size distribution, resistance of its transcription to small doses of actinomycin, etc.) that distinguish it from the bulk product of intact cells.

We have chosen, for purposes of these experiments, nucleate and enucleate half-eggs of *Arbacia punctulata*. These can be prepared in

is at *m*. Cortical organization is typical of that seen in the intact unfertilized egg. The scale line represents 10 μ.

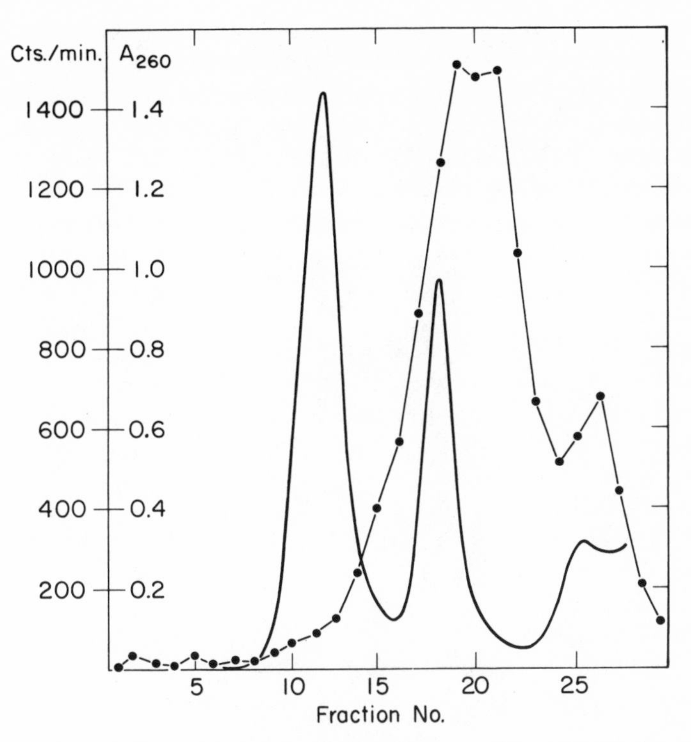

FIG. 2. RNA synthesized by red halves. Red halves were activated and incubated in 100 μCi uridine-5-^{3}H per milliliter, MFSW + ps for 2 hours. Extracted RNA was centrifuged (sedimentation right to left) for 18 hours at 23,000 rpm. ●——●, Radioactivity; ——, A_{260}.

a reproducibile way and in good yields. They are easy to score for contamination by other fragment types. Our approach has been to allow activated or fertilized half-eggs (and control intact cells) to incorporate RNA precursors into polymers. The incorporation was then characterized by autoradiography—optical and electron microscopic—and through biochemical study of the products.

The data show that activated enucleate half-eggs can synthesize high molecular weight RNA, and that this material is distinguishable, in some of its properties, from the total transcription product of nucleate fragments and of normal early embryos. The synthesis is clearly cytoplasmic, but no conclusions are as yet possible as to the actual sites of transcription. Nor is it possible to estimate what fraction of the normal zygotic RNA synthesis takes place in the cyto-

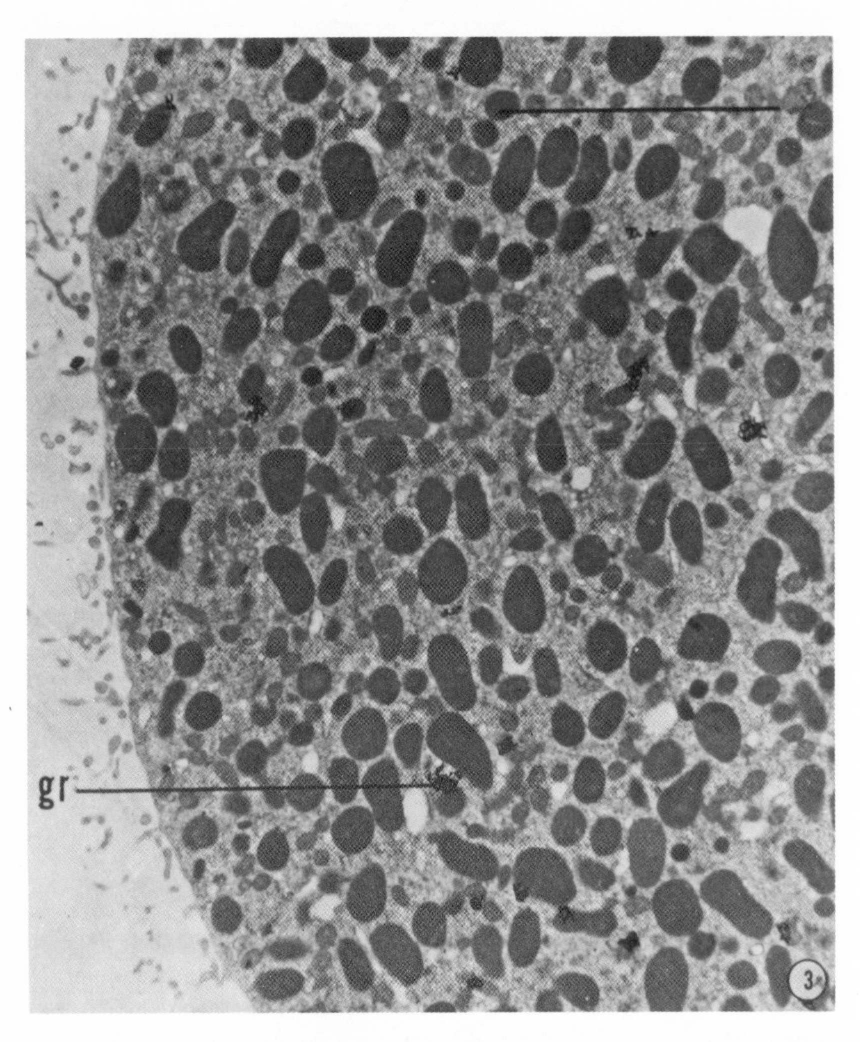

FIG. 3. Autoradiographic electron micrograph of the yolk region of a red half. Labeled with uridine-^{3}H as described in the text. *gr* identifies one of the twenty silver grains visible over this section. Mitochondria and a few vacuoles are seen interspersed among the dense yolk granules. The grains show no association with particular organelles. Note that the cortical granules have vanished, replaced by a microvillous surface typical of that seen on normally fertilized eggs. The scale line represents 5 μ.

plasm. The results do, however, demonstrate the reality of cytoplasmic RNA synthesis, and they suggest that such synthesis may contribute in normal circumstances to the total.

MATERIALS AND METHODS

Collection of gametes. Arbacia punctulata were obtained from Mr. Glendle Noble, Panama City, Florida. Ovaries were removed from the females and placed in Millipore-filtered sea water containing penicillin-streptomycin (MFSW×ps) 50 units/ml (Microbiological Associates, Inc., Bethesda, Maryland). The eggs were filtered through cheesecloth and washed repeatedly by sedimentation through MFSW+ps. Sperm were collected from excised testis and stored cold in a Pasteur pipette. An aliquot from each batch of eggs was fertilized. Only those eggs giving better than 90% fertilization were selected for experiments.

Preparation of half-eggs. Half-eggs were prepared following a procedure described earlier (Malkin *et al.*, 1965). Three milliliters of eggs were layered on a 27-ml linear gradient made from 1.2 *M* sucrose and MFSW+ps:1.2 *M* sucrose, 4:1. Three such gradients were placed in a Spinco SW 25.1 rotor and spun 12 minutes at 12,000 rpm in a Spinco L-2 Ultracentrifuge. The banded white halves were collected with a Pasteur pipette. Banded red halves were collected by piercing the bottom of the tube and applying air pressure to the top. Collected egg fragments were washed three times by sedimentation through MFSW+ps. They were then checked for contamination by careful surveys in the light microscope.

Activation of half-eggs. Half eggs were activated by the method of Harvey (1956). An equal volume of MFSW+ps with 60 g/l additional NaCl was added to the suspension of half-eggs. After 20 minutes, the cell fragments were collected by sedimentation and then resuspended in MFSW+ps.

Incorporation of radioactive precursors. Uridine-5-^{3}H 20 Ci/mmole and L-leucine-4,5-^{3}H, 30–50 Ci/mmole were obtained from New England Nuclear, Boston, Massachusetts. An additional 50 units/ml of penicillin-streptomycin were added to the incubation mixtures, and incubation flasks were kept dark (wrapped in aluminum foil) to prevent possible growth of algae during incubation.

For protein synthesis studies, leucine was added to give a concentration of 1–1.5 μCi/ml. After 15 minutes, an equal volume of 10% TCA was added, and the samples were chilled. Precipitated cells were pelleted at 1000 *g* for 15 minutes, resuspended in 5% TCA and

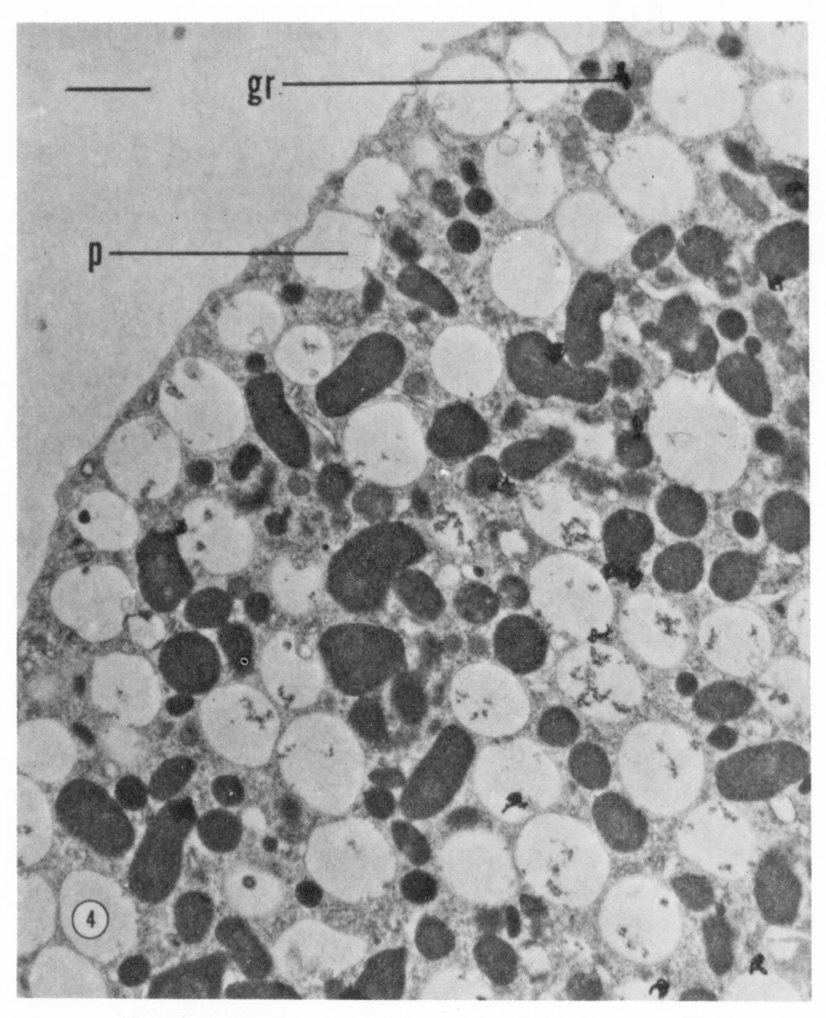

FIG. 4. Electron microscope autoradiogram of the centrifugal pole (Pigment pole) of a red half. Labeled as in Fig. 3; *p* identifies a pigment vacuole—remnant of a pigment granule. Note that the silver grain distribution is roughly the same (per unit area) as in the "yolky" region shown in Fig. 3. "Activated" surface. Scale line represents 1 μ.

incubated for 30 minutes at 80–90°C. The precipitates were cooled and washed twice with 5% TCA, then collected on Millipore filters and washed three times with 5% TCA. Dried filters were counted in a Beckman LS-250 liquid scintillation counter using toluene and PPO-POPOP scintillant (Spectrafluor, Nuclear Chicago).

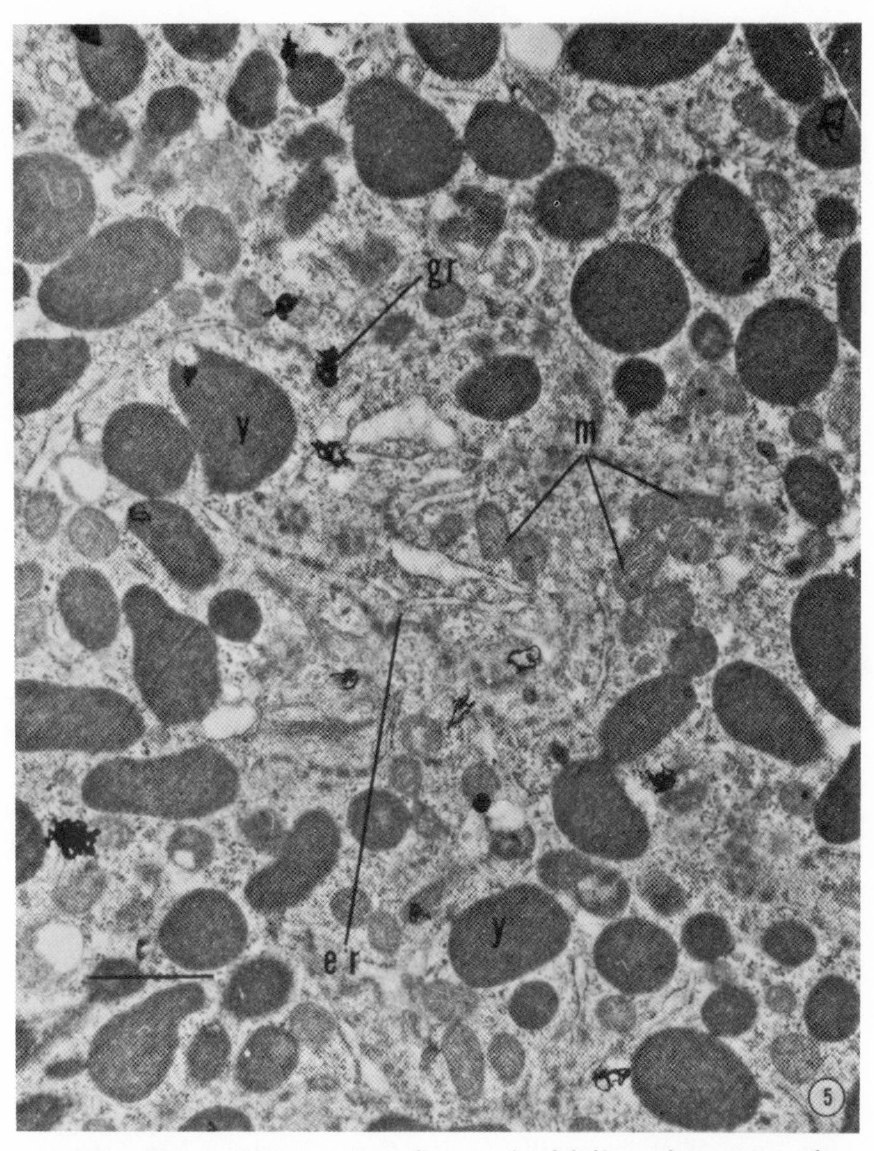

FIG. 5. Electron microscope autoradiogram of red half cytoplasm at somewhat higher magnification than in Figs. 1, 3, and 4, to show some ultrastructural features of the cytoplasmic matrix and the mitochondria. Labeling as in preceding figures. *er* identifies one of the rare local patches of rough endoplasmic reticulum. close to some vesicles and cisternae of smooth endoplasmic reticulum. *m* identifies a mitochon-

Uridine concentrations in incubation mixtures were 100 μCi/ml for red halves, 50 μCi/ml for white halves, and 10 μCi/ml for embryos. At the end of 2 hours, the incubation mixtures were chilled, and in some cases chilled unfertilized eggs were added as carrier. The material was pelleted, resuspended in chilled MFSW+ps, and repelleted. If RNA extraction did not immediately follow, the pellets were frozen in a dry-ice–acetone bath.

Extraction of RNA. RNA extraction followed the method of Brown and Littna (1964). The pellet, usually about 0.2 ml, was thawed by addition of 2 ml of homogenization medium: 0.1 *M* acetate buffer (pH 5.0) with 0.5% sodium dodecyl sulfate and 5 μg/ml polyvinyl sulfate. It was then homogenized at 4°C. An equal volume of water-saturated phenol was added, and the mixture was emulsified on a "Vortex" stirrer for 5–10 minutes at 4°C. The emulsion was broken by centrifuging at 12,000 *g* for 10 minutes. The phenol layer was removed and discarded, and the aqueous layer and interface were reextracted by this procedure two or three times. The aqueous layer was removed and made 0.1 *M* in NaCl. Nucleic acids were precipitated by adding 2.5 volumes of absolute ethanol and leaving the mixture at −20°C. Precipitates were pelleted and redissolved in the acetate buffer, and again precipitated with NaCl and ethanol.

Linear sucrose gradients, 15–30% (w/v) were prepared using the acetate buffer without polyvinyl sulfate in tubes for the Spinco SW 25.3 rotor. After centrifugation as described below, the gradients were collected by displacement pumping from the bottom of the tube. Optical density at 260 mμ was monitored continously in the 0.2-ml flow cell of a Gilford recording spectrophotometer and the gradient was divided into aliquots with the aid of a fraction collector. The aliquots were collected in tubes containing 100 μg of bovine serum albumin, and an equal volume of 15% TCA was added. Precipitates were collected on type HA Millipore filters and washed with 6% TCA, then counted as described above.

DNase treatment. Samples of purified RNA were dissolved in TNM buffer [tris(hydroxymethyl)aminomethane, pH 7.4, 0.01 *M*; NaCl, 0.01 *M*; $MgCl_2$, 0.005 *M*] containing 60 μg/ml DNase (RNase-free, from Worthington Biochemical Corp., Freehold, New Jersey).

---

drial cluster, with these organelles showing their typical (for this cell) organization; very finely granular to amorphous matrix and small, internal cristae. Note that the mitochondria in this cluster are not radioactive, and that no matrix structures inside them are like the abundant cytoplasmic ribosomes. Scale line represents 1 μ.

Control samples were dissolved in the buffer alone. Incubation was at 37°C for 5 minutes, following which the nucleic acids were precipitated with NaCl and ethanol as described. The precipitates were then redissolved, and their solutions were studied by sedimentation in sucrose gradients.

Actinomycin treatment. Actinomycin was used in the form of Cosmagen (Merck, Sharpe and Dohme) at concentrations indicated below.

Microscopy and autoradiography. Egg and egg fragment samples were fixed in 3% glutaraldehyde–0.01 *M* phosphate buffer (pH 7.5) with 2% NaCl and 0.5% $MgCl_2$. Fixation was generally allowed to continue overnight in the cold. Fixed material was rinsed very thoroughly with 2% NaCl and then with 4% NaCl. Postfixation was carried out for 1 hour in 1% OsO_4 in phosphate buffer with 4% NaCl. The osmium fixative was replaced by 4% NaCl, then the specimens were dehydrated and embedded in Epon by routine methods.

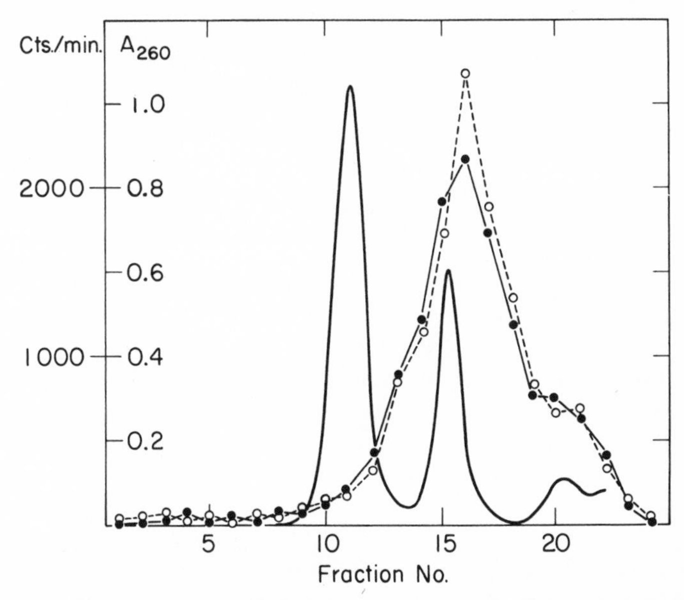

FIG. 6. DNase treatment of RNA from red halves. Red halves were activated and, after 90 minutes, incubated in 100 μCi uridine-5-^{3}H per milliliter, MFSW+ps for 2 hours. Extracted RNA was divided in half. One part was treated with DNase as described and the other with buffer. Samples were then centrifuged (sedimentation from right to left) for 17.5 hours at 24,000 rpm. O---O, control; ●——●, DNase treated; —— A_{260}.

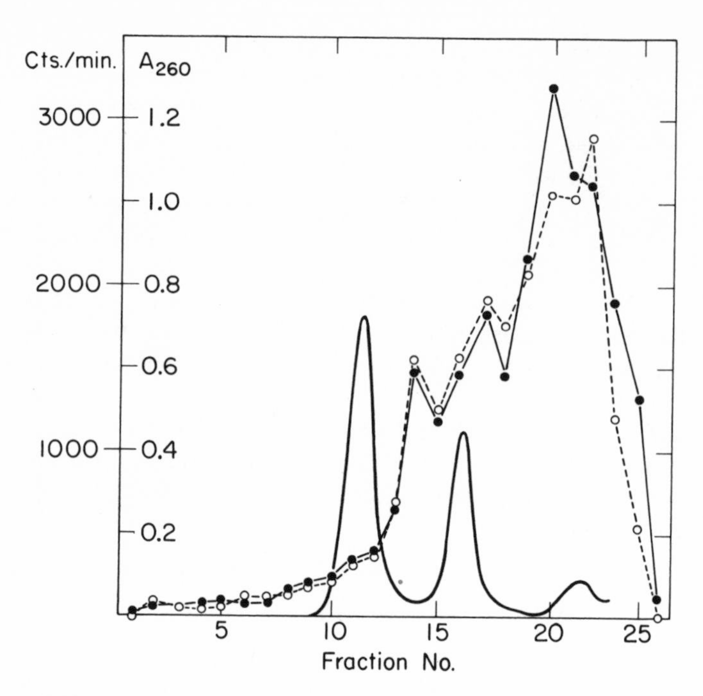

FIG. 7. DNase treatment of RNA from white halves. White halves were fertilized and, after 90 minutes, incubated in 50 μCi uridine-5-^{3}H per milliliter, MFSW+ps for 2 hours. Extracted material was treated as described in Fig. 6, except that centrifugation was for 17 hours. O---O, control; ●——●, DNase treated; ——, A_{260}.

Blocks were sectioned on a Porter-Blum MT-2 ultramicrotome. Thick sections of 1 μ served for optical microscopy and autoradiography, while thin sections were mounted on unfilmed clean copper grids for electron microscopy and electron microscope autoradiography. For the latter, a thin carbon film was deposited in the sections, and the grids were attached to glass microscope slides. Ilford L-4 emulsion was applied with a wire loop. Autoradiographic exposures of 1 month were required for usable numbers of silver grains over the half-egg sections. The backgrounds were negligible. Photographic development was for 4 minutes at 20°C in Kodak Dektol. Fixation and washing followed, after which the slides were dried. The grids were removed and stained for 15–20 minutes with lead acetate. Electron microscopy was done with a Hitachi HU 11a instrument operated at 50 kV.

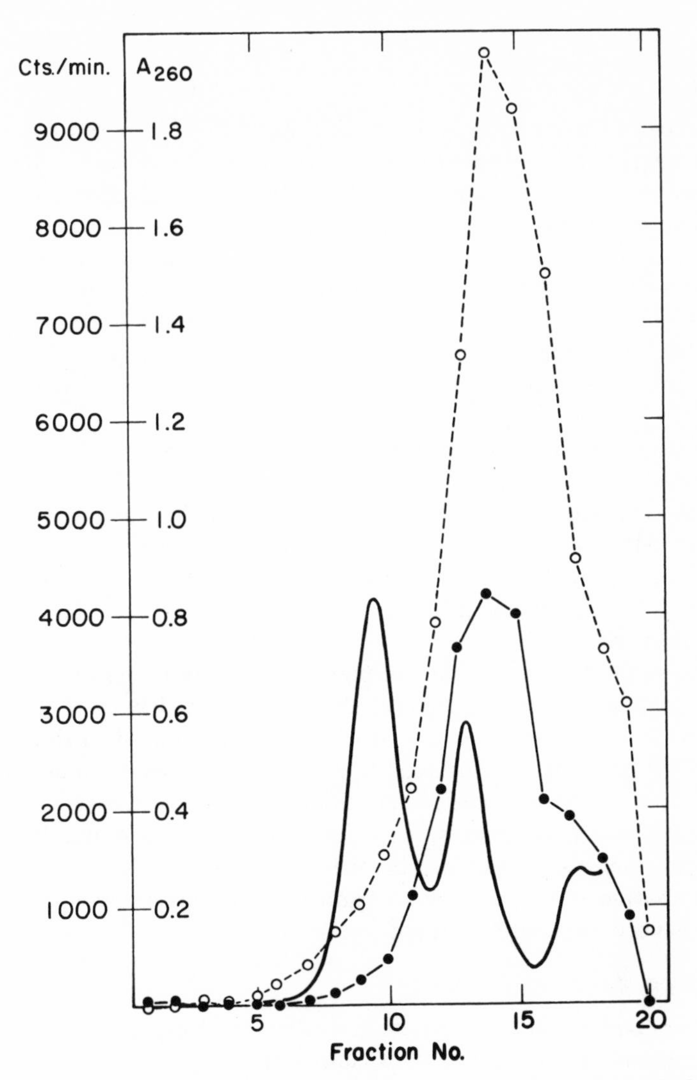

FIG. 8. RNA from red halves with and without actinomycin treatment. Red halves were activated, and the culture was divided in half. One portion was incubated in MFSW+ps, the other in MFSW+ps with 50 μg of actinomycin D per milliliter. After 90 minutes, both mixtures were made 100 μCi/ml in uridine-5-^{3}H and incubated for 2 hours. Extracted material was centrifuged (right to left) for 19 hours at 24,000 rpm. O---O, control; ●——●, actinomycin treated; ——, A_{260}.

RESULTS

Half-Egg Preparation

The particulate contents of *Arbacia* eggs stratify in the following order, centripetal to centrifugal pole: a cap of fat droplets, the pronucleus, a hyaline layer (ground cytoplasm containing annulate lamellae and some additional membranous elements), mitochondria, yolk granules, and the pigment vacuoles. The centrifuged egg develops a sulcus, then a deep constriction, and finally divides into approximately equal-sized halves. The differential content of particles causes the net densities of these halves to differ significantly, hence their isopycnic levels in the gradient are well separated (more than 1 cm).

Mitochondria remain in the centripetal halves, many of them associated with the oil droplets in clusters, but the centrifugal halves certainly contain a significant fraction of them. Figure 1 is a photomontage electron micrograph of a part of a centrifugal or "red" half, at low magnification, fixed shortly after completion of the centrifugation. The mitochondrial layer included in this egg fragment is easily identified.

Centrifugal halves contain the pigment and are therefore red. The centripetal halves, which contain the nuclei, are devoid of pigment and are white. Red and white bands are easily located in the gradients. Egg quarters are formed in a nonsynchronous manner during the centrifugation, so that preparations may vary somewhat in their content of these fragments, but the "red half" band usually contains no nucleate fragments, be they halves or quarters. Likewise, contamination of a "white half" band by enucleate fragments of any origin is very small, although not zero.

Activation of the Red Halves

The success of parthenogenetic activation treatments was estimated from leucine incorporation data, although in a number of experiments these conclusions were checked with the aid of the microscope. This is possible because of the cortex of activated merogones undergoes a typical cortical reaction (as in fertilization), including breakdown of the cortical granules, formation of a hyaline layer, and appearance of a fertilization membrane. These processes are not as easy to identify as they are in the intact egg, and hence electron microscopy of sections is more appropriate than optical microscopy of whole mounts.

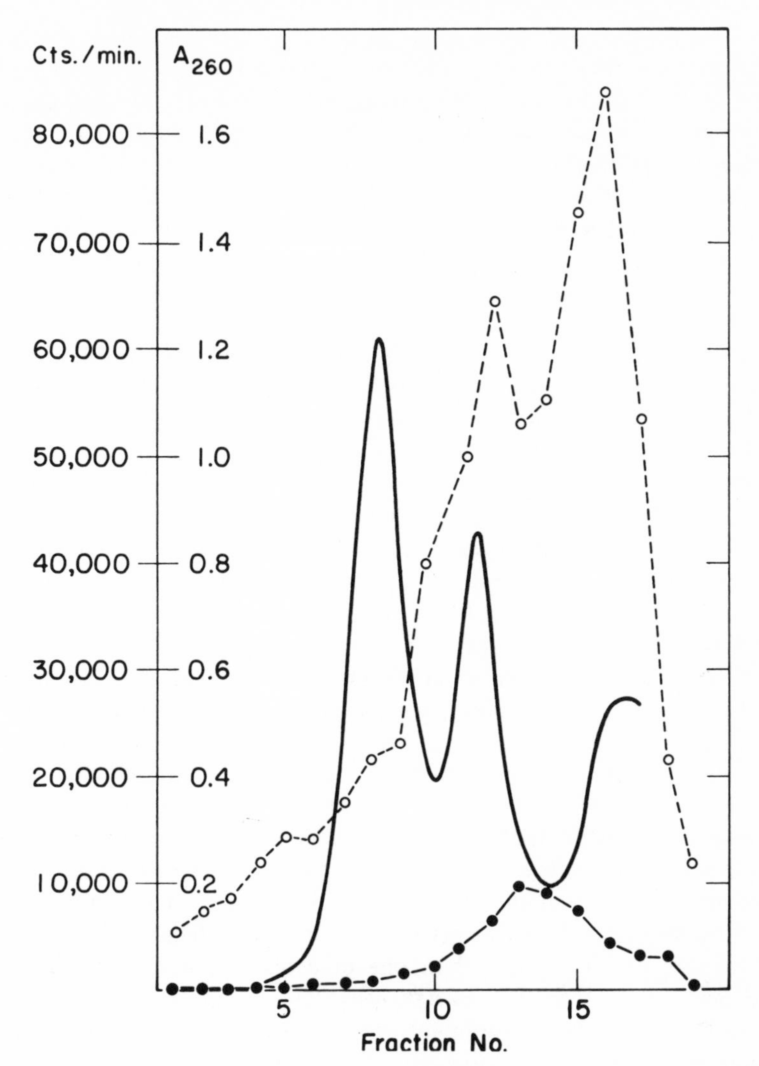

FIG. 9. RNA from whole embryos, with and without actinomycin treatment. Eggs were fertilized and after 3 hours the culture was divided and treated as described in Fig. 8, except that uridine-5-^{3}H concentration was 10 μg/ml. ○---○, control; ●——●, actinomycin treated; —— A_{260}.

Attention will be called to the ultrastructural data in legends for the accompanying micrographs. As to leucine incorporation into proteins: the treatment with hypertonic sea water increases it manyfold—up to 18 times in successful experiments—but it is clear from the biochemical and the morphological data that centrifugation and breakage are themselves responsible for activating a significant fraction of the red halves. Our experience may be summarized in the following way: between 5 and 20% of the red halves prepared as described are activated by the manipulations needed to obtain them. We have found no way to avoid this difficulty but, as we shall see, the presence of a cortical reaction, detectable with special clarity in the electron microscope (compare Fig. 1 with Figs. 3 and 4), is always correlated with significant incorporation.

Light microscope autoradiograms were made from thick sections of the material fixed for electron microscope autoradiography. With these, it was possible to survey relatively large numbers of sections in order to assess the uniformity of incorporation patterns. The results were consistent with those obtained from incorporation studies and with the membrane elevation seen *in vivo*: control unfertilized eggs showed very little radioactivity (at the photographic exposure times used, except for occasional very radioactive oocyte sections). Control half-eggs, red and white, were heterogeneous in respect to silver grain distribution. Most sections were no more radioactive than the control unfertilized eggs, but some, as many as 20%, showed significant accumulations of silver grains over them. The activated half-eggs, on the other hand, were quite uniformly radioactive, and resembled in their grain density sections of fertilized intact eggs.

These results offer no impediment to further study of the RNA synthesis in enucleate half-eggs, but they do indicate, as have other data (Gross, 1967), that quantitative statements about the rates of macromolecule synthesis in unfertilized or "nonactivated" eggs and fragments must be made with great caution. The results also indicate that the RNA synthesis of activated half-eggs is reliably an activity of the population, rather than one of contaminant oocytes, microorganisms, or interstitial ovarian cells. This is a simple, but at the same time important, statement: without the assurance of the autoradiographic result, the biochemical ones would be of uncertain significance.

Pattern of RNA Synthesis and Localization in the Red Half

RNA extracted from activated red halves gave the sucrose gradient pattern shown in Fig. 2. This pattern is typical of incorporation in both activated and nonactivated red halves: a broad peak with its modal point somewhat more slowly sedimenting than the 18 S ribosomal RNA peak, and a small peak or shoulder sedimenting at about 4 S.

Electron microscope autoradiograms from another experiment of the same kind are shown in Figs. 3–5. Grains appear distributed

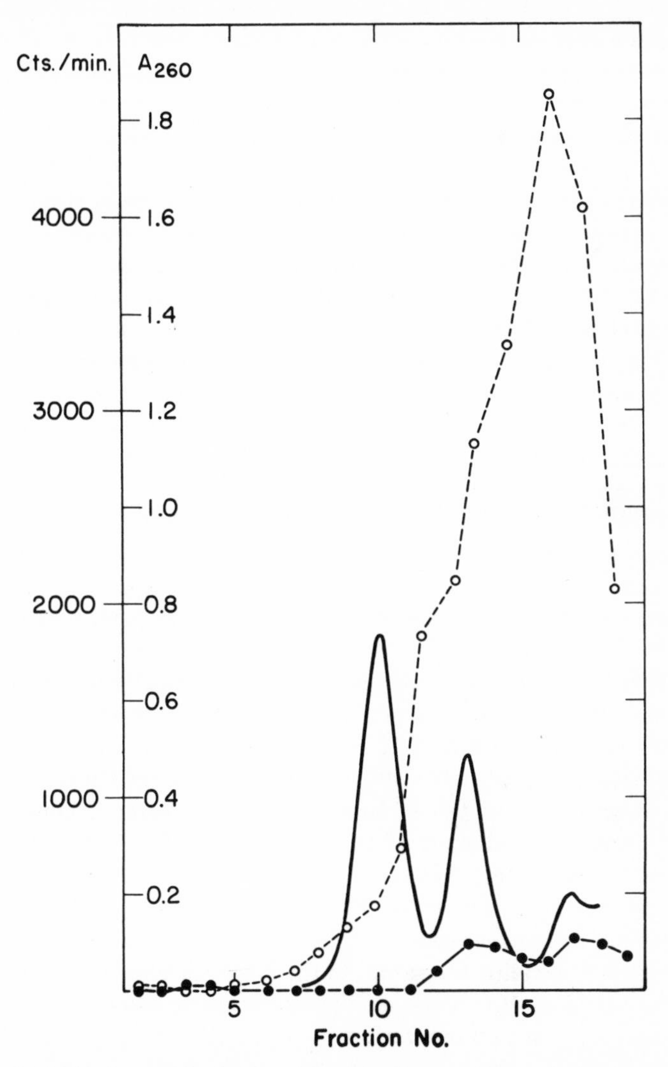

FIG. 10. RNA from white halves with and without actinomycin treatment. White halves were fertilized and treated as described for Fig. 8, except that uridine-5-^{3}H concentration was 50 μg/ml. O---O, control; ●——●, actinomycin treated; ——, A_{260}.

throughout the section. Since the labeling period was 2 hours, the radioactive molecules may not remain associated over such a long interval with the structures upon which they originate. Thus no conclusions ought to be drawn from these data alone as to the sites of transcription. In any event, the mitochondria are not especially radioactive after a 2-hour labeling period.

DNase Sensitivity of the Products

Activated red halves and fertilized white halves were incubated for 90 minutes without precursor, and then allowed to incorporate radioactive uridine for 2 hours. Incorporation was then stopped, and the RNA was extracted. The distribution of radioactivity in sucrose gradients was not altered by DNase treatment, as is shown in Figs. 6 and 7, whereas radioactivity and optical density of material sedimenting faster than 4 S were alkali-labile. These results show that under the conditions of these experiments, the radioactive product is RNA alone, and that contamination by DNA made radioactive through incorporation of deoxynucleotides bearing the label is insignificant.

Actinomycin Treatment of Red and White Halves and of Embryos

Figures 8–11 show the effect of actinomycin D on red halves, white halves, and embryos. Eggs were fertilized at the beginning of the experiment, and 3 hours later the red halves were activated and the white halves were fertilized. At this time the embryos were at 16-cell stage. The cultures were divided, and half of each was treated with 50 μg/ml actinomycin D. The other half was kept as control. After 90 minutes, uridine-5-H^3 was added to each incubation mixture so that the red half culture contained 100 μCi/ml, the white halves 50 μCi/ml and the embryos 10 μCi/ml. After 2 hours of exposure to the labeled nucleoside, incorporation was terminated and the RNA was extracted.

As is shown in Fig. 8, about 30% of red half incorporation remains with actinomycin treatment of the dose used, and the pattern of sedimentation remains the same. In the embryos (shown in Fig. 9) and white halves (Fig. 10), however, incorporation is much more severely depressed, and the sedimentation pattern appears altered. Incorporation is reduced by the actinomycin treatment to 8% in embryos, and to 7% in the white halves. In Fig. 11, the "normal" red half sedimentation pattern is plotted with that of the residual product from actinomycin-treated embryos. The radioactivity profiles are very

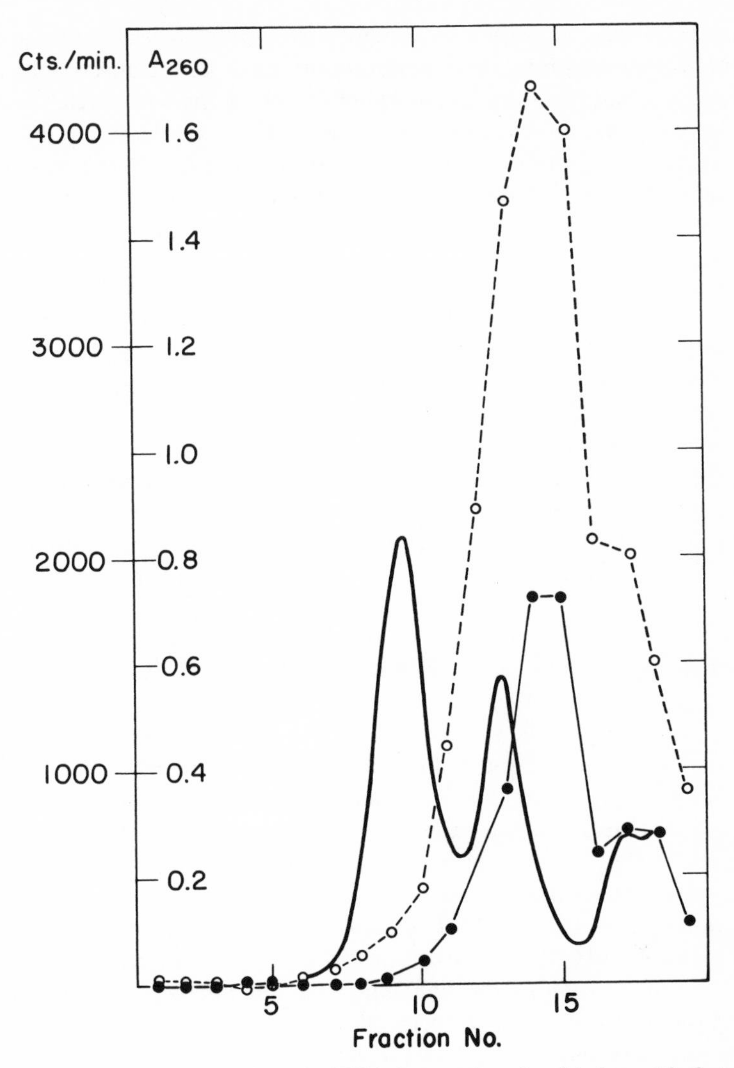

FIG. 11. Comparison of radioactive RNA from activated red halves with that from actinomycin-treated whole embryos. Red half RNA as in Fig. 8, and actinomycin-treated embryo RNA as in Fig. 9. O---O, embryos, ●——●, red halves; ——, A_{260}, which is appropriate for both radioactivity profiles. Shapes of the radioactivity profiles can be compared, but absolute radioactivities should not be taken as proportional to rates of synthesis (see text).

similar, suggesting that the small "actinomycin-resistant" component of embryos may be the same product as that transcribed in the enucleate half. The term "actinomycin-resistant" is a relative one, since we have no assurance that larger doses would not reduce the incorporation further. It is worth noting that, with the precursor, schedules, and techniques used in these experiments, end-labeled transfer RNA does not contribute greatly to the radioactivity of the purified RNA.

DISCUSSION

An RNA precursor is incorporated into molecules of moderate and high molecular weight by enucleate half-eggs of the sea urchin. This incorporation is observed in preparations that have been activated by treatment with hypertonic sea water, but some incorporation is also detected in populations not so treated. It is evident, from morphological examination of the egg fragments before and after the treatment, that some activation results from the preparation technique itself. Leucine incorporation experiments confirm this result. Autoradiography of populations indicated that the treatment with hypertonic sea water activates most egg fragments, whereas untreated populations have only 5–20% activated. In all cases, the RNA product is heterodisperse and noncoincident in sedimentation with any of the bulk species except the 4 S material, with which a portion (ca. 10%) of the radioactive product sediments. The labeling of RNA in enucleate halves cannot, therefore, be attributed solely to end-labeling of transfer RNA.

The labeled product is not DNA, since it is neither attacked nor altered in its sedimentation behavior by an enzyme preparation that degrades authentic DNA. The product has other properties, including alkali lability, that identify it as RNA.

Autoradiography at the light and electron microscope levels shows that the incorporation is intracellular. The autoradiographic backgrounds were negligible, and virtually all silver grains in the specimens represented by Figs. 1, 3, 4, and 5 were over the interior parts of cells. Light autoradiograms showed that in activated suspensions of red halves most of the fragments were radioactive. Hence the incorporation is not attributable to the presence in such preparations of contaminating oocytes, interstitial cells, or microorganisms.

There is no indication from these experiments of a particulate or membranous site for the cytoplasmic transcription. The red halves contain abundant mitochondria, but the silver grains show no tendency to locate specifically on or near mitochondria. There are mitochondrial clusters of sufficient size (e.g., Fig. 5) and in sufficient numbers in the specimens studied to have allowed such specificity in labeling to be detected. Rarely, silver grains *are* found on mitochondria. The resolution of the method is not good enough to permit the conclusion that such associations reflect the presence of labeled RNA in, rather than near, them.

It is well to be cautious about interpreting this result. It does *not* mean that the sites of synthesis are components of the cytoplasmic matrix. Quite possibly, transcription and release are very rapid, with the ultimate sites of function (and destruction) of the cytoplasmic RNA different from its site of synthesis. If this were so, it would be necessary to use pulse exposures to labeled uridine of very short duration, in order to detect the products at sites of transcription. The 2-hour exposures used here were as brief as was consistent with the incorporation of sufficient radioactivity to make possible autoradiographic detection. The problem of the site of cytoplasmic RNA synthesis must now be attacked with other methods than were available for these studies.

RNA synthesized in enucleate halves appears to differ in a number of respects from that produced in fertilized nucleate halves or intact embryos. This statement is necessarily less than absolute, because it is not easy to compare the three types of system in a biologically meaningful way. The 2-hour pulse label of a fertilized nucleate half may not be comparable to the 2-hour label of an activated enucleate half, and it is clear that a 16-cell normal embryo is not at all comparable with either, at least so far as the timing and synchrony of cleavages is concerned.

With this qualification, we may call attention to the fact that the enucleate half makes radioactive RNA that appears somewhat less heterodisperse than that of the nucleate half or the cleaving embryo. RNA of the enucleate half has a modal sedimentation of about 17 S, with a distinct shoulder at 4 S, and very little material sedimenting faster than 28 S. The nucleate systems both make RNA that include significant radioactivity with sedimentation greater than 28 S, and they label relatively much more 4 S. With the dosage and treatment schedules for actinomycin employed in these experiments, the enucleate half RNA synthesis is significantly less affected than is synthesis in either nucleate system. It is possible that this difference is not a real difference in resistance to actinomycin: e.g., the red half might, in consequence of its preparation, be much less permeable to the drug than is the zygote or the nucleate half fertilized. Yet the sedimentation behavior of that small fraction of RNA synthesis that escapes inhibition by actinomycin (at this dose) in the embryos is almost identical to that of the total enucleate half product. This suggests, although it certainly does not prove, that the synthesis of RNA in enucleate halves represents at least a component of the total normal transcription in early cleavage.

An interesting question that follows from the above concerns the fraction of normal RNA synthesis, e.g., in a zygote, that takes place in the cytoplasm. Experiments of the kind reported here cannot, in our opinion, answer such a question. They might be answered by suitable autoradiographic experiments on normal zygotes and embryos, but such experiments will be difficult. They will require either much higher specific activities in precursors or much longer photographic exposures than have been practical.

As to the enucleate halves: we do not know the composition or sizes of their RNA precursor pools, and it is not entirely justified to assume that these quantities are the same per unit of DNA and of polymerase as they are in the normal egg. We do not know whether the merogones have a greater, a lesser, or the same permeability to labeled uridine and to actinomycin as do nucleate halves or zygotes. We do not know, finally, whether or not the mechanical stress of breaking the egg (with the possibility of a subactivating influx of sea water ions) affects the rate or pattern of RNA synthesis. Thus it is not justified to speculate at this time about the absolute rates of synthesis in nucleus and cytoplasm of normal early embryos. Hybridization results (Hartman and Comb, 1969) cannot provide quantitative estimates, due to uncertainties in equating gene numbers as between cytoplasmic and nuclear DNA preparations, and for other technical reasons. To the extent that those results have any quantitative implications, they suggest that new gastrula RNA has a much larger contribution from cytoplasmic genes than does new blastula RNA, yet it is clear from autoradiography and biochemistry that the bulk of the posthatching incorporation is nuclear. Furthermore, there is a delay of about 15 minutes before significant radioactivity enters the cytoplasm at any stage beyond the first few cleavages (Kijima and Wilt, 1969; Aronson and Wilt, 1969; Kedes and Gross, 1969).

The experiments with half-eggs do demonstrate that the cytoplasm of an egg contains transcribable templates plus all of the enzymatic machinery and raw materials to accomplish the transcription. Such transcription might be a significant fraction of the total synthesis carried out during the earliest period of cleavage.

SUMMARY

Autoradiographic and biochemical studies on enucleate halves of sea urchin eggs (*Arbacia punctulata*) show that RNA synthesis can take place in the cytoplasm. The products transcribed in activated enucleate fragments are heterodisperse in sedimentation and of high molecular weight. After a 2-hour period of labeling, the new RNA molecules are distributed throughout the cytoplasm, as determined by autoradiographic detection of their radioactivity. No class of organelles or particles is strongly labeled. The transcription of cytoplasmic templates may be somewhat more resistant to actinomycin treatment than is RNA synthesis of the embryo or fertilized nucleate half-egg. No conclusions are possible in respect to relative rates of cytoplasmic and nuclear transcription during early development, but these results support the possibility of an important cytoplasmic contribution during early cleavage.

Note added in proof: A recent report on RNA synthesis in enucleate sea urchin egg fragments (Craig, S. P., *J. Mol. Biol.* **47,** 615–618, 1970) includes data from molecular hybridization experiments. The results are consistent with transcription of the RNA on mitochondrial DNA templates. Should this interpretation prove to be correct, our own results would constitute evidence that in this system the mitochondrial RNA leaves the sites of its synthesis quite rapidly. This is in contrast to what now appears to be the behavior of mitochondrial RNA in cultured mammalian cells (Zylber, E., and S. Penman, *J. Mol. Biol.* **46,** 201–204, 1969). The function of RNA synthesized by enucleate egg fragments remains obscure, since we have thus far been unable to obtain evidence of its presence in polyribosomes synthesizing proteins.

Further autoradiographic studies (Selvig, Hunter, and Gross, unpublished) concern the localization of pulse-labeled (20 mins) RNA in very early stages of normal sea urchin development. The results are relevant to issues raised in the discussion of this report, for they show that while cytoplasmic transcription accounts for a significant fraction of new RNA in zygotes and during the first two cleavages, it is the nuclei that are the major sites of synthesis at the eight cell stage and thereafter.

REFERENCES

ARONSON, A. I., and WILT, F. H. (1969). Properties of nuclear RNA in sea urchin embryos. *Proc. Nat. Acad. Sci. U.S.* **62,** 186–193.

BALTHUS, E., HANOCQ-QUERTIER, J., and BRACHET, J. (1968). Isolation of deoxyribonucleic acid from the yolk platelets of *Xenopus laevis* oocytes. *Proc. Nat. Acad. Sci. U.S.* **61,** 469–476.

BROWN, D. D., and LITTNA, E. (1964). RNA synthesis during the development of *Xenopus laevis*, the South African clawed toad. *J. Mol. Biol.* **8,** 669–687.

CHAMBERLAIN, J. (1968). Extranuclear RNA synthesis in sea urchin embryos. *J. Cell Biol.* **39,** 23a.

DAWID, I. B. (1965). Deoxyribonucleic acid in amphibian eggs. *J. Mol. Biol.* **12,** 581–599.

DAWID, I. B. (1966). Evidence for the mitochondrial origin of frog egg cytoplasmic DNA. *Proc. Nat. Acad. Sci. U.S.* **56,** 269–276.

GIUDICE, G., MUTOLO, V., and DONATUTI, G. (1968). Gene expression in sea urchin development. *Wilhelm Roux, Arch. Entwicklungsmech. Organ.* **161,** 118–128.

GROSS, P. R. (1964). The immediacy of genomic control during early development. *J. Exp. Zool.* **157,** 21–38.

GROSS, P. R. (1967). The control of protein synthesis in embryonic development and differentiation. *Curr. Top. Develop. Biol.* **2,** 1–46.

GROSS, P. R. (1968). Biochemistry of differentiation. *Ann. Rev. Biochem.* **37,** 631–660.

GROSS, P. R., KRAEMER, K., and MALKIN, L. I. (1965). Base composition of RNA synthesized during cleavage of the sea urchin embryo. *Biochem. Biophys. Res. Commun.* **18,** 569–575.

GROSS, P. R., MALKIN, L. I., and HUBBARD, M. (1965). Synthesis of RNA during oogenesis in the sea urchin. *J. Mol. Biol.* **13,** 463–481.

HARTMAN, J. F., and COMB, D. (1969). Transcription of nuclear and cytoplasmic genes during early development of sea urchin embryos. *J. Mol. Biol.* **41,** 155–158.

HARVEY, E. B. (1956). "The American Arbacia and Other Sea Urchins." Princeton Univ. Press, Princeton, New Jersey.

KEDES, L. H., and GROSS, P. R. (1969). Synthesis and function of messenger RNA during early embryonic development. *J. Mol. Biol.* **42,** 559–575.

KIJIMA, S., and WILT, F. H. (1969). Rate of nuclear ribonucleic acid turnover in sea urchin embryos. *J. Mol. Biol.* **40,** 235–246.

MALKIN, L. I., GROSS, P. R., and ROMANOFF, P. (1964). Polysomal protein synthesis in fertilized sea urchin eggs: The effect of actinomycin treatment. *Develop. Biol.* **10,** 378–394.

MALKIN, L. I., MANGAN, J., and GROSS, P. R. (1965). A crystalline protein of high molecular weight from cytoplasmic granules in sea urchin eggs and embryos. *Develop. Biol.* **12,** 520–542.

PIKÓ, L., BLAIR, D. G., TYLER, A., and VINOGRAD, J. (1968). Cytoplasmic DNA in the unfertilized sea urchin egg: Physical properties of circular mitochondrial DNA and the occurrence of catenated forms. *Proc. Nat. Acad. Sci. U.S.* **59,** 838–845.

RINALDI, A. M., and MONROY, A. (1969). Polyribosome formation and RNA synthesis in the early post-fertilization stages of the sea urchin egg. *Develop. Biol.* **19,** 73–86.

SPIRIN, A. S. (1966). On "masked" forms of messenger RNA in early embryogenesis and in other differentiating systems. *Curr. Top. Develop. Biol.* **1,** 1–38.

DNA Synthesis in Early Sea Urchin Embryos

INTRACELLULAR MIGRATION OF DNA POLYMERASE IN EARLY DEVELOPING SEA URCHIN EMBRYOS

L. A. LOEB AND B. FANSLER

INTRODUCTION

During the first 16 h of exponential cell division and rapid DNA synthesis after fertilization in *S. purpuratus* embryos, there is no significant change in the whole embryo DNA polymerase activity[1]. However, we have shown a marked change in the intracellular localization of polymerase activity. After each division cycle, progressively more activity is recovered in isolated nuclei while the corresponding cytoplasmic fractions show a concomitant loss of activity[1,2].

We now consider the change in localization of DNA polymerase in relation to the extent of synthesis and turnover of the enzyme. Evidence indicates that during early development of sea urchin embryos only a fraction of the total protein undergoes turnover[3]. We find polymerase is not an exception. It must have been synthesized during oogenesis and stored in the cytoplasm of the egg. During early development the enzyme migrates to the newly forming nuclei, thus maintaining a special relationship between nuclear DNA and polymerase.

MATERIALS AND METHODS

With the exception of those described below, all other procedures used in this investigation are identical with those previously reported[1,2].

Sea urchin nuclear DNA polymerase was purified by the method of LOEB[4] and the steps indicated here are consistent with that procedure. Chromatography on hydroxylapatite was omitted since it did not increase the purity of the enzyme. All steps are carried out in the presence of 20 % (w/v) glycerol or 1 M dextrose: the enzyme requires these or other polyglycols for stability[4]. DNA polymerase was assayed throughout the purification by using "activated" calf thymus DNA as a primer[4]. One unit of activity is defined as the amount of activity required to convert 1 mμ-mole of labeled deoxynucleoside triphosphate into acid-insoluble material in 10 min at 37°. Based on initial rate of incorporation, a unit of sea urchin nuclear polymerase is equivalent to about 1.2 units of *Escherichia coli* DNA polymerase[5] or 6.0 units of calf thymus DNA polymerase[6]. Deoxyribonuclease activity was assayed with partially degraded sea urchin ^{3}H-labeled DNA as a substrate[4]. The amount of labeled amino acid incorporated into protein was determined by precipitating the protein with 1 M $HClO_4$, reprecipitating from 0.2 M NaOH and dissolving in 1 ml of Hyamine chloride (Packard Instrument Co.). Thereafter, a toluene-phosphor solution was added and radioactivity determined[4]. A standard solution of [^{3}H]toluene was added to each vial and results are given in disint./min.

RESULTS

In order to measure the extent to which DNA polymerase was synthesized during early development the enzyme was purified from hatched *S. franciscanus* embryos which had been grown in the presence of L-[^{3}H]leucine (Table I, Expt. A). If polymerase was preferentially synthesized, relative leucine incorporation (^{3}H disint./min per mg protein) would increase as each fraction was selectively enriched for the enzyme. Even though the extent of polymerase purification was over 300-fold, the amount of [^{3}H]leucine which had been incorporated into the most purified fraction (17 900 disint./min per mg protein) was not significantly different from that of the unfractionated embryonic proteins (19 500 disint./min per mg protein). Furthermore, there was little variation in extent of leucine labeling among the various fractions. These results indicate that during early development DNA polymerase does not undergo selective turnover. Of note is the increased leucine incorporation into certain proteins of the nuclear fraction; these do not copurify with the polymerase but are excluded during phase separation. Studies of NEMER AND LINDSAY[7] indicate that these proteins may be histones which are preferentially synthesized during the 32–128-cell period.

The preceding experiment provides evidence of nonpreferential turnover (synthesis and degradation) of polymerase during the first 16 h of development. We now ask, what is the rate of synthesis of the enzyme compared to the bulk protein at the time immediately prior to hatching? *S. purpuratus* embryos at this stage were exposed to ^{3}H-labeled amino acids for 30 min and then harvested. DNA polymerase was purified from these embryos and the incorporation of ^{3}H into the different protein fractions was determined (Table I, Expt. B). As a source of precursors for protein synthesis, an algal hydrolysate of labeled amino acids offered two distinct advantages: since many amino acids were labeled, there was less selection for proteins which have an exceptionally high composition of a single amino acid. Second, a greater amount of radioactivity was incorporated for a given amount of protein synthesized. Amino acid incorporation in the most purified fraction is only a little greater than

TABLE I

PURIFICATION OF ^{3}H-LABELED SEA URCHIN NUCLEAR DNA POLYMERASE

In Expt. A, 0.5 mC of L-[^{3}H]leucine was added to a culture 15 min after fertilization which contained 800 ml of embryos and the culture was harvested at hatching (16 h). In Expt. B, 0.5 mC of an algal hydrolysate of [^{3}H]amino acids (Schwarz BioResearch) was added to a culture containing 200 ml of embryos 30 min prior to harvesting at the hatching stage.

Fraction and step	*Expt. A (S. franciscanus)*		*Expt. B (S. purpuratus)*	
	Polymerase activity (units/mg)	L-[^{3}*H*]*leucine incorporated (disint./min per mg* $\times 10^{-2}$)	*Polymerase activity (units/mg)*	[^{3}*H*]*Amino acids incorporated (disint./min per mg* $\times 10^{-2}$)
I Whole embryo	0.88	195	1.20	61*
II Nuclei	2.84	342	7.45	76
III Phase separation	7.60	116	10.75	67
IV $(NH_4)_2SO_4$	12.2	139	18.4	119
Acid precipitation	12.7	156	21.8	49
V Phosphocellulose	73.2	320	85.0	53
VI DEAE-cellulose	187.0	187	288.0	111
VIII Sephadex	270.0	179	425.0	110

* Incorporation of [^{3}H]amino acids into whole embryos was determined in two separate experiments. 3 ml of hatched embryos were exposed to 0.1 mC of the same algal hydrolysate in 100 ml of sea water for 30 min. Embryo and nuclear fractions were obtained. The average incorporation into whole embryos and nuclei was $69 \cdot 10^3$ and $86 \cdot 10^3$ disint./min per mg protein, respectively. The incorporation in the whole embryo fraction was taken to be 80 % of the nuclear incorporation that was obtained in this purification procedure.

that of the nuclei, indicating that during the purification there was no selection for a rapidly synthesized protein.

The final step in the purification of the polymerase labeled with [^{3}H]amino acids (Table I, Expt. B) is shown in Fig. 1. A complete separation of deoxyribonuclease (endonuclease) activity and DNA polymerase activity was achieved. The spe-

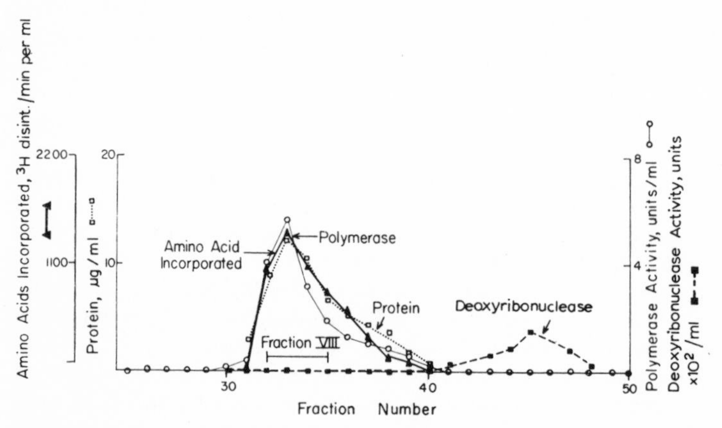

Fig. 1. Gel filtration of sea urchin nuclear DNA polymerase in Sephadex G-100. 200 units of Fraction VI were applied and eluted from the column[4].

cific activity of the polymerase (units/mg protein) was nearly constant over the major part of the peak. Furthermore, the ratio of amino acid incorporation (^3H disint./min) to μg protein did not significantly vary among the fractions of the peak.

DISCUSSION

Early development of sea urchin embryos is characterized by exponential cell division. The cells divide in 2 h or less, and the DNA, amounting to $1.8 \cdot 10^{-12}$ g per diploid nucleus, can double in about 10–12 min at 15° (ref. 7). This high rate of DNA synthesis is accompanied by an exceptionally high level of DNA polymerase activity *in vitro*[8], the specific activity being greater than other reported eucaryotic tissues. We have shown previously that the total DNA polymerase activity per embryo remains nearly constant throughout the first 16 h of development[1]. Further studies demonstrated that the great majority of the polymerase activity was found in the cytoplasm of the egg[1]. As the embryo proceeded through successive cell divisions, progressively more polymerase activity was found in the nuclear fraction with a concomitant loss of activity in the cytoplasm. By hatching, 200–400-cell stage, up to 95 % of the polymerase activity was recovered in nuclei[1,2]. The ratio of DNA polymerase activity to the DNA content in nuclei isolated at intermediate stages before hatching was nearly constant, *i.e.* the ratio of activity to DNA was 35, 39, 31 and 25 units per mg DNA in 2-, 8-, 32- and 128-cell embryos. This translocation of polymerase activity could have resulted from either a breakdown in the cytoplasm and preferential synthesis of polymerase in nuclei, extremely rapid turnover in cytoplasm with some transfer to nuclei, or a physical migration of a stored polymerase from the cytoplasm to the nucleus.

To distinguish between these possibilities we have looked for evidence of selective synthesis of DNA polymerase. If there is a rapid breakdown and resynthesis of the enzyme, it is reasonable to assume that during resynthesis the enzyme would incorporate radioactive amino acids added to the culture. The embryos are readily permeable to exogenous amino acids and we find no evidence for special pools of precursors destined for the synthesis of particular proteins. If polymerase is selectively synthesized in embryos grown in the presence of labeled amino acids, the enzyme would contain proportionally more radioactivity than the bulk of the cell proteins. Upon purification the ratio of disint./min per mg protein would increase with greater enrichment of the polymerase. The alternative to selective synthesis in the nucleus or cytoplasm during early development involves the migration of a preformed cytoplasmic enzyme. In this case, one would expect the enzyme to undergo little protein turnover at a rate no greater than the bulk embryo protein. During purification of the polymerase the ratio of radioactivity to protein would not increase. This latter result was observed. The ratio of ^3H disint./min per mg protein in the most purified polymerase fraction was not markedly different from that in the bulk of the nuclear protein.

Evidence indicates that there is little protein turnover in early developing *S. purpuratus* embryos[3]. The amount of protein (40 μg) in the unfertilized egg is no different than that of the 400-cell embryo[8]. Upon fertilization there is an immediate increase in amino acid incorporation[9]; thereafter the rate of protein turnover remains constant[3]. During the first 16 h of development there is only 20 % turnover of the total protein from the unfertilized egg (calculated from leucine and valine incorporation data in ref. 3). Since polymerase is not selectively synthesized, no more than 20 % would undergo turnover. This is insufficient, in terms of breakdown and resynthesis, to account for the 90 % change in localization of polymerase activity from the cytoplasm to the nucleus. Therefore, the observed change in the localization of DNA polymerase activity reflects a physical migration of the enzyme. Polymerase must have been synthesized during oogenesis and stored in the cytoplasm of the egg. With successive replications after fertilization, the polymerase is quantitatively transferred to newly formed nuclei.

Some attention must be given, at this point, to the composition of the most pure fraction (Fraction VIII) from the purification. As shown in Fig. 1, this final step, Sephadex gel filtration, yields a single symmetrical retarded peak of protein coincident with polymerase activity. The ^{3}H disint./min indicative of amino acid incorporation give a remarkably unvarying ratio (^{3}H disint./min per μg protein) in the fractions comprising the peak. In addition to evidence reported in this paper, chromatographic, chemical and electrophoretic studies indicate that the most purified fraction of the enzyme is homogeneous[4]. Rechromatography of the most purified fraction by a variety of methods yielded a single symmetrical retarded peak. Chemical analysis indicated that if DNA was present, it is less than 0.5 %. Fraction VIII exhibited one distinct migrating protein band after electrophoresis in polyacrylamide gel. We conclude that the major part, if not all, of the protein in the most purified fraction is DNA polymerase.

The only other purified proteins which uniquely participate in cell division are those of the mitotic apparatus. Studies on the synthesis of these proteins during the first cell division of sea urchin embryos indicate that they are not synthesized at a rate exceeding that of the bulk cell proteins[10]. If these studies can be generalized, a molecular catalog of an egg would include most of the enzymatic and structural components necessary for rapid replication for a number of cell generations.

It is possible that the intracellular migration of DNA polymerase is limited only to very early development. Alternatively, this migration may occur prior to DNA synthesis in all eucaryotic cells, but the requirements for rapid DNA synthesis during early development in these embryos has given us the opportunity to detect it. The results with partially synchronized L-cells reported by LITTLEFIELD *et al.*[11] and GOLD AND HELLEINER[12] support the latter concept; during DNA synthesis a decrease in the polymerase activity of the supernatant fraction and an increase in the particulate fraction were noted.

ACKNOWLEDGMENTS

This investigation was supported by grants from the American Cancer Society (E-483) and from the Stanley C. Dordick Foundation. Support was also derived from grants to this institute: U.S. Public Health Service grants CA-06927 and FR-05539, American Cancer Society grant IN-49, and an appropriation from the Commonwealth of Pennsylvania. We thank Drs. Daniel Mazia, Martin Nemer and Jack Schultz for generous counsel.

REFERENCES

1 B. Fansler and L. A. Loeb, *Exptl. Cell Res.*, 57 (1969) 305.
2 L. A. Loeb, B. Fansler, R. Williams and D. Mazia, *Exptl. Cell Res.*, 57 (1969) 298.
3 B. J. Fry and P. R. Gross, *Develop. Biol.*, 21 (1970) 125.
4 L. A. Loeb, *J. Biol. Chem.*, 244 (1969) 1672.
5 C. C. Richardson, C. L. Schildkraut, H. V. Aposhian and A. Kornberg, *J. Biol. Chem.*, 239 (1964) 222.
6 M. Yoneda and F. J. Bollum, *J. Biol. Chem.*, 240 (1965) 3385.
7 M. Nemer and D. T. Lindsay, *Biochem. Biophys. Res. Commun.*, 35 (1969) 156.
8 L. A. Loeb, D. Mazia and A. D. Ruby, *Proc. Natl. Acad. Sci. U.S.*, 57 (1967) 841.
9 D. Epel, *Proc. Natl. Acad. Sci. U.S.*, 57 (1967) 899.
10 F. H. Wilt, H. Sakai and D. Mazia, *J. Mol. Biol.*, 27 (1967) 1.
11 J. W. Littlefield, A. P. McGovern and K. B. Margeson, *Proc. Natl. Acad. Sci. U.S.*, 49 (1963) 102.
12 M. Gold and C. W. Helleiner, *Biochim. Biophys. Acta*, 80 (1964) 193.

DNA Synthesis and Development in Reciprocal Interordinal Hybrids of a Sea Urchin and a Sand Dollar[1]

JOHN W. BROOKBANK

INTRODUCTION

Echinoid hybrids have been extensively used in the past as sources of information about interactions between nucleus and cytoplasm during embryonic development, and as favorable objects for the study of specific interactions between egg and sperm at fertilization. The present report is intended as a contribution to the former category, as well as a contribution to the developmental history of hybrid embryos. In particular, the crosses to be described were made to provide material for studies of RNA, DNA, and proteins during early development. The hybrid, *Strongylocentrotus purpuratus* female × *Dendraster excentricus* male (SD), and the reciprocal cross (DS) have been previously reported by Moore (1957, SD and DS), and by Flickinger (1957, SD only). Flickinger described these embryos as strictly maternal in terms of appearance and rate of development, before and after gastrulation. Moore described the SD embryos as exhibiting biparental inheritance. The present author is in agreement with the description of SD offered by Moore. In the same paper, Moore briefly alludes to the DS cross, saying that the plutei are "not very satisfactory." The data presented here extend the observations of Moore on SD and DS, and include new information on DNA synthesis during development of SD and DS embryos, as well as parallel observations on the embryos of both parent species (SS and DD).

[1] This research was supported in part by a grant (GM 04659) from the National Institutes of Health.

MATERIALS AND METHODS

SD and DS embryos. Adult animals were obtained from live boxes maintained at the Friday Harbor Laboratories, and, on a few occasions, by air express to the Gainesville campus from the Pacific Bio-Marine Supply Co., Venice, California. Spawning was induced by injection of 0.5 *M* KCl. The eggs were collected in filtered seawater. Sperm were collected "dry" in embryological watch glasses. Hybrid fertilization was effected by mixing 5 ml of a 10–20% suspension of sperm with each 100 ml of 50% suspension of unfertilized eggs. The suspension was monitored for membrane elevation, and, after a maximum of the eggs showed membranes (5 minutes or less), exhaustive washing was begun to remove the excess sperm. Washing was continued until the supernatant was clear. Homospermic fertilizations were accomplished in the same manner, using 1–5 drops of 1% sperm suspension per 100 ml of 50% egg suspension. Hybrid fertilization percentages ranged from 10 to 60, as judged both by membrane elevation and ensuing cleavages. Cultures showing 20% fertilization or better were used. Control (homospermic) cultures were 90% (or better) fertilized. Unfertilized eggs were removed using a 67 μ mesh Nitex filter (SD) or by allowing them to settle to the bottom of the cultures after hatching. Cultures were maintained at 13°C, at 0.2% concentration, with constant gentle agitation. Swimming embryos were harvested daily, using a Lucite plankton centrifuge, and transferred to fresh seawater. Cultures of SD from which unfertilized eggs were removed prior to hatching did not differ from those containing unfertilized eggs until hatching. Trypsin treatment (Bohus Jensen, 1953; Tyler and Metz, 1955) was not employed. Trypsin treatment of the unfertilized eggs prevents elevation of the fertilization membrane (S eggs) and allows the eggs to clump tightly upon settling. This clumping is to be avoided if uniform cultures are to be obtained. Most descriptive and biochemical work to date has concerned cultures of SD (and SS, DD), produced at Friday Harbor and at Gainesville. All critical observations made at Friday Harbor have been repeated and confirmed on the cultures produced under conditions in Florida, using air-expressed animals. The DS data derives from two cultures produced in the Gainesville laboratories.

All possible precautions against homospermy were taken. The syringe used for KCl injection was rinsed with tap water before use on the next animal, and males and females of the same species were spatially separated during spawning to prevent contamination. All "unfertilized" eggs were examined before addition of sperm as a further check on uncontrolled fertilizations. Polyspermy (as evidenced by multipolar divisions) was not observed.

DNA analyses were done on lyophilized embryos using the diphenylamine reaction (Burton, 1956); total nitrogen of digests of unfertilized eggs and later stages was estimated iodometrically following steam distillation. DNA synthesis has also been assayed by measurement of incorporation of thymidine-^{14}C (sp. act. = 60 mC/mmole) by a 10% suspension of fertilized eggs in seawater (Millipore filtered) containing per milliliter 200 μg of streptomycin sulfate and 0.5 μC thymidine. Samples taken at 10-minute intervals after fertilization and addition of isotope were collected on glass fiber filters. Samples were then washed with cold 5% trichloroacetic acid (TCA) and absolute ethanol, air dried, and counted in a liquid scintillation system. The data are corrected for quenching and background to disintegrations per minute (dpm) per 10^5 eggs. Aliquots of the supernatant were also counted. The unfertilized eggs were not preloaded with thymidine-^{14}C, mainly because the "hybrid" eggs could not be so treated due to the exhaustive rinsing necessary to remove the excess sperm used in cross-fertilizations. Labeled thymidine was added no later than 10 minutes before the first S period, and earlier if possible. Equilibration of the eggs and label occurs within 10 minutes, as indicated in homospermic (SS and DD) situations where rinsing of the fertilized eggs is more rapidly accomplished. Egg counts were made on dilutions of small aliquots of the 10% suspensions. At least 200 eggs were enumerated in each sample. Four separate counts of a diluted aliquot were made, the results were averaged, and the final number of eggs per unit volume was calculated from dilution factors. Percentage fertilization and cleavage were also confirmed during egg counts. Samples for autoradiography were fixed in Carnoy (acetic alcohol) fluid at the same time samples were taken for incorporation estimates. These samples were rinsed in cold 70% ethanol, dehydrated, paraffin embedded, and sectioned at 10 μ. After coating with photographic emulsion (Kodak NTB-3), and exposure (48 or 96 hours), the slides were developed and stained with gallocyanin. Some slides were subjected to RNase digestion prior to staining.

All magnifications given in the figures refer to the initial magnification by the microscope. All photos were originally enlarged from 35-mm film by a factor of approximately 4.5.

RESULTS

Table 1 indicates the schedule of development for SS, DD, SD, and DS embryos at 13°C. It is evident that the rate of development of hybrid embryos, as evidenced by time of cleavages, hatching, and formation of primary mesenchyme, is strictly maternal. Delay in development is observed after the beginning of gastrulation, in the hybrid embryos (SD and DS). At this time, the SD embryos form a dense layer of cells (mesenchyme?) at the gastral plate, from which the archenteron emerges. The archenteron invaginates toward the region of the future stomodeum. Control embryos exhibit bilaterality during the final phases of invagination, by the bending of the archenteron toward the future oral side. The SD embryos also show this bending of the archenteron, but the invagination proceeds more slowly, and the tip of the archenteron may make contact with the wall of the blastocoel at a point closer to the vegetal pole, resulting in a shorter endodermal tube. The control SS embryos (Fig. 1a) expand in volume during and following gastrulation, and the epidermal wall of the blastocoel becomes quite thin, except at the points where the oral and anal arms of the pluteus are destined to grow out. The SD embryos also expand, and ultimately exhibit a thinning of the blastocoel wall. Control embryos (SS) develop a prism shape following gastrulation, and elaborate triradiate spicules as the prism develops the typical arms of the pluteus. The SD larvae assume various shapes, some reminiscent of SS, others, DD, and some simply spheroidal shapes. These hybrids elaborate tetraradiate spicules

TABLE 1
SUMMARY OF DEVELOPMENT[a]

Cross	First division (min)	S_1 (min)	S_2 (min)	Hatching	Gastrulation, start-end (hours)	Early pluteus (hours)
SS	135 (50%)	55–70	130–150	22–23	35–45	72
DD	90 (50%)	35–50	95–115	17–18	23–32	48
SD	135 (50%)	35–70	100–155	22–23	35[b]	96[c]
DS	90 (50%)	35–70	100–155	17–18	23[b]	60[c]

[a] Times are minutes or hours post-insemination, at 13°C.

[b] Hybrid development delayed from this point on, with variable morphology.

[c] Time when pigment and skeleton (of some larvae) are similar in amount and length, respectively, to maternal control embryos.

(Fig. 1d) reminiscent of the paternal (DD) form (Fig. 1b). Figures lc and d illustrate the development of SD through 96 hours. By 5 days (Figs. 1, e, f, and g) the SD embryos show the beginning of enterocoelic pouches (sometimes unilateral). These sacculations resemble the prominent pouches of the DD embryos, rather than the solid outgrowths characteristic of the later development of maternal SS embryos. The skeleton continued to be elaborated (Fig. 1g), and the gut may become well differentiated (Fig. 1f) and show peristalsis. The DS embryos (Fig. 2) are slow to develop bilaterality and a skeleton, though they eventually do so (Fig. 2e). The DS hybrids show two types of gastrulae (Figs. 2, b and c), one with a relatively thick archenteron (Fig. 2c). The skeleton which forms at 60–70 hours (Fig. 2f) is reminiscent of the complex and variable structures observed in SD larvae, and may begin as a Tri-or tetraradiate spicule. Similar (quantitatively) skeletal development is evident by 48 hours in DD.

Synthesis of DNA as a function of total nitrogen [nitrogen content is constant to the pluteus stage (Ballentine, 1940; and others)] is shown in Table 2, for SS, DD, and SD embryos. The amount of DNA present by the mesenchyme blastula stage of SD is approximately 1.3 times the amount present in maternal controls (SS). The amounts presented for DD controls cannot be directly compared with SS and SD on a per egg basis, since data giving nitrogen content per egg are lacking. The results of thymidine-^{14}C incorporation studies are shown in Fig. 3. The incorporation by SD and DS is represented by solid bars, representing two experiments (SD), and a single experiment (DS). The solid curves (SS and DD) are averages of two separate experiments, and for the sake of clarity they are presented without points. Samples were taken at 10-minute intervals in all cases. The standard counting error was limited to $\pm 5\%$ during S_1, and to $\pm 2\%$ for subsequent samples. The error in egg counts, as stated above, is approximately $\pm 7\%$, giving a total standard error of from ± 9 to $\pm 12\%$. The counts of supernatants are not plotted, but they consistently mirrored the incorporation data. The presence of saturating levels of labeled thymidine at the end of the experiment was also confirmed from the supernatant samples. The results confirm the published information of Hinegardner *et al.* (1964) on SS, and extend the observations of Simmel and Karnofsky (1961), made on *Echinarachnius parma*, to another sand dollar. The hybrid data are previously unreported.

The DNA:N values (Table 2) presented for SD hybrids, though

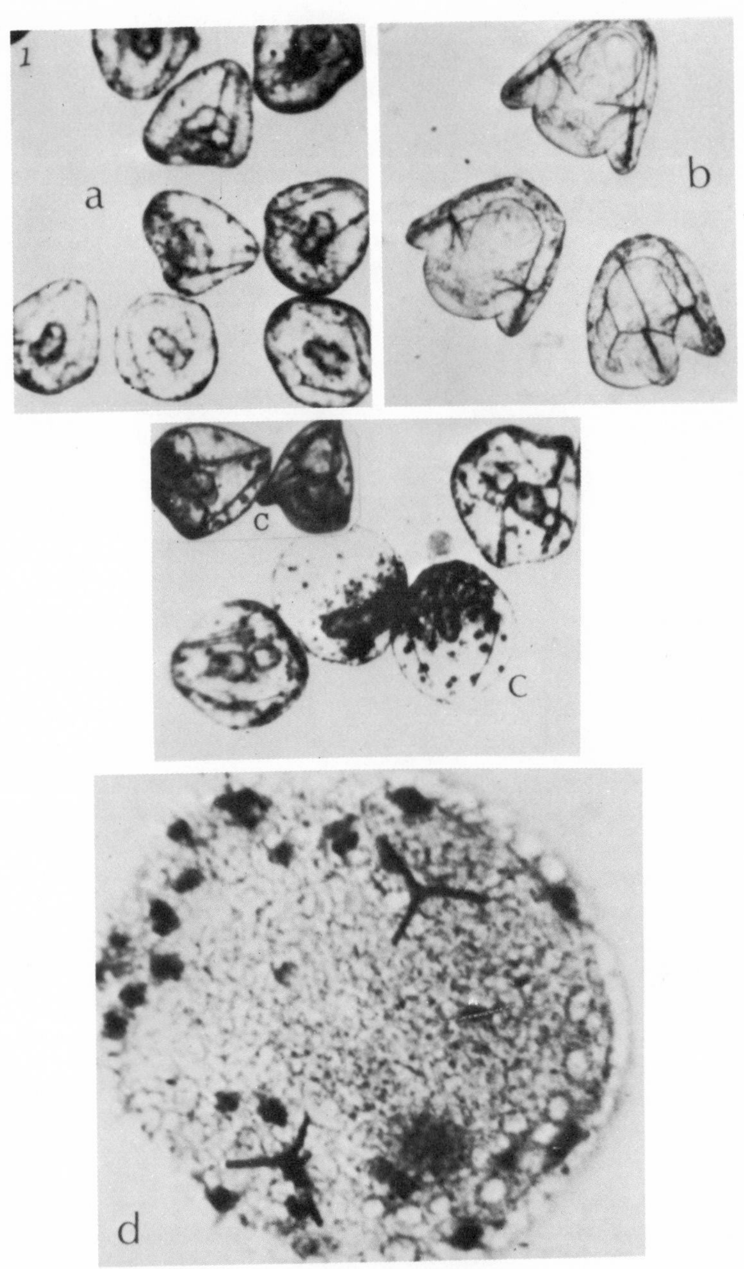

FIG. 1. a–d

FIG. 1. (a) Early plutei of SS, 72 hours. Single skeletal rods. Distinct oral and anal arms absent. Enterocoelic pouch absent. × 100. (b) Early plutei of DD, 50 hours. Anal arms with fenestrate rods present. Enterocoelic pouches present. × 100. (c) "Plutei" of SD, 96 hours. Variability in length of archenteron and skeletal elements

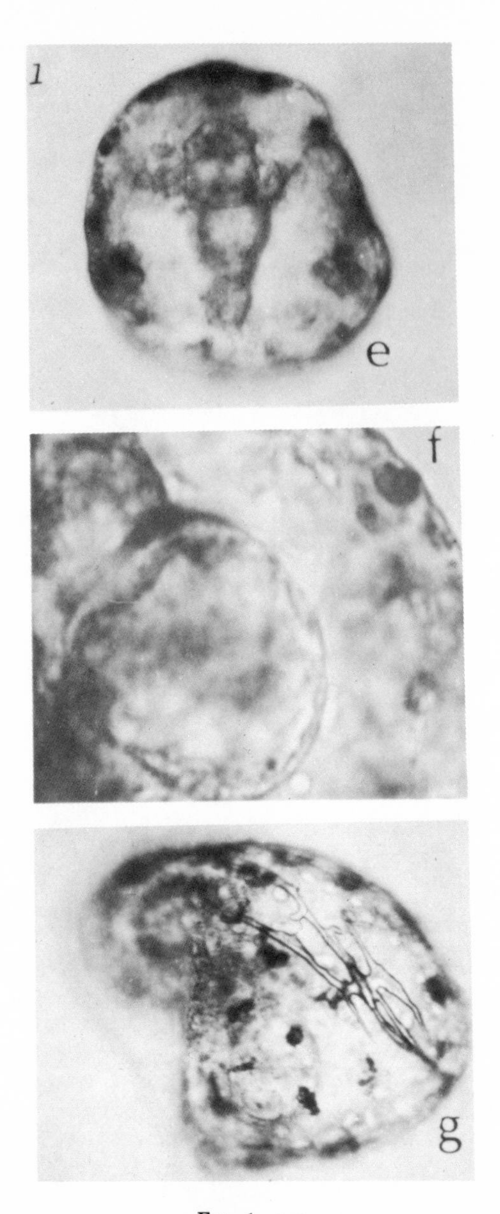

FIG. 1. e–g

evident. Some skeletal rods fenestrate. External form variable. × 110. (d) Early (70 hours) pluteus of SD, compressed under coverslip. Tetraradiate spicule evident (lower left). × 400. (e) SD pluteus, 5 days. Mouth at top. Archenteron with enterocoelic pouch (on left of embryo) evident. × 250. (f) SD pluteus, 7 days. "Stomach" portion of archenteron shown. × 500. (g) SD pluteus, 8 days. Extensive skeleton, irregular in shape. Echinochrome granules, anal arms, and oral field (out of focus) in evidence. × 250.

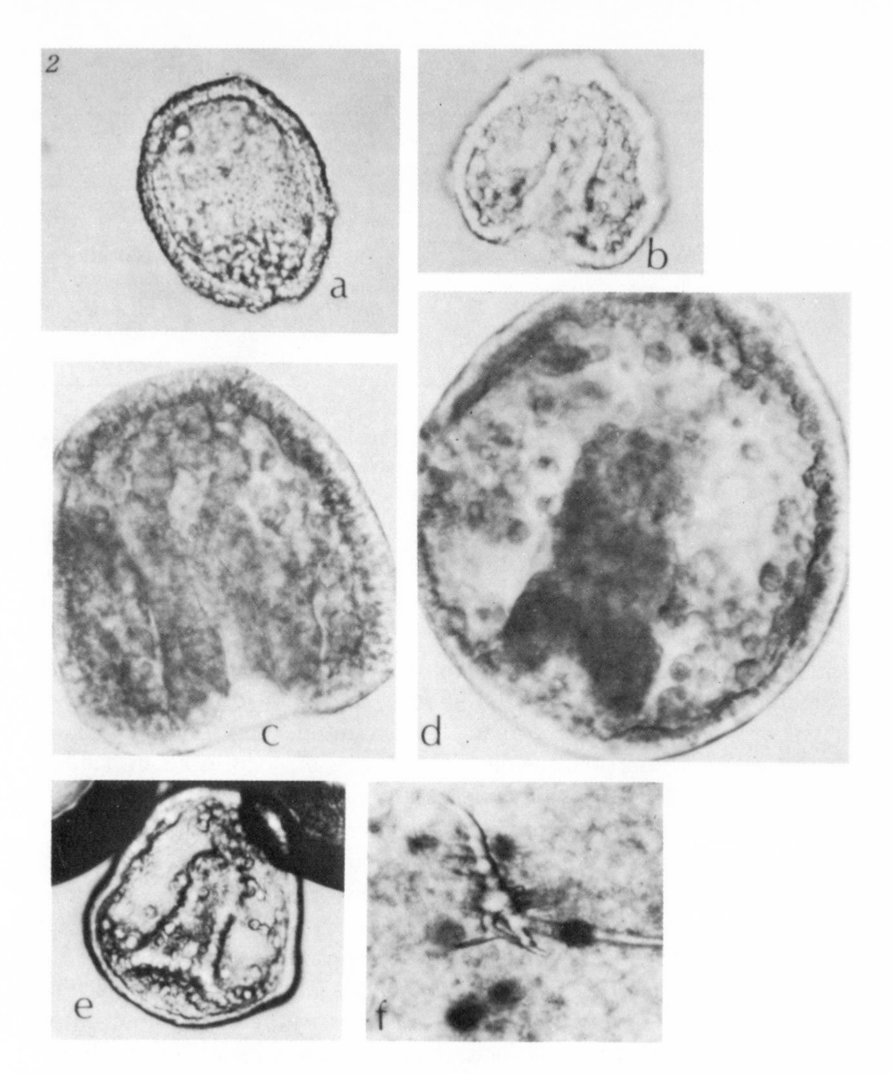

FIG. 2. (a) Blastula, with primary mesenchyme, of DS; 2 hours post-hatching (19 hours). × 164. (b) Gastrula of DS. 36 hours. × 164. (c) Gastrula of DS, 44 hours; stained with acetocarmine, and flattened. × 328. (d) Later gastrula, or early "pluteus" of DS, 49 hours; flattened, and stained with acetocarmine. × 328. (e) Early "pluteus" of DS, 50 hours. Anal arm protuberances evident. × 164. (f) Skeletal spicule from 65-hour embryo of DS. An example of the more complex forms observed. × 328.

TABLE 2
TOTAL DNA[a] DURING DEVELOPMENT OF SS, SD, AND DD EMBRYOS[b]

Stage	SS	SD	DD
Unfertilized	1	—	3
Mesenchyme blastula	45	60	51
	(25 H)	(27 H)	(20 H)
Gastrula	53	58	49
	(42 H)	(50 H)	(29 H)
Pluteus	80	57	81
	(72 H)	(71 H)	(51 H)

[a] As micrograms of deoxyadenosine per microgram of nitrogen $\times 10^{-3}$.

[b] H = hours, at 12°C, after fertilization. Numbers represent averages of at least two separate determinations, also made in duplicate. Maximum range between the lowest and highest averaged values is 7.

initially higher than SS controls, are not as high as one might expect on the basis of the S_1 and S_2 incorporation data of Fig. 3. Accordingly, incorporation measurements were made on 16-hour blastulae of SS, DD, and SD. These embryos were allowed to incorporate thymidine-^{14}C for 1 hour (0.5 μC/ml), and were then concentrated, washed with TCA and ethanol, and counted. Egg (embryo) counts were made on the suspensions. The results are expressed in Table 3 as dpm per 10^5 embryos. Supernatant samples, taken at the beginning and end of the experiment, confirm the data presented. It is apparent that the SD embryos, by 16 hours, show a marked decrease in incorporation as compared with SS controls.

Autoradiographs confirm the nuclear location of the incorporated thymidine, and the incorporation before pronuclear fusion in the DD series (Fig. 4). The preparations do not allow precise cytological (chromosomal) observations, nor are they suitable for quantitative comparisons. RNase treatment improves visualization of the nuclei and chromosomes without noticeable loss of radioactivity. Figure 5a shows (qualitatively) the precocial (with respect to SS) label in SD, and the added intensity of the label, compared to maternal controls, in the later phases of synthesis of both (Figs. 5a and b) SD and DS hybrids. This observed added intensity of label is assumed to result from the "precocial" replication of D chromosomes in the hybrids (see below).

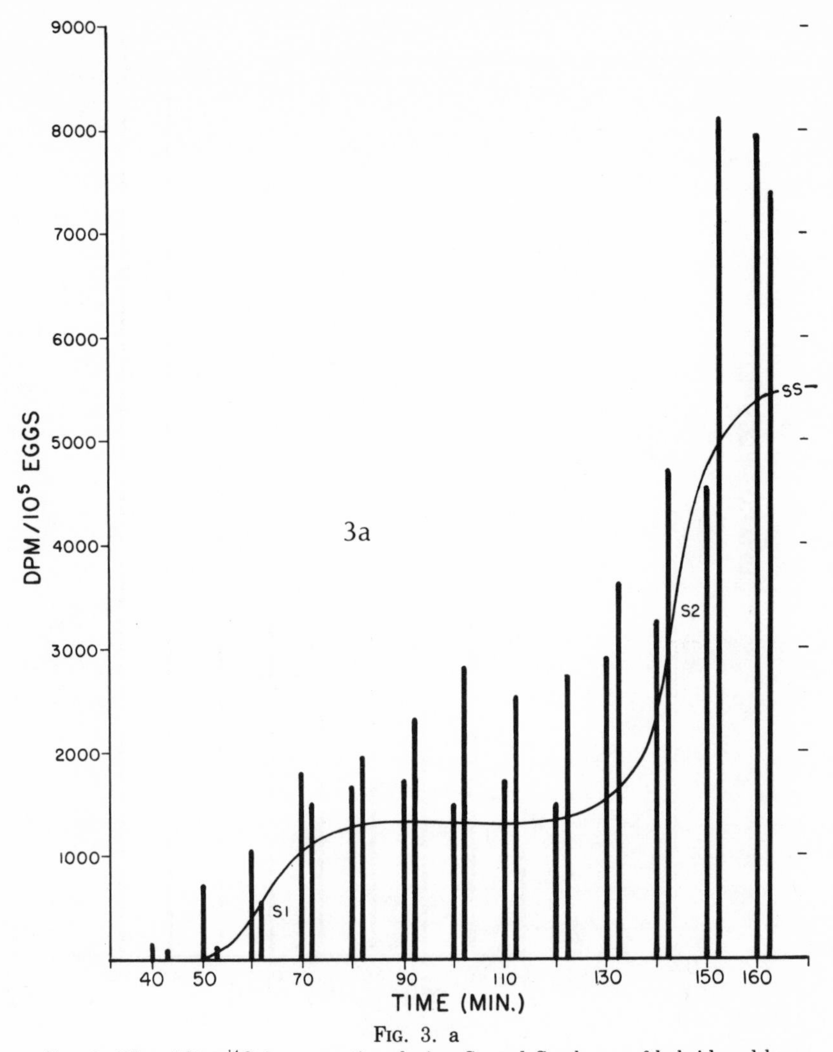

FIG. 3. a

FIG. 3. Thymidine-^{14}C incorporation during S_1 and S_2 phases of hybrid and homospermic embryos. Thymidine added (0.5 μC/ml) at 20 minutes, post-insemination). S_1 and S_2 = incorporation phases of SS or DD. (a) Black bars: SD (two experiments, sam-

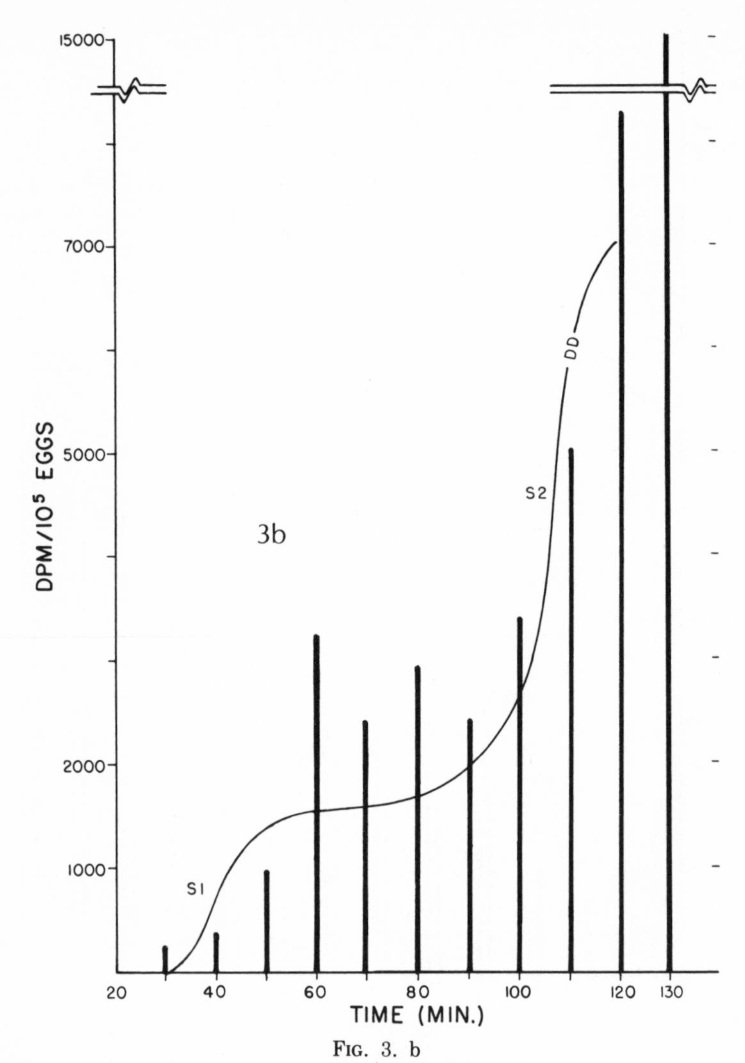

FIG. 3. b

ple times 3–5 minutes apart). Solid line: SS (maternal) curve. Sampled at 10-minute intervals, with a standard error of ±5% during S_1, and ±2% subsequently. (b) Black bars: DS (single experiment). Solid line: DD (maternal) curve. Other conditions as in 3a.

TABLE 3

INCORPORATION OF THYMIDINE-^{14}C (DURING 60-MINUTE PULSE) BY 16-HOUR BLASTULAE OF SS, SD, AND DD AT 13°C[a]

SS	SD	DD
38,249 (40,678; 35,821)	23,825 (23,942; 23,709)	118,969 (121,738; 116,200)

[a] Data are expressed as dpm/10^5 embryos. DD embryos beginning hatching. Values in parentheses are the averaged duplicates.

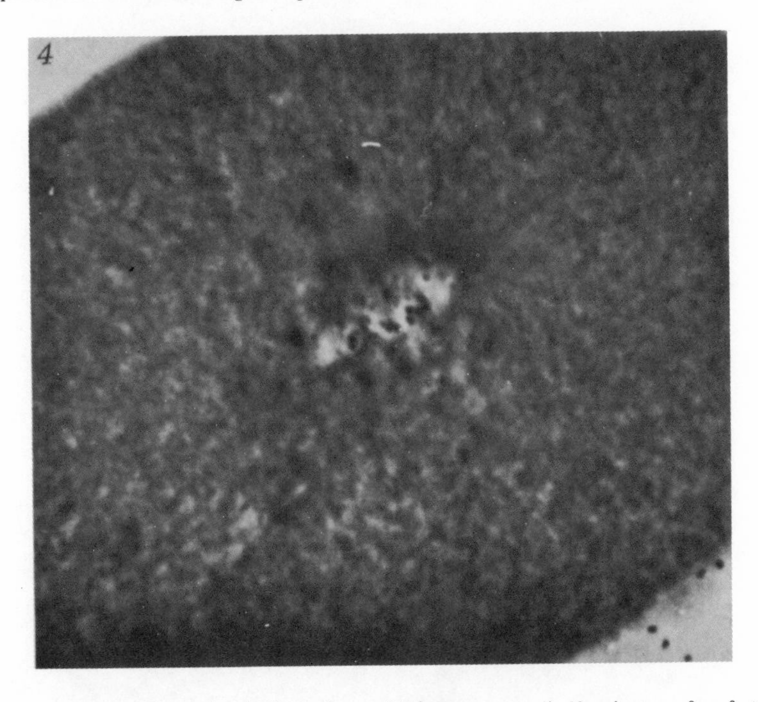

FIG. 4. Autoradiograph of DD embryos (72-hour exposure) 40 minutes after fertilization. Thymidine-^{14}C (0.5 μC/ml) was added at 15 minutes. Acetic acid–alcohol fixation, 10 μ section. Gallocyanin stain. Focus is on grains immediately above the pronuclei. × 500.

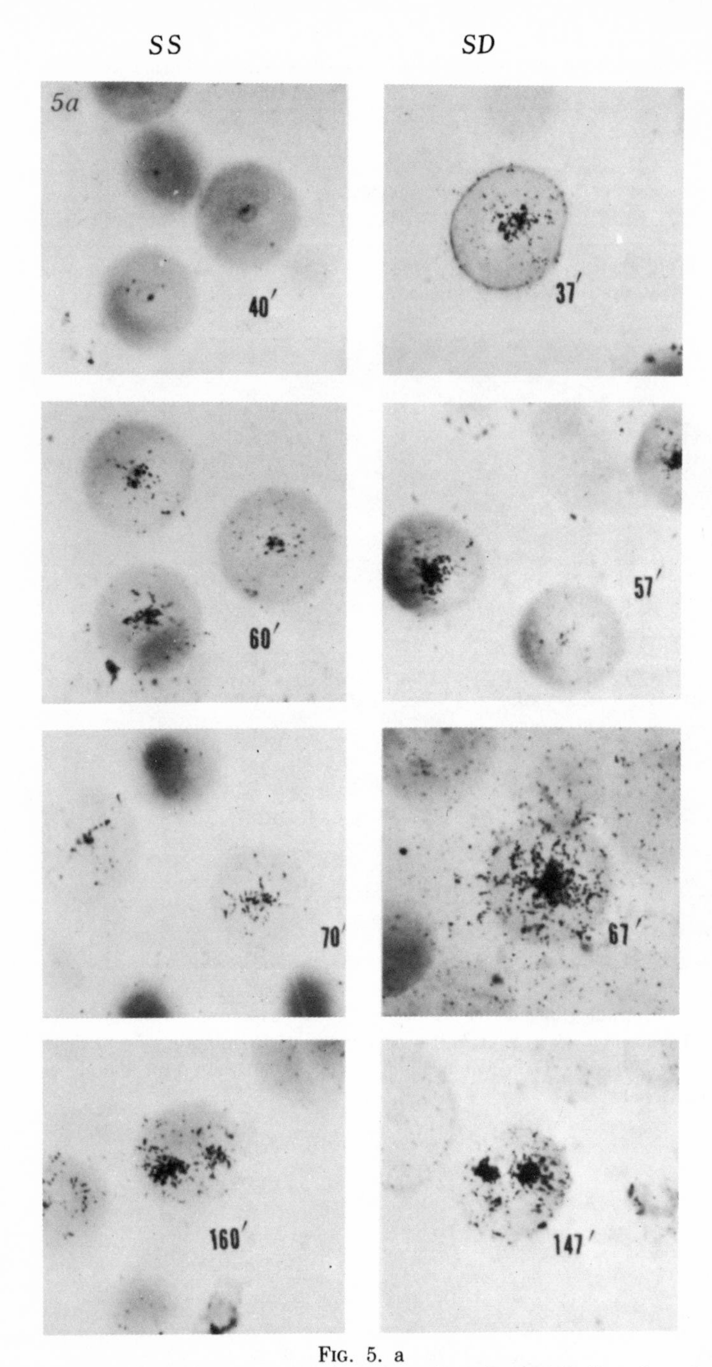

FIG. 5. a

FIG. 5. Autoradiographs of eggs at times (post-insemination) indicated on the photographs. Exposure time = 96 hours. Sectioned at 10μ. Gallocyanin stain. (a) SS and SD. $\times$ 170. Note precocial label at 37 minutes in SD. Black dot in SS at 40 minutes

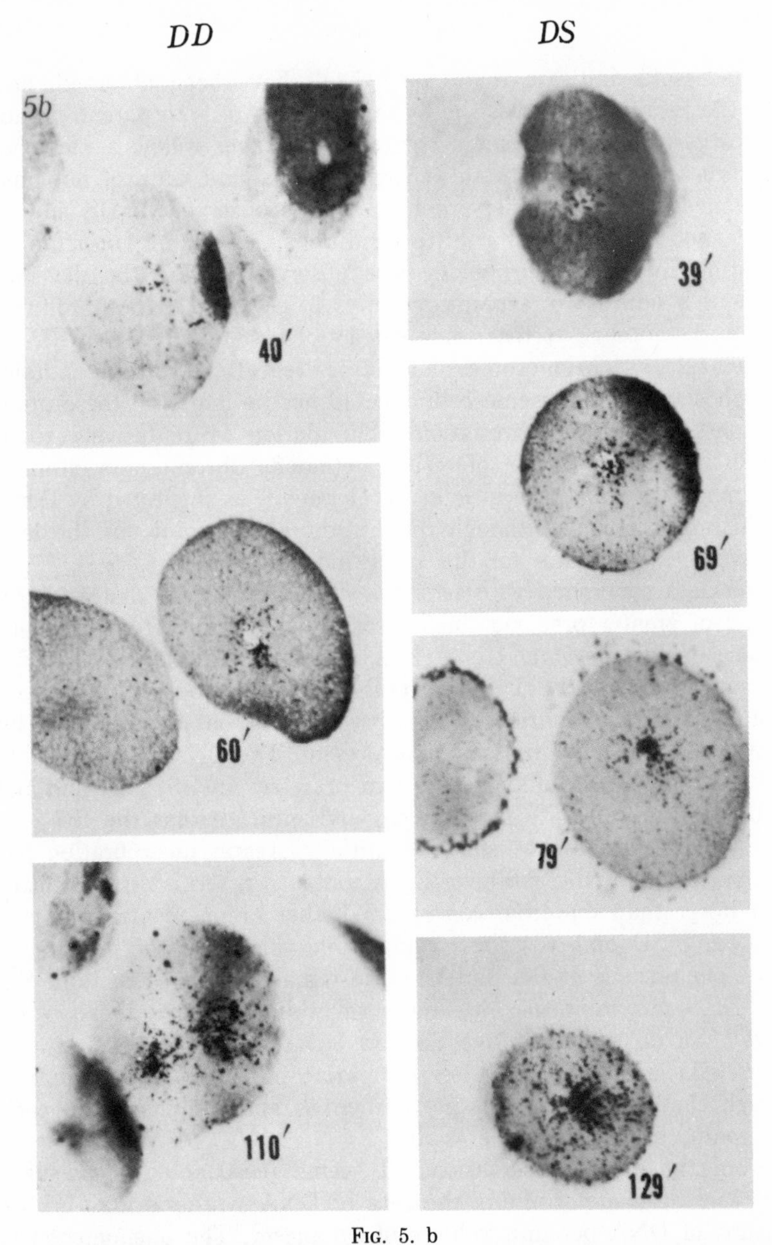

FIG. 5. b

is the (unlabeled) sperm pronucleus. Also, note heavier labeling at 57 minutes through 147 minutes in SD, in accord with results of Fig. 3a. Sections treated with RNase, 100 μg/ml, 30 minutes, 37°C. (b) DD and DS. × 200. Early label (40 and 39 minutes) comparable in DD and DS. At 69 and 79 minutes (DS), the heavier label is due to the S_2 of D, occurring with the S_1 of S. Label at 129 minutes of DS includes contribution from the S_3 of D. RNase not used.

DISCUSSION

The results indicate that true hybrids have been obtained, and that the resulting embryos show characteristics of both parent species in postgastrular development. Pregastrula stages follow a maternal schedule insofar as the rate of segmentation and time of hatching are concerned. The variety of larval forms observed in DS and in SD is not ascribable to culture conditions. The same spectrum of variation in larval morphology was observed during the development of a number of separate cultures of SD, and in both cultures of DS. Temporary acetocarmine smears of first metaphase of SD do not reveal any chromosomes outside or free of the mitotic spindle, though actual chromosome counts could not be made on these preparations. Therefore, chromosome elimination is tentatively ruled out in the interpretation of results obtained. Chromosome elimination may play some role in later development, as suggested by Denis and Brachet (1969), although these authors also point out the lack of cytological evidence for this in echinoid hybrids.

The data concerned with synthesis of DNA indicate that an early period of greater precursor incorporation, or of more frequent replications of chromosomal DNA (Figs. 3 and 5), is followed by a decreased incorporation (Table 3), both of these phenomena being accentuated in the hybrid (SD) embryos. Data on incorporation by early blastulae of DS are lacking at present. The DNA:N data (Table 2) reflect the net result of these two processes in SS, DD, and SD embryos from the late blastula on, and indicate that the SD embryos, even though they show an initially greater incorporation per unit time, eventually exhibit a DNA content (in terms of total nitrogen) lower than maternal controls. Whether or not control SS embryos and SD embryos show comparable values for the amount of DNA per nucleus at the pluteus stage remains to be seen. Both SD and DS embryos incorporate more thymidine through the two-cell stage than do the respective SS and DD controls. However, by 16 hours, SD is incorporating less thymidine per hour pulse than SS, though the number of cells per embyro is about the same in each (by count, in sectioned material).

From the above observations, it seems reasonable to speculate that the factor determining the rate of DNA synthesis may be the amount of DNA per unit volume of cytoplasm. The phenomenon is thus tentatively regarded as a negative feedback system. The diminished synthesis of nuclear DNA during late cleavage would be a consequence of the exceeding of a fixed proportionality between the amount of nuclear DNA and the volume of cytoplasm. One may, of course, be dealing with the concentration of a molecular species in the cytoplasm, rather than volume per se. A gradual decrease in mitotic index would be predicted from this assumption, with the hybrids ultimately showing a mitotic index and DNA content lower than control embryos by late gastrula and later stages. Mitotic index data for SD are not available as yet, but the lower DNA content of SD embryos (compared to SS) is evident by the third day of development (Table 2). Such a slowdown of DNA synthesis (following a period of greater thymidine incorporation) might also explain the lower DNA content of other echinoid hybrids, when these are compared to maternal controls (Whiteley and Baltzer, 1958; Chen and Baltzer, 1964; Denis and Brachet, 1969). An early period of greater thymidine incorporation would seem possible if the $S_{1,2}$, replication periods occur at different times in the parent species. The hybrid *Paracentrotus* ♀ × *Arbacia* ♂, used by the above-named authors, may meet this requirement, since the first and second cleavages occur earlier, by 20–30 minutes, in the *Paracentrotus* parent. It is important to note that the above interpretation assumes that differences between SS and SD in the size of the unlabeled precursor pool are negligible at 16 hours, insofar as their effect on the final level of incorporation of labeled thymidine is concerned. This assumption, while not unreasonable, remains unsubstantiated.

Although alternative explanations exist that are compatible with the data presented (e.g., elimination of paternal chromosomes following cleavage), the above interpretation seems direct, and to offer opportunities for future experimentation. Furthermore, no cytological evidence of chromosome loss in SD was found by Moore (1957) or in the present study.

The measurements of thymidine-^{14}C incorporation in SS and DD, during the S_1 and S_2 time periods, are in agreement with previously mentioned reports. The data for SD and DS hybrids suffer from the following limitations: (1) Since the fertilization percentage is lower, the counts observed during the initial minutes of the S_1 incorporation period are reliable only to $\pm 12\%$. (2) The longer time required for fertilization (5 minutes in hybrids, vs. 1–2 minutes in controls, for membrane elevation), and the subsequent extensive washing period to remove the excess sperm (during which time additional eggs may become fertilized), result in greater asynchrony at first cleavage in the hybrids as opposed to controls. The precise degree of asynchrony, as determined by Hinegardner *et al.* for SS, is not known. Estimates at first cleavage of SD and DS indicate that the initial furrowing occurs at the same time as in maternal control cultures, but that the time of 50% cleavage (an arbitrary end point) is approximately 15% later in the hybrids (compared to maternal controls). It is therefore difficult to determine from these data precisely when thymidine incorporation periods begin, particularly during the first 60 minutes. Fortunately, the data from scintillation counting are augmented by the autoradiographs.

With these limitations in mind, the following interpretation of the data is suggested. In both SD and DS, the first event to occur is the S_1 of the D complement, which occurs (in DD) prior to nuclear fusion. This event is assumed to occur in both SD and DS during the same time period as in DD (see Fig. 5). Following this S_1 of D, pronuclear fusion occurs, followed in turn by the S_1 of the S complement of chromosomes, during the period expected from the SS data. This S_1 replication of S chromosomes in turn induces (allows) a precocial S_2 in the D chromosomes (in both SD and DS) and implies a lack of species specificity in the conditions favoring replication. After completion of pronuclear fusion (ca 45 minutes in DD, 60 minutes in SS) and the attendant replications, subsequent S periods ensue as dictated by the egg cytoplasm. In DS, one therefore observes an S_3 in the D chromosomes, and an induced S_2 of the S chromosomes at the normal S_2 period of DD, beginning at 100 minutes. At about 140 minutes (S_3 of DD, and S_2 of SS) the DS embryos undergo what amounts to an S_4 of the D set, and an S_3 of the S set. Similarly, in SD hybrids, pronuclear fusion and concomitant replications are followed by normal S_2 of SS, which begins at 140 minutes and results in an S_3 of D as well as the scheduled S_2 of S. If one follows through this interpretation, and substitutes incorporation values (dpm/10^5 eggs) for haploid sets of S and D chromosomes of 600 (S) and 800 (D), one arrives at theoretical cumulative totals of SD = 8800 (S_3 of D + S_2 of S), and DS = 17,600 (S_4 of D + S_3 of S). The observed values from the plots in Fig. 3 are, SD = 8000 (150 minutes), and DS = 15,080 (130 minutes), which are close to expectation. In calculating the expected results, the quantitative relationship between S_1 and S_2 reported here for SS and DD, and by Hinegardner *et al.* for SS, have been used. These observations indicate that the S_2 is approximately $4(S_1)$, S_3 is $8(S_1)$, and S_4 is thus

$16(S_1)$. In theory, one would expect $S_2 = 3(S_1)$. It is probable that the S_1 incorporation levels observed are low due to initial dilution of labeled thymidine by the unlabeled precursor pool of the eggs. In any event, the observed differences between the SD and DS curves (with respect to the levels of incorporation at a particular time) reflect cytoplasmic influences.

One could speculate further that that pronuclei possess a supply of DNA polymerase sufficient for S_1, but must synthesize new enzyme, for subsequent S periods, at the direction of the cytoplasm. This notion is supported by the work of Black *et al.* (1967), who find that puromycin has little effect on the S_1 incorporation period of *Arbacia* eggs, whereas subsequent incorporation periods are completely inhibited by this suppressor of protein synthesis.

In addition to the pattern of DNA synthesis in the hybrids, the variability of hybrid morphology is noteworthy. It could be argued that precocial replications (as well as possible differential slowing of DNA synthesis later on) result in heteroploidy and chromosomal imbalance in the hybrid embryos. The analysis of this aspect of the results, along with verification of the precocity of the early replication periods, awaits thymidine-^{3}H autoradiographs, and critical cytological work on chromosome behavior and morphology during mitosis in hybrids. A serological analysis of proteins produced during hybrid development (SD) has been completed (Badman and Brookbank, 1970).

SUMMARY

Reciprocal, interordinal crosses are described. The hybrids seem to be examples of biparental inheritance, rather than maternal dominance.

The variability in hybrid morphology (plutei) is ascribed to chromosomal imbalance (pending further investigation), rather than to external influences.

Thymidine-^{14}C incorporation data are interpreted in terms of initial (pronuclear) replication periods which are independent of the cytoplasm and catalyzed by enzymes (DNA polymerase) carried by the gamete nuclei. Subsequent replication periods are presumed to be dependent on maternal (cytoplasmic) timing, and on protein synthesis.

The observed decrease in nuclear DNA synthesis during the early blastula stage of SD is discussed in terms of negative feedback control through a specific relationship between DNA content and cytoplasmic volume (or concentration of a cytoplasmic constituent(s)).

REFERENCES

BADMAN, W. S., and BROOKBANK, J. W. (1969). Serological studies of two hybrid sea urchins. *Develop. Biol.* **21,** 243–256.

BALLENTINE, R. (1940). Total nitrogen content of the Arbacia egg. *J. Cellular Comp. Physiol.* **15,** 121–122.

BLACK, R. E., BAPTIST, E., and PILAND, J. (1967). Puromycin and cycloheximide inhibition of thymidine incorporation into DNA of cleaving sea urchin eggs. *Exptl. Cell Res.* **48,** 431–439.

BOHUS JENSEN, A. (1953). The effect of trypsin on the cross fertilizability of sea urchin eggs. *Exptl. Cell Res.* **5,** 325–328.

BURTON, K. (1956). A study of the conditions and mechanism of the diphenylamine reaction for colorimetric estimation of deoxyribonucleic acid. *Biochem. J.* **62,** 315–323.

CHEN, P. S., and BALTZER, F. (1964). Further morphological and biochemical studies on normal and hybrid embryos of sea urchins. *Experientia* **20,** 236–240.

DENIS, H., and BRACHET, J. (1969). Gene expression in interspecific hybrids. I. DNA synthesis in the lethal cross *Arbacia lixula* ♂ × *Paracentrotus lividus* ♀. *Proc. Natl. Acad. Sci. U.S.* **62,** 194–201.

FLICKINGER, R. A. (1957). Evidence from sea urchin-sand dollar hybrid embryos for a nuclear control of alkaline phosphatase activity. *Biol. Bull.* **112,** 21–27.

HINEGARDNER, R. T., RAO, B., and FELDMAN, D. E. (1964). The DNA synthetic period during early development of the sea urchin egg. *Exptl. Cell Res.* **36,** 53–61.

MOORE, A. R. (1957). Biparental inheritance in the interordinal cross of sea urchin and sand dollar. *J. Exptl. Zool.* **135,** 75–83.

SIMMEL, E. B., and KARNOFSKY, D. A. (1961). Observations of the uptake of tritiated thymidine in the pronuclei of fertilized sand dollar embryos. *J. Biophys. Biochem. Cytol.* **10,** 59–65.

TYLER, A., and METZ, C. B. (1955). Effects of fertilizin-treatment of sperm and trypsin-treatment of eggs on homologous and cross-fertilization in sea-urchins. *Pubbl. Staz. Zool. Napoli.* **27,** 128–145.

WHITELEY, A. H., and BALTZER, F. (1958). Development, respiratory rate, and content of desoxyribonucleic acid in the hybrid Paracentrotus ♀ × Arbacia ♂. *Pubbl. Staz. Zool. Napoli* **30,** 402–457.

SEA URCHIN NUCLEAR DNA POLYMERASE

I. *Localization in Nuclei during Rapid DNA Synthesis*

L. A. LOEB, B. FANSLER, R. WILLIAMS and D. MAZIA

The expectation that DNA polymerase should be localized in nuclei follows from general experience with the occurrence of enzymes at sites of function. It could have a deeper meaning only if a great deal more was known about the mechanisms of DNA synthesis in chromosomes—but a knowledge of such mechanisms first requires that we know whether or not the enzymes are associated structurally with nuclei or more specifically with chromosomes.

The initial investigations of eucaryotic cells which followed the discovery of the DNA polymerase system by Kornberg and his collaborators did not fulfill this expectation. The DNA polymerase activity of eucaryotic cells was found predominantly in the soluble supernatant fraction after high speed centrifugation [1–6]. However, evidence obtained by isolating nuclei in non-aqueous solvents indicates that the enzyme may be concentrated in the nucleus [6, 7].

Mazia & Hinegardner [8] reported a high DNA polymerase activity in nuclei isolated from sea urchin embryos in an aqueous medium, but were unable to estimate what proportion of the total enzyme content of the embryos was so localized. In this communication, we show that a very large proportion of the total DNA polymerase activity of rapidly dividing sea urchin embryos can be recovered in isolated nuclei. In comparing these results with those in the literature, two qualifications must be made. We are dealing with a DNA polymerase which is characterized by a strong preference for native DNA as a primer [9] and we are dealing with an organism in early development, at a stage when cell division with its necessary DNA synthesis is the main activity, and when there is relatively little protein synthesis.

In the second communication, evidence will be presented to indicate that DNA polymerase is a cytoplasmic component in the egg, and the activity becomes progressively localized in the nucleus during early development.

MATERIALS AND METHODS

Materials

Sea urchins were collected along the northern coast of California or purchased from Pacific Bio-Marine Supply Company, Venice, Calif. [α-^{32}P]-thymidine triphosphate was obtained from the International Chemical and Nuclear Corp. and unlabeled deoxynucleotides from California Foundation for Biochemical Research. Calf thymus DNA, crystalline pancreatic DNase and RNase were acquired from Worthington Biochem. Corp. "Activated" DNA was prepared by subjecting DNA to a limited digestion with deoxyribonuclease by a modification of the procedure of Aposhian & Kornberg [10]. Sea urchin nuclear DNA polymerase was purified from nuclei of hatched *Strongylocentrotus franciscanus* embryos [11]. In all assays the most purified fraction, VIII, was used (specific activity, 270 units/mg protein).

Preparation of cell fractions

Eggs of the sea urchins *S. purpuratus* and *S. franciscanus* were fertilized and grown to the hatching stage [12]. After at least 95 % hatched, the embryos were allowed to settle and were washed with sea water by low speed centrifugation.

To prepare homogenates for the determination of total DNA polymerase activity, the embryos were washed twice with 1 M dextrose, after which they were suspended in 5 vol of solution A: 1 M dextrose, 0.02 M potassium phosphate buffer (pH 7.4), 0.004 M reduced glutathione, 0.0004 M potassium versenate. No DNA polymerase was lost during the washing procedures. The suspension was then homogenized by intermittent sonication at 0–1°C until there were no intact embryos. There was no detectable loss of polymerase activity during sonication or storage at −40°C for up to two months. The DNA polymerase activity of these preparations is considered to represent the total activity in the embryos under the conditions used in the assay. This preparation is subsequently referred to as the "embryo fraction."

Nuclei were isolated from the same batches of hatched embryos. These were washed by centrifugation at 1200 g for 10 min, once in sea water, twice in 1 M dextrose and once in a hypotonic solution of 0.15 M sodium chloride with 0.015 M sodium citrate. The pellet was suspended in 4 vol of the NaCl-citrate solution which ruptured the cells. They were then dispersed by forcing the suspension twice through a number 20 hypodermic needle. An equal volume of 2 M sucrose was thoroughly mixed with the suspension, after which it was centrifuged at 14,000 g for 30 min. The pellet, which consisted of rather pure nuclei, was dispersed in 5 vol of solution A and sonicated as described above. This will subsequently be referred to as the "nuclear fraction."

The cytoplasmic supernatant fraction was prepared from hatched embryos by a method which is described here for the first time. Embryos were washed by suspending in three vol of 1.0 M dextrose and centrifuging at 2000 g for 10 min. The embryos, already breaking up into individual cells, were then resuspended in three vol of 0.02 M potassium phosphate buffer, pH 7.4, 0.003 M $MgCl_2$ and 1.0 M dextrose. Triton X-100 (Rohm & Haas) was added to a final concentration of 0.1 %, the suspension mixed gently for 1 min and placed in an ice-water bath for an additional 4 min. Centrifugation at 9500 g for 5 min yielded a pellet of nuclei and the remaining unbroken cells. The supernatant was carefully decanted and centrifuged at 100,000 g for 45 min. This supernatant is termed the "cytoplasmic fraction" and can be assayed directly for DNA polymerase activity.

Triton X-100 at twice the concentration used for isolation did not inhibit sea urchin DNA polymerase activity, while a tenfold increase (1.0 %) resulted in a 15 % loss of activity. It should be mentioned that another detergent, deoxycholate, was unacceptable in this procedure because concentrations as low as 0.01 % produced an 85 % loss of polymerase activity. The concentration of Triton X-100 and the length of time the embryos are exposed to it were found to be critical in obtaining only cytoplasmic material. By direct phase microscopic observation, this method gives approx. 50 % cell breakage with essentially no nuclear rupture. The latter observation was confirmed by preparing cytoplasmic supernatant fractions from embryos which had incorporated ^{14}C-thymidine into their DNA. Less than 1.0 % of the total incorporated radioactivity was found in the cytoplasmic fraction.

DNA polymerase activity

The assay measures the initial rate for the incorporation of [α-^{32}P]-thymidine triphosphate into an acid-insoluble product. In order to assay the DNA polymerase activity of the embryo fraction, it is necessary to minimize the effects of destructive enzymes. This was accomplished by using high concentrations of added native or "activated" DNA, deoxynucleoside triphosphates, and especially by the employment of a very short period of incubation. The incubation mixture (0.3 ml) contained 25 μmoles Tris-maleate buffer, pH 7.4; 4.5 μmoles $MgCl_2$; 0.75 μmoles KCl; 0.3 μmoles 2-mercaptoethanol; 25 mμmoles each of dATP, dCTP, dGTP and [α-^{32}P]-dTTP (about 2×10^4 dpm/mμmole); 530 mμmoles native calf thymus DNA and 0.01–0.05 ml of enzyme preparation. Incubation was at 37°C and all determinations were performed simultaneously in triplicate. Incorporation was determined by precipitation and washing as previously described [9]. Under these conditions incorporation was proportional to enzyme concentration.

Protein was determined by the method of Lowry et al. [13] after precipitation and washing with cold perchloric acid. Deoxypentose was determined by the diphenylamine procedure of Dische [14], after preparation of cell extracts by the procedure of Schneider [15].

RESULTS

Characteristics of sea urchin DNA polymerase

The ability of sonically disrupted embryos, nuclei and purified sea urchin nuclear DNA polymerase [11] to catalyze the incorporation of labeled thymidine triphosphate into acid-insoluble material was compared under a variety of conditions (table 1). Whatever the source of enzyme, for maximal incorporation the presence of all four deoxynucleotide triphosphates, Mg^{2+} and added DNA primer were required. The small but definite activity of the embryo and nuclear fractions in the absence of added primer is attributed to the presence of DNA in these fractions. In the nuclear and embryo fractions, considerable incorporation is shown in the absence of the complete complement of the four deoxynucleoside triphosphates. This lack of dependency has been reported in many crude preparations of eucaryotic DNA polymerases and is usually attributed to the presence of a terminal DNA polymerase [6]. However, even purified sea urchin nuclear DNA polymerase [11] and calf thymus DNA polymerase [16] do not stringently require all four deoxynucleoside triphosphates. An even less stringent requirement in the unpurified dialyzed cellular fractions might be attributed to the sequential action of DNase and deoxynucleotide kinases which could provide the required substrates; the latter activities are known to be present in sea urchin embryos. Irrespective of the source of DNA polymerase activity, the product appears to be easily digested with deoxyribonuclease. The main conclusion from table 1 is that the properties of this enzyme are characteristic of DNA polymerizing systems and that under a variety of conditions whole embryo sonicates, nuclei and purified sea urchin nuclear DNA polymerase exhibit similar requirements. It has been reported that native DNA is a more effective primer than equivalent amounts of heat-denatured DNA for the DNA polymerase activity in isolated sea urchin nuclei [9], as well as for purified sea urchin nuclear DNA polymerase [11]. Table 2 shows that a similar primer preference is exhibited by sonicated preparations of whole embryos. The activity without added DNA is minimal and presumably dependent on the endogenous DNA present in the preparations.

Table 1. *Requirements for deoxynucleotide incorporation into DNA*

	dTM^{32}P incorporated (mμmoles/mg protein)		
Reaction mixture	Embryo fraction	Nuclear fraction	Purified enzyme
Complete	0.99	8.50	280.0
Minus native DNA primer	0.06	0.10	<0.1
Minus $MgCl_2$	<0.01	<0.01	<0.1
Minus dGTP	0.45	1.62	28.5
Minus dGTP, dCTP	0.41	1.32	20.8
Minus dGTP, dCTP, dATP	0.42	0.78	18.0
Minus enzyme	<0.01	<0.01	<0.1
Complete+2.5 μg DNase[a]	<0.01	0.10	1.2
Complete+2.5 μg RNase[a]	1.05	8.04	280.0

Sonicated preparations of whole embryos and nuclei were dialyzed against 1000 vol of solution A at 0°C and assayed for DNA polymerase activity (see Methods) with native DNA as primer. In assays with purified enzyme, 177 mμmoles of "activated" DNA were used as primer. Reaction mixtures containing the embryo fraction were incubated for 5 min, all other incubations were for 10 min. In certain tubes, as designated,[a] after incubation, the reaction was terminated by heating at 60°C for 10 min. After cooling, 0.025 ml of either DNase (0.1 mg/ml) or RNase (0.1 mg/ml) was added and the tubes were incubated an additional 30 min at 37°C. The reaction was then stopped as in the standard DNA polymerase assay. All determinations were performed in triplicate.

Localization of DNA polymerase activity in nuclei at hatching

A comparison of the activity of sonically disrupted whole hatched embryos with that re-

Table 2. *Effectiveness of native and heat-denatured DNA as primers for DNA polymerase in sonically disrupted sea urchin embryos*

Expt	dTM^{32}P incorporated (mμmoles/mg of protein/4 min) No added DNA	Native DNA	Heat-denatured DNA
1	0.080	0.880	0.236
2	0.088	0.805	0.223
3	0.070	0.693	0.148
4	0.038	0.316	0.058
5	0.037	0.507	0.098
6	0.030	0.354	0.088

The DNA polymerase activity of the embryo fraction was determined in the presence of 354 mμmoles of native or heat-denatured DNA or in the absence of added DNA. Details of the assay are given in "Methods." In experiments 1 and 2, 288 and 296 μg of protein from *S. purpuratus* embryos and in experiments 3, 4, 5 and 6, 288, 395, 437, and 126 μg of protein from *S. franciscanus* embryos were added to each reaction mixture. Heat-denatured DNA was prepared by keeping a solution of DNA (1 mg/ml) in 0.01 M KCl at 100°C for 10 min and then immediately cooling in an ice-water bath.

covered in the nuclei isolated from samples of the same populations is shown in table 3. In each experiment the conditions of the assay were those which were optimal for the embryo fraction. The optimal Mg^{2+} concentration, which occasionally varied, was determined individually for each embryo fraction. It is shown in table 3 that the specific activity of the nuclear preparations, on the basis of protein, was 2.2 to 6.0 times greater than that of the embryo fractions. In all, we have directly compared activities in 29 corresponding preparations and without exception, the activity in the nuclear preparations was greater. Calculated on the recovery of DNA, the nuclear fraction contains 56 to 95 % of the total DNA polymerase activity present in the whole embryos at the time of hatching.

The additivity of synthesis with different cell fractions

The possibility that the measured activity of the embryo fraction is low because of inhibitors or the action of destructive enzymes even in the present assays was examined by studying the additivity of the synthesis in embryo fractions and purified sea urchin

Table 3. *Localization of DNA polymerase activity in nuclei*
Polymerase activity, dTM^{32}P incorporated

Expt	Recovery of nuclei (%)	Fraction: Embryo (mμmoles/mg protein)	Nuclear (mμmoles/mg protein)	Embryo (mμmoles/mg DNA)	Nuclear (mμmoles/mg DNA)	Activity in nuclei (%)
1	92	0.81	2.89	25.8	17.7	69
2	94	0.82	1.79	23.1	12.9	56
3	91	0.35	1.89	15.2	11.1	73
4	69	0.55	1.83	27.0	16.2	60
5	59	0.49	2.96	22.9	21.7	95
6	88	0.55	1.61	21.1	16.5	78

In experiments 1 and 2, corresponding sonicated preparations of whole embryos and nuclei were obtained from the same batches of hatched *S. purpuratus* embryos, and in experiments 3 through 6 from hatched *S. franciscanus* embryos. The yield of nuclei was determined as the per cent of the total DNA recovered in the nuclear fraction. We made these calculations on the assumption that at this stage of development cytoplasmic DNA is but a small fraction of the total DNA of the embryo; if not, the per cent activity in isolated nuclei would be greater. The amount of the embryo fraction added to each tube in experiments 1 to 6 was 289, 296, 395, 312, 437, 289 μg of protein or 9.0, 10.4, 9.1, 6.3, 9.6, 7.1 μg of DNA, respectively. The amount of nuclear fraction added to each tube in experiments 1–6 was 194, 190, 240, 205, 270, 432 μg of protein or 30.6, 26.4, 37.8, 23.2, 34.8, 42.2 μg of DNA, respectively. In experiment 1, the Mg^{2+} concentration was 10 mM for the determination of activity of the embryo fraction, in all other assays the concentration was 15 mM. In all experiments native DNA was used as a primer and incubation was for 4 min.

Table 4. *The additivity of synthesis with cell fractions from* S. franciscanus

Expt	Polymerase activity, $dTM^{32}P$ incorporated (mμmoles/5 to 4 min)		
	Embryo fraction	Nuclear fraction	Embryo + Nuclear
1	0.15	0.37	0.66
2	0.22	0.52	0.82
3	0.16	0.45	0.67

DNA polymerase was determined as given in "Methods" with native DNA as primer. Homogenates of whole embryos and nuclei were first dialyzed against 1000 vol of solution A for 18 h at 0–1°C. The enzyme fractions which were incubated together were mixed before addition to the reaction mixture in the same proportions as used singly. Enzyme was added after the reaction mixture had equilibrated to 37°C and incubation was for 4 min.

nuclear DNA polymerase. DNA polymerase activity was determined using cell fractions from *S. franciscanus* embryos with native calf thymus DNA as primer (table 4). No inhibitory effect of the cytoplasmic constituents is indicated; if anything, there is a tendency for the assays to be more than additive. This finding could be explained by the previous observations that a brief exposure of DNA to DNase increases its priming ability [8]. The ratio of DNase to DNA polymerase activity is greater in the embryo fraction than in the isolated nuclei [11]. In table 5 assays were carried out with cell fractions derived from *S. purpuratus* embryos with "activated" DNA as a primer. "Activated" DNA was prepared by treating native DNA with DNase until it exhibited maximal priming for purified sea urchin nuclear DNA polymerase. This prior treatment appears to mitigate the tendency for the combined fractions to be more than additive. Additivity of synthesis is also exhibited when the cellular fractions are assayed together with the purified enzyme.

Table 5. *Lack of apparent inhibitor in additivity studies*

Enzyme fraction	Polymerase activity, $dTM^{32}P$ incorporated (μμmoles)	
	Experimental	Calculated[a]
Embryo	12	
Nuclear	17	
Cytoplasmic supernatant	14	
Purified	9	
Embryo + Nuclear	30	29
Embryo + Cytoplasmic	26	26
Embryo + Purified	24	21
Nuclear + Cytoplasmic	35	31
Nuclear + Purified	21	26
Supernatant + Purified	28	23

Methods were as given in table 4 except that crude enzyme fractions were from *S. purpuratus* embryos and "activated" DNA was used as a primer. Bovine serum albumin (40 μg) was added to the reaction mixture containing only purified enzyme.
[a] Calculated values represent the sum of syntheses by the individual fractions.

Heat denaturation studies

The rate of loss of DNA polymerase activity upon heating different cell fractions could indicate whether the same enzyme catalyzed the reaction. These measurements are particularly exacting for sea urchin nuclear DNA polymerase since the purified enzyme is extremely heat sensitive. In this experiment, embryo, nuclear and cytoplasmic fractions were simultaneously heated at 42°C in the presence of 1 M dextrose (fig. 1). When indicated, the heat-denaturation was terminated and the amount of DNA polymerase activity remaining was determined. The rate of loss of DNA polymerase activity is approximately the same: evidence indicating that the same enzyme is present in the three cellular fractions.

DISCUSSION

Sea urchin embryos are particularly well suited for in vivo and in vitro studies of DNA

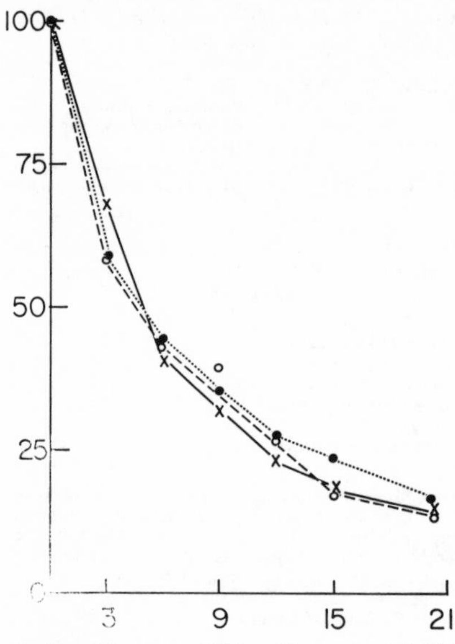

Abscissa: Minutes at 42°C; *ordinate:* % of initial activity.

Fig. 1. Loss of DNA polymerase activity in different cell fractions upon heating. Cell fractions obtained from the same batch of hatched *S. purpuratus* embryos were simultaneously incubated at 42°C in Solution A for the times indicated. Thereafter, DNA polymerase activity was determined as given in "Methods" with 354 mμmoles of "activated" DNA as primer. Incubation was for 4 min at 37°C and incorporation without preincubation (100 %) was 0.46, 0.53 and 0.16 for the whole embryo, nuclear and cytoplasmic fractions, respectively. ●---●, sonicated embryo; ×—×, nuclear fraction; ○--○, cytoplasmic fraction.

synthesis. After fertilization, large batches of these embryos divide synchronously for several cell generations. The permeability to many nucleic acid precursors permits one readily to monitor periods of DNA synthetic activity. In the early cleavage stages, the number of nuclei doubles in 1–2 h and the DNA, amounting to 1.8×10^{-12} g per diploid nucleus, doubles during an S-period of 10 to 12 min at 16°C [17, 18]. This rate of DNA synthesis is about 60 times greater than in most other eucaryotic cells. The unusually high DNA synthetic activity in the embryos is accompanied by a corresponding high level of DNA polymerase activity in vitro. In fact, the polymerase activity per mg of cell protein is considerably higher than that of other investigated animal cells, being comparable to that in *Escherichia coli.* One can imagine that a DNA polymerase which may be intimately involved in DNA replication is more massively or more firmly bound in nuclei of embryos which are so active in DNA synthesis.

The localization of the DNA polymerase in nuclei reported here must be considered against the background of nucleo-cytoplasmic relations in the embryos at the stage examined. The embryos reach the hatching stage in about 18 h after fertilization and have produced approx. 300–400 nuclei. During this time there has been no significant net increase in protein content per embryo. At hatching we do find that 56–95 % of the enzyme is in the nuclear fraction, but still a substantial amount is not recovered with it. We do not attribute this to experimental error. In fact, evidence will be presented in the succeeding paper to indicate that the bulk of the measured DNA polymerase is found in the cytoplasm of the egg and is transferred to nuclei as they are formed in successive divisions.

We thank Dr Robert Perry for suggesting the use of "Triton–X" for isolating cytoplasmic fractions and Grace Ziegler for skilled assistance.

This investigation was initially supported by USPHS grant no. GM-13882 to Dr D. Mazia, Department of Zoology, University of California, Berkeley. Subsequent support came from the Stanley C. Dordick Foundation, the American Cancer Society (grant E-483), and by grants to this Institute: NIH grants CA-06927 and FR-05539, American Cancer Society grant IN-49, and an appropriation from the Commonwealth of Pennsylvania.

REFERENCES

1. Davidson, J N, Smellie, R M S, Keir, H M & McArdle, A H, Nature 182 (1958) 589.
2. Bollum, F J, J biol chem 235 (1960) 2399.

3. Montsavinos, R & Canellakis, E S, J biol chem 233 (1959) 635.
4. Furlong, N B, Arch biochem biophys 87 (1960) 154.
5. Bach, M K, Biochim biophys acta 91 (1964) 619.
6. Keir, H M, Progress in nucleic acid research (ed J N Davidson & W E Cohn) vol. 4, p. 81. Academic Press, New York (1965).
7. Behki, R M & Schneider, W C, Biochim biophys acta 68 (1963) 34.
8. Mazia, D & Hinegardner, R T, Proc natl acad sci US 50 (1963) 148.
9. Loeb, L A, Mazia, D & Ruby, A D, Proc natl acad sci US 57 (1967) 841.
10. Aposhian, H V & Kornberg, A, J biol chem 237 (1962) 519.
11. Loeb, L A, J biol chem. In press.
12. Mazia, D, Mitchison, J M, Medina, H & Harris, P J, J biophys biochem cytol 10 (1961) 467.
13. Lowry, O H, Rosebrough, N J, Farr, A L & Randall, R J, J biol chem 193 (1951) 265.
14. Dische, Z, The nucleic acids (ed E Chargaff & J N Davidson) vol. 1, p. 289. Academic Press, New York (1955).
15. Schneider, W C, J biol chem 161 (1945) 293.
16. Yoneda, M & Bollum, F J, J biol chem 240 (1965) 3385.
17. Nemer, M, J biol chem 237 (1962) 143.
18. Hinegardner, R T, Rao, B & Feldman, D E, Exptl cell res 36 (1964) 53.

SEA URCHIN NUCLEAR DNA POLYMERASE

II. *Changing Localization during Early Development*

B. FANSLER and L. A. LOEB

Early development in sea urchin embryos is characterized by exponential cell division with accompanying DNA synthesis. In the previous paper, we showed that at hatching, a stage of rapid DNA synthesis, the enzyme which catalyzes this synthesis is localized in the nucleus. In this communication we will consider the amount and localization of polymerase in the unfertilized egg and at embryonic stages prior to hatching. The natural synchrony of these embryos has permitted us to examine DNA polymerase during the cell cycle. It will be shown that when development begins DNA polymerase activity is predominantly in the cytoplasm of the embryo and is progressively localized in the nuclear fraction with each successive division cycle.

MATERIALS AND METHODS

With the exception of those described below, all other procedures used in this investigation are identical to those in the preceding article.

Deoxynucleotide kinase activity

The rate of phosphorylation of ^{14}C-labeled deoxynucleoside monophosphates into the corresponding di- or tri-phosphates was determined by a modification of the method of Furlong [1]. The optimum concentrations of all components in the reaction mixtures were determined individually for each of the kinase assays. Activity was measured at kinase levels where the rate of reaction was proportional to enzyme concentration. A unit is defined as the amount of activity required to convert one mμmole of the deoxynucleoside monophosphate into the corresponding di- or tri-phosphate in the assays described below:

The incubation mixture contained in a volume of 0.05 ml, in 6×50 mm culture tubes, the following: 2.15 μmoles of Tris HCl (pH 8.0), 0.64 μmole $MgCl_2$, 0.8 μmole ATP, 0.12 μmole of the indicated ^{14}C-labeled deoxynucleoside monophosphate and a kinase source from crude fractions of sea urchin embryos (2–50 μg of protein). In addition, 5 μmoles of KCl was present in assays of deoxyguanosine monophosphate kinase. In the thymidine monophosphate kinase assay 0.8 μmole of phosphoenolpyruvate and 2 μg of pyruvic kinase were present. After incubating for 10 min at 37°C, the reaction was stopped by heating 3 min at 95°C. After cooling, 40 μl of 0.4 M sodium acetate buffer, pH 4.9, and 10 μl of human semen phosphatase [2] (approx. 20 units) were added and the tubes were incubated an additional 3 min at 37°C. After centrifuging 10 min at 4000 g, aliquots of 50 μl of the clear supernatant were applied to DEAE disks, 2.2 cm diameter (DEAE paper was obtained from H. Reeve Angel & Co.). These were washed successively in 0.004 M ammonium formate, water and 95 % ethanol. Radioactivity was determined by scintillation counting.

Preparation of nucleate and non-nucleate halves of unfertilized sea urchin eggs

The separation of small quantities of sea urchin eggs into nucleate and non-nucleate halves by centrifugation was originally described by Harvey [3]. In order to accommodate larger quantities of eggs, the procedure was modified in that the separation was carried out in a stabilizing linear sucrose gradient. An aliquot of 1 ml of the suspension of eggs obtained from a single sea urchin (one volume of eggs in 4 vol of calcium^{2+}-free sea water) was carefully layered over a 35 ml linear sucrose gradient (0.5–1.0 M sucrose). Gradients were prepared by the method of Martin & Ames [4] using equal volumes of 1.0 M sucrose solution made up in distilled water and 0.5 M sucrose made up by diluting the former solution with artificial Ca^{2+}-free sea water. Centrifugation was performed in a swinging bucket rotor 12,000 g for 70 min. Generally the non-nucleate halves banded sharply about 5 cm from the bottom of the tube, while intact eggs were in the middle of the tube and the nucleate halves about 1 cm from the top of the tube. Fractions were collected by puncturing the bottom of the centrifuge tube with a number 20 needle. Samples were examined to ascertain the presence or absence of nuclei and to measure the diameters of the cell fragments. Only pure samples of nucleate or non-nucleate halves (of uniform size) were pooled and used for further studies. The nucleate and non-nucleate halves were diluted with 4 vol of cold 1 M dextrose and centrifuged at 1500 g for 10 min. The pellet was washed in 1 M dextrose by centrifugation and resuspended in 5 vol of solution A. Sonication was carried out as previously described for the preparation of whole embryo extracts.

Fractionation of embryos during early development

Fifty ml of *S. purpuratus* eggs were collected, washed and fertilized as described. Immediately thereafter the fertilization membranes were removed by the method of Mazia et al. [5]. If not removed, the fertilization membranes sediment with the nuclei and cause nuclear aggregation and trapping of cell debris. The embryos were grown at 16°C in 7.5 l of sea water. At 2, 4, 6, 9, 12 and 18 h after fertilization, samples were removed, centrifuged at 1500 g for 10 min and the embryos resuspended in 15 vol of 1 M dextrose. Because embryos in very early development have few nuclei, making them more difficult to isolate, it was necessary to take the largest sample at 2 h (1935 ml) and progressively smaller samples at subsequent times. From each dextrose suspension of embryos a 7.5 ml aliquot was taken for preparation of the cytoplasmic supernatant and a 6 ml aliquot was used to prepare the whole embryo fraction. The remaining suspension was used for nuclear isolation. These fractionation procedures are described in Methods of the first paper in the series.

RESULTS

The constancy of DNA polymerase activity during the cell cycle

The characteristic interval in the eucaryotic cell cycle during which DNA synthesis occurs is referred to as the S-period. In sea urchin embryos this interval is particularly short, beginning at mid-telophase and extending into early interphase [6, 7]. In fig. 1 we have taken advantage of the natural synchrony of early developing embryos to compare in the same cultures incorporation of labeled thymidine into DNA in vivo with DNA polymerase activity in vitro. Consistent with the studies of Hinegardner et al. [7] the first DNA synthetic period, concurrent with pronuclear fusion, occurs about 30 min after fertilization. The subsequent S-periods are associated with cell division, beginning at mid-telophase and lasting about 15 min. In contrast to the sudden turning-on and turning-off of DNA synthesis during the cell cycle the total level of DNA polymerase activity stays fairly constant.

The large increase in metabolic activity that occurs immediately after fertilization

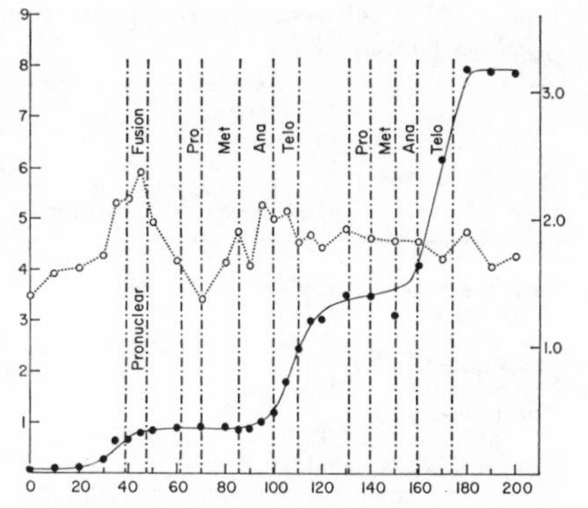

Abscissa: Minutes after fertilization; *ordinate:* (left) thymidine incorporation, cpm × 10^{-3}, ●—●; (right) polymerase activity, mμmoles/mg protein, ○----○.

Fig. 1. DNA polymerase activity of *S. purpuratus* embryos during the cell cycle. ^{3}H-Thymidine incorporation into synchronized embryos was determined by the method of Hinegardner et al. [7]. Immediately after fertilization 0.1 mC of ^{3}H-thymidine was added to 5 ml of embryos in 500 ml of sea water. Samples were taken when indicated for measuring thymidine incorporation and for determining DNA polymerase activity after washing and sonically disrupting the embryos.

Table 1. *DNA polymerase activity in embryos during early development*

Stage	Incorporation (mμmoles dTM^{32}P/mg protein/4 min)	
	S. franciscanus	*S. purpuratus*
Unfertilized egg	0.340	1.12
15-min embryo	0.535	1.20
2-cell embryo	0.505	0.96
4-cell embryo	0.392	1.28
8-cell embryo	0.545	1.52
Hatched embryo (18 h)	0.578	1.29
Gastrula (24 h)	0.635	1.61

Sea urchin embryos were grown to the indicated stages, sonically disrupted and assayed for DNA polymerase activity with 177 mμmoles of "activated" DNA as primer.

[8] is not accompanied by an increase in DNA polymerase activity. During the first division cycle there was sometimes observed (as shown in fig. 1) an increase in the level of polymerase at the time of fertilization as well as a variable decrease at early metaphase. These changes were peculiar to the first division cycle and not observed during subsequent division cycles.

The constancy of DNA polymerase activity during early development

The amount of DNA polymerase activity in unfractionated homogenates of developing sea urchin embryos is shown in table 1. There is little change in protein content during this time; the specific activity is thus considered to represent the total activity per embryo. The relative invariability in total embryo activity reflected against the rapid exponential synthesis of DNA during early development suggests that the total amount of polymerase necessary for DNA synthesis, at least through early gastrulation, is present in the egg before fertilization.

Table 2. *Deoxyribonucleotide kinase activity at different developmental stages*

Stage	Kinase activity (units/mg of protein)			
	dCMP	dAMP	dGMP	dTMP
Unfertilized egg	60	242	37	33
2-Cell embryo	77	256	45	51
100-Cell embryo	95	291	67	32

At the stage of development indicated, sonically disrupted *S. purpuratus* embryos were prepared as given in Methods of article I in this series. After centrifugation at 100,000 g for 1 h, the kinase activity in the supernatant was determined as given in Methods.

In other experiments, the level of deoxynucleotide kinase activity was determined at various stages during early development (table 2). These enzymes phosphorylate the deoxynucleoside monophosphates into corresponding di- and tri-phosphates and appear to function in the over-all DNA synthetic pathway. In contrast to the nuclear localization of the polymerase, the specific activity of the kinases in the nuclear fraction was slightly less than that of the embryo fraction. Some increases are to be noted in the activity of some of the kinases in the soluble cytoplasmic fraction, but the over-all pattern was one of constancy compared to the magnitude of the change in the in vivo rate of DNA synthesis.

The level of DNA polymerase activity at late gastrula (30–40 h), a time at which the rate of DNA synthesis is rapidly declining, is of special interest. Initial experiments revealed a striking decrease in the amount of DNA polymerase activity during this stage of development. However, when the polymerase of gastrula homogenates was assayed together with purified sea urchin nuclear DNA polymerase, the activities were not additive, suggesting that an inhibitor might be present in the gastrula.

Table 3. *Distribution of DNA polymerase and deoxynucleotide kinase activity in nucleate and non-nucleate halves of* S. purpuratus *eggs*

Enzyme	Specific activity (units/mg of protein)		
	Nucleate	Intact eggs	Non-nucleate
DNA polymerase			
Expt 1	2.68	3.66	1.42
Expt 2	3.09	—	1.30
Expt 3	0.95	1.42	1.10
dTMP kinase	53	51	36
dCMP kinase	102	71	51
dGMP kinase	65	55	50
dAMP kinase	268	428	460

Distribution of DNA polymerase activity in nucleate and non-nucleate halves of sea urchin embryos

Procedures which are successful for the isolation of nuclei from early embryos are not easily applicable to isolating nuclei from eggs. For the most part, this difficulty is explainable on quantitative considerations; the small amount of nuclear material compared to the much larger cytoplasm of the egg. In order to gain insight into the intracellular distribution of DNA polymerase activity, the activities of nucleate and non-nucleate halves were compared. These eggs can be split by centrifugation into halves. The less dense nucleate half is stratified into a lipid "cap" and a small nucleus above a large amount of cytoplasmic material. The non-nucleate halves are entirely free of nuclei. If polymerase activity is localized in egg nuclei, the non-nucleate halves would be free of activity. Table 3 presents representative experiments listing the polymerase activity of nucleate and non-nucleate halves of *S. purpuratus* eggs as well as the activity of eggs that have resisted separation on the same gradient. Unfortunately, centrifugation is also accompanied by stratification and rearrangement of most intracellular materials [3]. Perhaps this accounts for some of the variation between different batches of eggs and the difference between the average polymerase activity of the two halves compared to intact eggs of the same batch. In 9 of 11 experiments there is a tendency for localization of the polymerase in the nucleate half. A considerable portion of the polymerase activity appears in the non-nucleate half indicating that polymerase is not a nuclear component of the egg.

Similarly, the distribution of the four deoxynucleoside monophosphate kinases was determined in nucleate and non-nucleate halves (table 3). For three of the enzymes, a tendency towards concentration in the nucleate half is apparent. In contrast to DNA polymerase, the deoxynucleoside kinases were not found to be concentrated in the nucleus at hatching. However, the tendency of a greater activity in the nucleate half is apparent for DNA polymerase and the three kinases. This extent of partitioning between the halves might be that of any soluble enzyme that is not significantly localized in the egg nucleus.

Intracellular distribution of DNA polymerase during early development

Considering the relative constancy of DNA polymerase activity during the cell cycle and early development, as well as the low rate of protein synthesis in these embryos, it seems probable that the synthesis of DNA polymerase is not characteristic of the early sea urchin embryo. One possible explanation for these results would be the localization of the polymerase in the egg nucleus and subsequent partitioning with each division. The egg nucleus would have to accomodate adequate polymerase to satisfy the requirements of four hundred progeny nuclei. The distribution between nucleate and non-nucleate

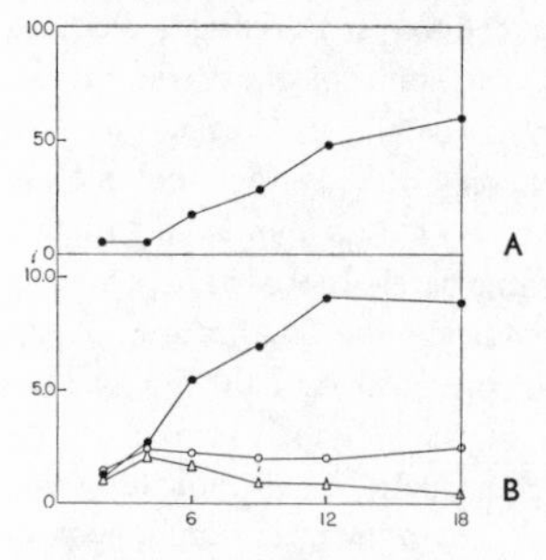

Abscissa: Hours after fertilization; *ordinate:* (*A*) % activity recovered in nuclear fraction; (*B*) polymerase activity mμmoles TM ^{32}P/mg protein.

Fig. 2. Activity of DNA polymerase from various fractions of *S. purpuratus* embryos during early development. The polymerase assay for all samples is that described in Methods for whole embryo sonicate. (*A*) Values represent the percentage of the whole embryo activity which is recovered in isolated nuclei. ●—●, nuclear fraction; ○—○, whole embryo; △—△, cytoplasmic supernatant.

halves does not support this concept. The simplest alternative is an intracellular migration of DNA polymerase from the cytoplasm to the nuclei during early development. In order to explore the latter possibility, samples of embryos were obtained during early development and fractionated into various cellular components to ascertain the distribution of DNA polymerase activity among them. Ideally, it would be desirable to isolate from one batch of embryos a pure cytoplasmic fraction (free of contamination from broken nuclei) and a pure nuclear fraction (free of cytoplasmic contamination). A method to accomplish this was not found. As an alternative, we isolated nuclei and cytoplasm by separate procedures. The cytoplasmic fraction was obtained by gently lysing the embryos with Triton-X, yielding as a by-product a pellet of disrupted cells and nuclei. As discussed in Methods, this cytoplasmic fraction was free of nuclear contamination. The DNA polymerase activity of the various fractions is shown in fig. 2*B*. The level of activity of the embryo fraction changes very little between 4 and 18 h while the soluble cytoplasmic fraction activity decreases about 75 % during the same period. As development proceeds there is an increase in the polymerase activity of the nuclear fraction with the contemporaneous decrease in the cytoplasmic fraction.

Table 4. *Ratio of DNA polymerase activity to DNA in isolated nuclei*

Stage of development	Activity mμmoles/mg DNA
2-cell embryo	35.0
8-cell embryo	39.0
32-cell embryo	31.0
128-cell embryo	25.0
300-cell embryo	13.5
400-cell embryo	11.5

DNA was assayed by the method of Giles & Myers [9].

It is more pertinent to calculate the per cent of DNA polymerase activity of the embryo that is recovered in the nuclear fraction (fig. 2*A*). At the 2-cell stage only 5 % of the total activity is present in isolated nuclei, while at the 400-cell stage (18 h) 60 % is recovered in the nuclear fraction. Furthermore, during early development the extent of localization of the polymerase in the nuclear fraction progressively increases at stages in which the total amount of polymerase activity in the embryo does not significantly change. The per cent of activity in isolated nuclei is probably less than that present in the embryo in vivo; corrections have not been made for recovery of nuclei in the isolation procedure, and localization here can only refer to retention under conditions of isolation. If polymerase activity in isolated nuclei is considered in relationship to the amount of DNA (table 4), the ratio of polymerase activity to DNA is relatively constant until

9 h after fertilization (about 64- to 128-cell stage). This constancy implies a stoichiometry between the amount of nuclear DNA and the DNA polymerase activity associated with it. The decrease in the ratio of polymerase to nuclear DNA after the 128-cell stage appears to be real and could be related to the in vivo decrease in the rate of DNA synthesis at this time.

DISCUSSION

The evidence presented shows that the egg starts with a large amount of DNA polymerase, the localization of which changes with development. It means that polymerase has been synthesized during oogenesis and is stored in the cytoplasm of the egg. This is not surprising, for if requirements exist for preformed enzymes, the egg would be a logical place for their storage.

The change in localization is consistent with the embryo's changing pattern of DNA synthesis. The first DNA synthetic period occurs prior to pronuclear fusion [10]; the DNA of the sperm pronucleus must be replicated in the egg cytoplasm. We have not detected DNA polymerase activity in sperm, thus polymerase must be present in the cytoplasm immediately after fertilization. Further development proceeds by a nearly exponential rate of cell division with accompanying DNA synthesis. From the evidence presented it appears that as the amount of nuclear DNA per embryo increases, DNA polymerase binds quantitatively to the newly formed templates. The ratio of polymerase activity to DNA in isolated nuclei during early development is constant. Nuclei isolated from sea urchin embryos in which tetraploidy has been induced by oestradiol have doubled the DNA polymerase activity of untreated embryos [11]. The complexing of polymerase to DNA could effectively withdraw the enzyme from the cytoplasm into the nucleus. The extent of changing localization is such that by hatching, a time of rapid DNA synthesis, polymerase activity is predominantly nuclear. It may be fortuitous that the rate of cell division markedly decreases after hatching or it could reflect the depleted reserve of polymerase in the cytoplasm. Further cell division would require the production of more polymerase. Evidence indicates that the rate of protein synthesis is increased at this time and that this increase is selective for particular protein species [12]. It is not known if polymerase is one of these proteins.

So far evidence in support of a change in the localization of DNA polymerase must refer only to activity, for this is the only property we have measured. It would be advantageous to equate the observed change in localization with a molecular migration of the polymerase. In a later paper some evidence will be presented to support such a model.

This investigation was supported by grants from the American Cancer Society (E-483) and from the Stanley C. Dordick Foundation. Support was also derived from grants to this Institute: USPHS grants CA-06927 and FR-05539, American Cancer Society grant IN-49, and an appropriation from the Commonwealth of Pennsylvania.

REFERENCES

1. Furlong, N B, Anal biochem 5 (1963) 515.
2. Lehman, I R, Bessman, M J, Simms, E S & Kornberg, A, J biol chem 233 (1958) 163.
3. Harvey, E B, The American *Arbacia* and other sea urchins. Princeton University Press, Princeton, N.J. (1956).
4. Martin, R G & Ames, B N, J biol chem 236 (1961) 1372.
5. Mazia, D, Mitchison, J M, Medina, H & Harris, P J, J biophys biochem cytol 10 (1961) 467.
6. Nemer, M, J biol chem 237 (1962) 143.
7. Hinegardner, R T, Rao, B & Feldman, D E, Exptl cell res 36 (1964) 53.
8. Epel, D, Proc natl acad sci US 57 (1967) 899.
9. Giles, K W & Myers, A, Nature 206 (1965) 93.
10. Simmel, E B & Karnofsky, D A, J biophys biochem cytol 10 (1961) 59.
11. Mazia, D, J cell comp physiol suppl. 1 62 (1963) 123.
12. Gustafson, T & Hasselberg, I, Exptl cell res 2 (1951) 642.

Experimental Morphology of Embryonic Development

GROWTH AND DEVELOPMENT OF THE LABORATORY CULTURED SEA URCHIN[1]

RALPH T. HINEGARDNER

The sea urchin embryo has played a key role in embryological studies for almost a century. Derbès first described the developmental process (though not with complete accuracy) in 1847. In 1876 the eggs were used in Hertwig's experiments demonstrating the role of sperm in fertilization, and in 1891, Driesch used sea urchin eggs in his experiments showing indeterminate cleavage. In recent years the sea urchin egg has become particularly prominent in the study of both the morphology and biochemistry of early development. (See reviews by Gustafson and Wolpert, 1963; Gross, 1967; Davidson, 1968.)

Probably more is now known about the early development of the sea urchin than about any other organism. The reasons for this, other than the fact that early echinoderm and vertebrate development are similar, are primarily technical. The animals are easy to obtain on almost any sea coast, they spawn readily and yield large numbers of eggs. The eggs complete meiosis in the ovaries and therefore can be fertilized immediately after spawning. Fertilization is easily accomplished simply by adding sperm to an egg suspension. Development usually follows with good synchrony and with minimum care. These eggs readily take up a large number of chemicals of biological interest, and they have the further advantage of being small and containing much less yolk than either frog or chicken eggs.

One major disadvantage in the use of sea urchins has been the extreme difficulty in raising the larvae beyond plutei to adults and thus obtaining a second generation. This has seldom been done, and never as a useable laboratory procedure. (See Harvey, 1956 for the relevant references.) If the larvae could confidently be raised, it would then be possible to apply genetic techniques to the study of sea urchin development. This, along with the other advantages these eggs offer, would make them almost ideal material to use in unraveling the developmental process.

This paper reports the first steps in that direction. It is an outline of the techniques for raising sea urchins in the laboratory, and a description of the general features of the developmental stages. The techniques have now been developed to the point where large numbers of urchins can be taken from egg to egg. Though the procedure is not yet as simple as raising *Drosophila,* it is practical.

[1] This research was supported by National Science Foundation grant GB-7984.

The life cycle of the sea urchin can be divided into six more-or-less distinct phases. (1) The fertilized egg, (2) development through blastula and gastrula to pluteus, at which time egg nutrients are usually consumed, (3) growth and development of the feeding pluteus to a mature larva, (4) development of the embryonic urchin inside the growing larva, (5) metamorphosis, and (6) growth of the young urchin to a reproductive adult. Most other Echinoderm groups develop in a similar way.

This sequence is not a smooth continuum with the structures of each stage giving rise to those of the next. The fourth one is particularly unusual. During this period the urchin, or more accurately the ventral half of the urchin, grows almost as a parasite within the larva. Most of the urchin is derived from a combination of the middle portion of the left hydrocoel and the overlying ectoderm (MacBride, 1903). Few, if any, of the larval organs give rise to comparable organs in the adult. The urchin develops its own mouth and anus, most if not all, of its internal organs, spines, *etc.* The larva comes close to being little more than a source of nutrient and protection. A similar type of embryological development is found in the insects and Nemertines. In both of these there are imaginal discs which give rise to portions of the adult. These discs also have close to an independent existence. The Echinoids differ, however, since one equivalent to a disc ultimately gives rise to the whole urchin.

The larvae

Culture Methods: The larvae of *Arbacia punctulata, Lytechinus pictus,* (Pacific Bio-Marine Supply Co., P. O. Box 536, Venice, California 90291), *Lytechinus variagatus,* (Gulf Specimen Co., Panacea, Florida 32346), *Strongylocentrotus purpuratus* and *Echinometra mathaei,* have all been raised. If several conditions are met, any of these larvae can readily be grown to maturity (*i.e.,* up to metamorphosis) in the laboratory. The proper food is most important. Larval concentration and agitation must also be controlled. Both larvae and urchins can be grown in either filtered sea water or a synthetic salt mixture such as *Instant Ocean,* (Aquarium Systems Inc., 1450 E. 289 Street, Wickliffe, Ohio), but growth has been consistently better in sea water.

The type of food organism is critical and out of 14 algal species tried, using either *Arbacia punctulata* or *Lytechinus pictus* as the test organism, only three were found to be satisfactory. These were species of *Dunaliella, Rhodemonas,* and *Pyranimonas.* All three are flagellated algae. The algae that would not serve as food were: *Amphidinium operculayum, Coccolithus huxleyi, Cryptomonas, Cyclotella nana, Cylindrotheca closterium, Eutreptiella, Isochrysis galbana, Melosira nummuloides, Monochrysis lutheri, Nitzschia brevirostris, Phaeodactylum tricornatum.* No diatom has been found that alone could serve as food. All algae were grown in pure culture using half strength Guillard's medium (Guillard and Ryther, 1962). The larvae of *Arbacia* grew well on *Dunaliella tertiolecta. Lytechinus pictus, L. variagatus, S. purpuratus* and *E. mathaei* developed better on a diet of an alga designated 3C by Guillard and tentativelly identified as a species of *Rhodemonas. Lytechinus pictus,* and possibly the other species as well, will also develop on a species of *Pyranimonas* (designated LB 997) that was obtained from Dr. John West, University of California, Berkeley. The concentration of algae used depends on the stage of larval development, with earlier stages being fed much less than older. The larvae were usually fed once a day and given the amount of algae they would consume in 24 hrs or, in the case of young plutei, about 3000 algae per ml. No attempt was made to illuminate the larval cultures or otherwise maintain the algae. Prior to use the algae were centrifuged from their culture media and resuspended in sea water.

If mixtures of algae were used a larger number of species would probably be applicable. The natural conditions in the ocean would also be more closely approximated. However, from an experimental point of view, a single species offers a number of advantages. Maintenance of the algae is easier, feeding is simplified and any nutritional studies or radioactive labeling experiments can be better controlled.

The maximum number of larvae that can be cultured in a given volume of water depends on their stage of development. One individual per milliliter represents a comfortable maximum for mature larvae.

Some form of agitation is usually necessary during larval growth. This prevents a number of potential ills. In still water the plutei of some species tend to remain near the surface and often stick there and die. *Lytechinus pictus* is particularly susceptible to this. About half way through development there is the opposite tendency and the larvae will stay near the bottom. Here they can become trapped in debris. All this can be prevented by some form of gentle stirring. A simple apparatus for doing this is illustrated in Figure 1.

This is essentially a magnetic stirrer, but instead of placing the stirring magnets above the drive magnet, they are placed laterally. This permits a large number of stirrers to be driven simultaneously by one motor. The apparatus illustrated in Figure 1 is 20 inches high and 18 inches in diameter. The bottom shelf rests on ball bearings which allow the shelves to be rotated. Three 6 by $\frac{3}{4}$ by $\frac{1}{4}$ inch Alnico V magnets are mounted on a center shaft that is driven by a 25 RPM motor. The larvae are grown in $3\frac{3}{4}$ inch diameter by $2\frac{3}{4}$ inch polystyrene dishes (no. 42F, Tristate Plastic Molding Company, Henderson, Kentucky). Two holes approximately $\frac{3}{32}$ of an inch in diameter are drilled in the lids of the dishes, one near the edge, the other in the center. Both serve to ventilate the culture. The center hole also holds the axle of the stirrer.

An assembled stirrer is illustrated in the insert at the bottom of Figure 1. The stirrer floats inside the culture dish and consists of a small circular ferrite magnet $\frac{1}{8}$ by $\frac{3}{8}$ inches in diameter (other small magnets can work equally well) cemented inside two lids of 35 × 10 mm polystyrene petri dishes (Falcon Plastics Company, Rochester, New York). A piece of monofilament nylon fish line (approximately 20 pound test) is cemented in the top to serve as an axle. Two paddles are attached to the bottom. The particular paddle design shown, tends to minimize contact between larvae and paddle. The paddles should have a total area of about 2 cm^2 for satisfactory operation.

An alternative method is to agitate the culture by gently bubbling air through it, but this generally slows development and reduces the length of larval spines. Of the species that have been raised, *Arbacia* is the only one that grows well without agitation. In fact, it can even be raised in test tubes.

FIGURE 1. Culture apparatus, with culture dishes. Insert shows construction of the magnetic stirrer. See text for description.

The five species that have been worked with are not all equally adaptable to laboratory conditions. *Strongylocentrotus purpuratus* requires a temperature of 15° C or lower and therefore necessitates a cold room or water bath. *Arbacia, L. pictus, L. variagatus* and *E. mathaei* will all grow at room temperature (22–24° C). *Lytechinus variagatus* is the least hardy of the four. *Arbacia* is probably the easiest to raise, but the young urchins tend to hold on to any surface with tenacity and are difficult to transfer without breaking off their tube feet. The urchins are also less hardy than their larvae. The larvae of *E. mathaei* are smaller than those of the other species and this creates some difficulties, particularly in handling. The young urchin is also difficult to raise. Therefore most of what will be reported here will be results obtained with *L. pictus.* These larvae are not difficult to raise and the young urchins grow well under laboratory conditions.

Larval Development: The eggs of *L. pictus* are obtained by injection of 0.1–0.2 ml of a fresh 0.1 molar acetylcholine-sea water solution (Hinegardner, 1967).

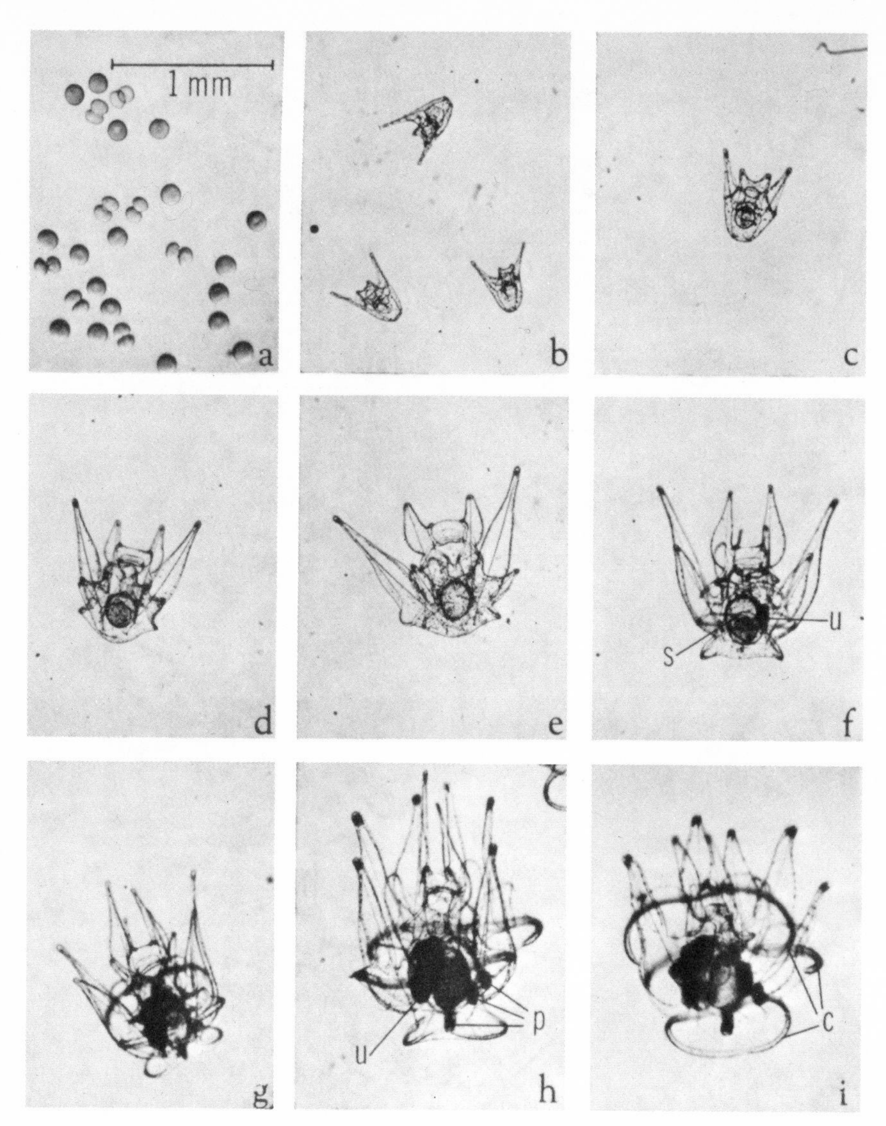

FIGURE 2. Development of the larva of *Lytechinus pictus* from fertilized egg to maturity at 24° C; (a) One and two cell embryos, (b) Pluteus, 2 days old, (c) 4 days, (d) 7 days, (e) 8 days, (f) 11 days (because of orientation the developing urchin appears on the right side), (g) 13 days, (h) 19 days, (i) 26 days. Abbreviations are: c—ciliary bands or epaulets, p—pedicillaria, s—stomach, u—urchin. All pictures are at the same magnification.

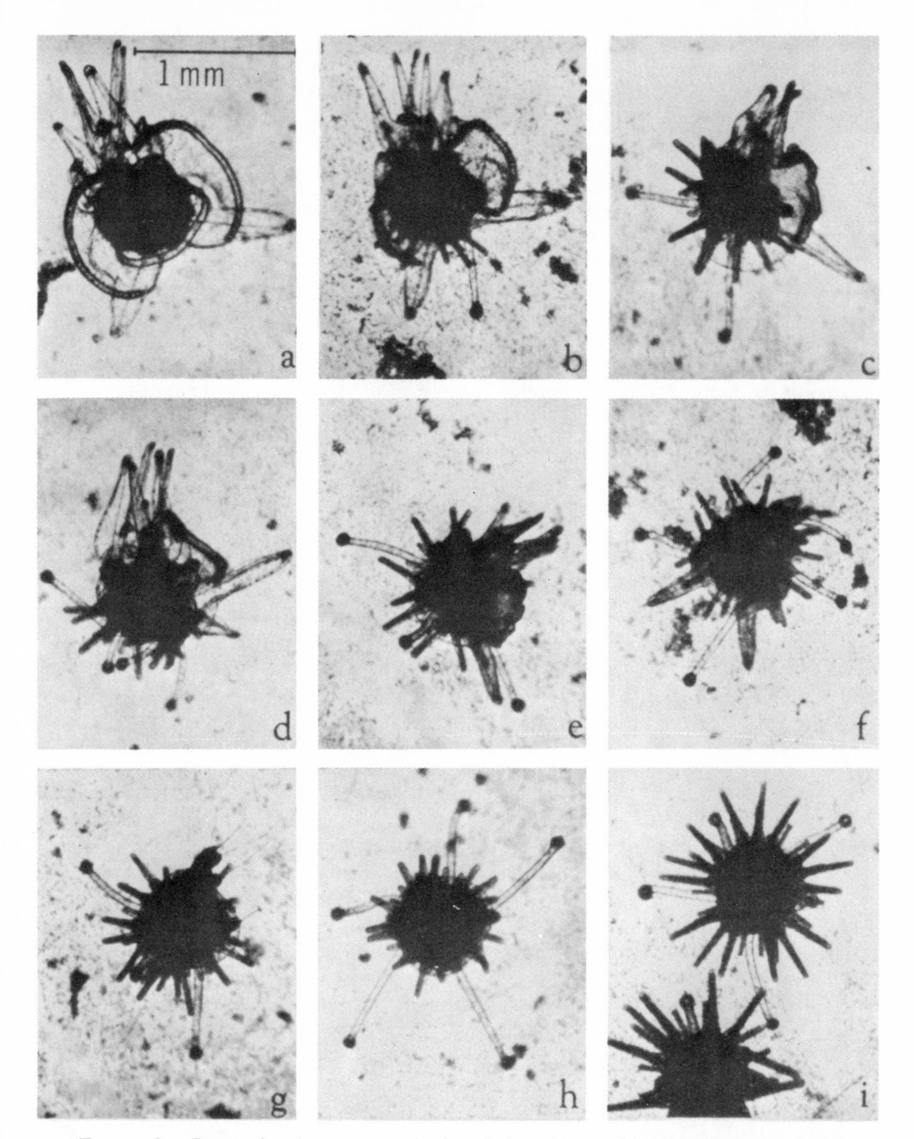

FIGURE 3. Stages in the metamorphosis of *Lytechinus pictus* larvae; (a) 1 minute after adding to the appropriate substrate, (b) 6 minutes, (c) 9 minutes, (d) 11 minutes, side view showing relation of larval structures to the emerging urchin, (e) 12 minutes, (f) 15 minutes, (g) 37 minutes, larval spines are visible in upper right, (h) 80 minutes, (i) 27 hours. All pictures are at the same magnification.

The adult urchins are more apt to survive after this treatment than after the usual 0.5 molar KCl injection or electrical stimulation. The eggs are fertilized and developed at room temperature to early pluteus in the usual way (Costello *et al.,* 1957; Tyler and Tyler, 1966; Hinegardner, 1967). In about two days the plutei are able to feed and, if conditions are close to optimum, they will reach maturity in a month or less.

Figure 2 illustrates larvae at various stages of development. The major morphological changes occurring during this period, aside from overall increase in size, are: (1) The appearance of four additional arms for a total of eight (compare Fig. 2c with Fig. 2h). (2) The formation of heavy ciliary bands which have been called epauletes, and which are most easily seen in Figure 2i. (3) The differentiation of the left hydrocoel and overlying ectoderm into the urchin tube feet primordia (Fig. 4a-c). This is followed by (4) the appearance of tooth and spine primordia and (5) urchin growth and development, primarily of the ventral half. Along with this, there is (6) the development of the three pedicellariae, one posterior and two on the right side of the larva (Fig. 2h). Finally (7) there is metamorphosis of the larva into a small urchin (Fig. 3). The development of the dorsal half and adult internal organs then begins.

Morphological details of the various stages up to metamorphosis have been described by MacBride (1903) for *Echinus esculentus.* A general description of larval development is also given in Kumé and Dan (1968).

Providing the larvae are given resonable care, about 80% of the plutei can be raised to mature larvae. It is difficult to determine what fraction of the non-survivors died from genetic or congential defects, but it would seem that the survival rate is close to maximum.

Metamorphosis

Loss of larval form. The physiological aspects of metamorphosis are not yet well undertood, but the visible changes that occur during this process have been followed in detail and they are illustrated in Figure 3. The first event is the settling of the larva on its left side on to an appropriate substrate. A surface covered with a mixture of algae and bacteria can induce this response. Within a minute, the anterior portion of the larva, and the arms, flex sharply toward the larva's right side (compare Fig. 3a to the normal larva in Fig. 2i). This exposes the left side of the larva and the tube feet of the young urchin to the substrate.

The tissue surrounding the urchin is then drawn up and the urchin spines appear. This is accompanied by the collapse of the larval tissue on to the top of the urchin (Fig. 3b-3g). Within about an hour (3g and h), all that remains of the larva is a lump of tissue on top of the young urchin. A few naked larval spines may extend out from it (Fig. 3g) but these are lost in a few hours. Over the next 24 hours the spines of the urchin greatly elongate (Fig. 3i).

Metamorphosis is not an obligatory stage. If the larvae are kept in clean containers they will usually not metamorphose. Instead, they continue to feed, but they grow little if at all. After two months evidence of deterioration becomes apparent and eventually they die. The maximum life span of well fed larvae is probably about four months. Their ability to metamorphose is almost com-

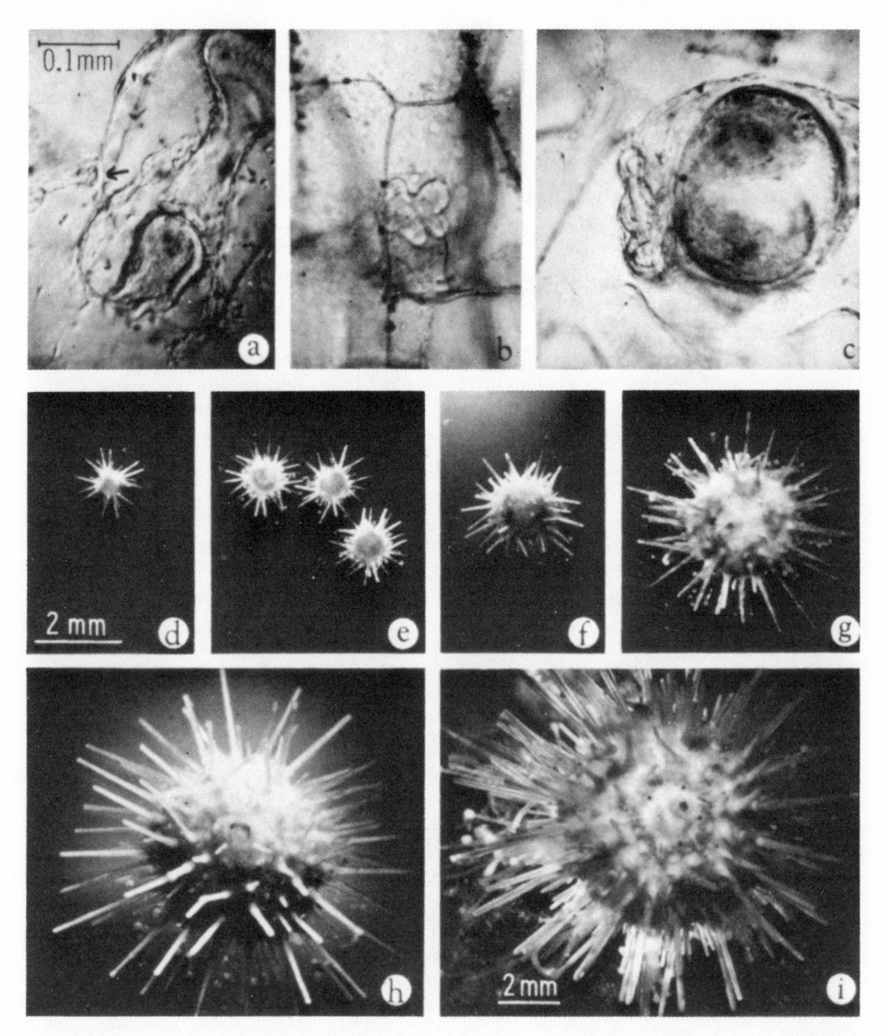

FIGURE 4. Stages in the development of the urchin; (a) 7 day old larva, arrow indicates the invaginating ectoderm; the left hydrocoel is just to the right of arrow and the gut is in the right center of the field; (b) A 9 day larva, showing the beginning of the five tube feet; this larva is rotated 90° to (a) and (c), (c) lateral view of the developing urchin in an 11 day larva; (d) 1 day old urchin; (e) 9 day urchin; (f) 27 day; (g) 71 day; (h) 117 day; (i) 208 day. (a) to (c) are taken through Nomarski optics and are at the same magnification, (d) to (h) are also at the same magnification.

pletely lost at the end of about two months. However if the larvae are underfed, this slows development and the life span can be greatly increased.

The events described here, apply to all the urchin species that have been examined. In a more general way they also apply to other echinoderm classes. Mature starfish and brittle star larvae have been obtained from plankton tows. Their metamorphosis has also been observed and it appears to follow the basic features of the pattern outlined here.

Development of the urchin. Figures 3i and 4d illustrate the young urchin. With one exception, the urchins of all the species that have been examined are very similar. However, the young *Arbacia,* which belongs to a different suborder, looks quite different (see Harvey (1956). It has 15 paddle-like spines and looks almost like a flower. The other species have 20 spines. These are cylindrical and arranged in five groups of four. In all species there are three pedicellaria which developed in the larvae and are now on the urchin's dorsal surface.

The young urchin has neither a mouth nor anus and according to MacBride (1903) no gut either. In terms of formal organs, it is little more than half an urchin; the ventral half. There is half a test, 20 ventral spines, five ventral tube feet and the five teeth. At this stage the dorsal half is essentially a rounded lump of larval tissue punctured by the three pedicellaria. The dorsal organs appear to develop out of this tissue. For the first two days the larval tissue can easily be picked off the urchin. If this is done the urchin will still continue to crawl about for a number of days afterward. The digestive system, and probably other internal organs appear at about four or five days. The urchin then begins to feed. This marks the end of the metamorphic period.

The urchin

Growth: A number of stages during the growth of *Lytechinus pictus* are shown in Figure 4. The young urchin begins to grow after it is 8–10 days old. Along with size increase, there is an increase in the number of spines and tube feet. At a shell diameter of 2 mm the madreporite begins to develop. The gonadopores appear at an age of about two months and a shell diameter of 3.2 mm. Gonads also begin to develop at this time, starting as a single lobe near the gonadopore. They grow ventrally from the gonadopores and contain some ripe gametes when the urchin reaches a diameter of 6 mm, and an age of four to five months.

Culture method: The nutritional requirements of the urchin seem to be more complex or more restricted than those of the larva and it has taken more than a year to find an appropriate algal food that could be reliably cultured in the laboratory. At present a nonsterile surface dwelling diatom is used. This has been identified as a species of *Nitzschia.* Whether or not any of the bacteria in the culture are necessary, has not been determined. Cultures of this algae are maintained in Guillard's media.

The urchins are raised in plastic petri dishes (100 or 150 mm × 25 mm). The algae is first grown in these dishes until the bottom is covered (lightly for young urchins, heavier for older urchins), then the medium is diluted 50% with sea water and the urchins introduced. The dishes are kept in an illuminated incubator at 22–24° C. Under these conditions the algae growth tends to counterbalance consumption. Conditions for a balanced ecology are hard to establish and ultimately either the algae is consumed or it grows so thick it begins to die off. In either situation the urchins are transfered to a fresh dish. If the young urchins are properly maintained they increase in diameter at a rate of approximately one millimeter in 18 days.

No real attempt has been made to determine urchin survival rates. This is because most of the urchins now being raised have been subjected to some experimental treatment. However, a rough estimate, based on experience, can be made. Survival depends, among other things, on the control of disease, maintenance of proper feeding conditions and on the particular male and female the gametes come from. Some matings appear to yield a hardier line of urchins than do others. When all the variables are taken into account, including survival of the plutei, somewhere around 50% of the young plutei can probably be grown to mature urchins. As more crosses are made within the laboratory stock, survival should improve.

Since some mature gametes can be obtained from urchins about four or five months old, a second generation can then be started. Therefore, the generation time is six months or less. Of course, no more than a few hundred eggs can be obtained at this stage, but these should be enough for genetic tests, or to establish a particular genetic line. Six months is significantly longer than the generation time of *Drosophila* but not so different from corn or mice, both of which have been used extensively in genetic studies. Therefore, if urchins are susceptible to genetic analysis, it should now be both possible and practical to use genetic techniques in the study of sea urchin development.

I wish to thank Mrs. Saundra Parra for her skilled technical assistance and Dr. R. R. L. Guillard of the Woods Hole Oceanographic Institute for kindly supplying a large number of algal species. I would also like to thank Dr. John West, University of California, Berkeley, for algae identification and for providing several algae, and Dr. Robert Kane, University of Hawaii, for sending specimens of *Echinometra.*

Summary

A method is described for raising sea urchins from egg to egg in the laboratory. The larvae are raised on flagellated marine algae and the young urchins on a substrate-dwelling diatom. The major features in the developmental process are: growth of the larva, development of the urchin inside the larva, metamorphosis and growth of the young urchin to sexual maturity. The entire life cycle takes about six months.

LITERATURE CITED

Costello, D. P., M. E. Davidson, A. Eggers, M. H. Fox and C. Henley, 1957. *Methods for Obtaining and Handling Marine Eggs and Embryos.* Marine Biological Laboratory, Woods Hole, Massachusetts, 247 pp.

Davidson, E., 1968. *Gene Activity in Early Development.* Academic Press, New York, 375 pp.

Derbès, M., 1847. Observations sur le méchanisme et les phénomènes qui accompagnent la formation de l'embryon chez l'oursin comestible. *Ann. Sci. Natur. Zool.*, **8**: 80–98.

Driesch, H., 1891. Entwicklungs-mechanische Studien. I. Der werth der beiden ersten Furchungszellen in der Echinodermentwicklung. Experimentelle Erzeugung von Thiel- und Doppelbildungen. *Z. Wiss. Zool.*, **53**: 160–178.

Gross, P., 1967. The control of protein synthesis in embryonic development and differentiation, pp. 1–46. *In:* A. Monroy and A. A. Moscona, Eds., *Current Topics in Developmental Biology*. Academic Press, New York.

Guillard, R. R. L., and J. H. Ryther, 1963. Studies on marine planktonic diatomes. I. *Cyclotella nana* Hustedt and *Detonula confervacea* (Cleve) Gran. *Can. J. Microbiol.,* **8**: 229–239.

Gustafson, T. and L. Wolpert, 1963. The cellular basis of morphogenesis and sea urchin development. *Int. Rev. Cytol.,* **15**: 139–214.

Harvey, E. B., 1956. *The American Arbacia and other Sea Urchins.* Princeton University Press, Princeton, New Jersey, 298 pp.

Hertwig, O., 1876. Beiträge zur Kentniss der Bildung, Befruchtung und Theilung des thierischen Eies. *Morphol. Jahr.,* **1**: 347–434.

Hinegardner, R. T., 1967. Echinoderms, pp. 139–155. *In:* F. H. Wilt and N. K. Wessells, Eds., *Methods in Developmental Biology.* Thomas Y. Crowell Co., New York.

Kumé, M. and K. Dan, 1968. *Invertebrate Embryology.* Nolit Publishing House, Belgrade, Yugoslavia, 605 pp.

MacBride, E. W., 1903. The development of *Echinus esculentus,* together with some points in the development of *E. miliaris* and *E. acutus. Phil. Trans. Roy. Soc. London, Series B,* **195**: 285–327.

Tyler, A. and B. S. Tyler, 1966. The gametes; some procedures and properties, pp. 639–682. *In:* R. Boolootian Ed., *Physiology of Echinodermata.* Interscience Publishers, New York.

OOCYTE DIFFERENTIATION IN THE SEA URCHIN, *ARBACIA PUNCTULATA*, WITH PARTICULAR REFERENCE TO THE ORIGIN OF CORTICAL GRANULES AND THEIR PARTICIPATION IN THE CORTICAL REACTION

EVERETT ANDERSON

INTRODUCTION

The eggs of echinoderms have claimed the attention of many cell biologists. Whereas investigators have recognized the unique value of eggs from organisms within the entire phylum, it seems that the eggs of the Echinoidea, the class to which sea urchins belong, possess characteristics that are particularly useful for obtaining answers to specific questions. Harvey (38) states: "the experimental work on sea urchin eggs has included every line of approach, cytology, embryology, physiology and biochemistry, and has been concerned with the solution of many fundamental problems." While a great deal has been learned about the eggs of organisms throughout the animal kingdom, much cytological information is still needed in an effort to understand further certain events during and immediately after oogenesis. For example, as oocytes of many organisms develop there appears, within the ooplasm, a population of bodies of varied sizes and internal configurations. Initially these structures are randomly distributed; however, as the oocyte approaches maturity they come to lie in the peripheral ooplasm. Because of the position they occupy in the mature egg, these structures have been called cortical granules both in invertebrates and in vertebrates. In the piscine egg they have been referred to as cortical alveolae (see reference 87).

Harvey (39) was the first to call our attention to the cortical granules of the eggs of *Arbacia*. The granules were later described in cytological preparations by Hendee (40), and they have since been a topic of interest and controversy among cytologists, embryologists, and physiologists. The function of cortical granules in oocytes of some organisms remains unexplained (7, 36, 42). However, it is the consensus that the cortical granules of oocytes of a wide variety of organisms are involved in the initial phase of the multistep phenomenon of fertilization (see references 2, 4, 11, 32, 33, 76, 87).

Notwithstanding the numerous papers dealing with these structures, our knowledge concerning the origin of cortical granules is somewhat nebulous and fragmentary. In our efforts to understand those structural changes occurring during differentiation of oocytes and associated structures, we have found it desirable to inquire further into the genesis and fate of cortical granules in *Arbacia*. Therefore this paper explores the origin and ultrastructure of these ooplasmic structures and their participation in the cortical reaction.

MATERIALS AND METHODS

The principal organism used in this study was the sea urchin, *Arbacia punctulata*. For comparison, more limited observations were made on the origin of cortical granules in the following species of echinoderms: *Asterias forbesi* (starfish), *Ophioderma bievispinum* (brittle star), and the sea urchins *Echinarachnius parma* and *Strongylocentrotus purpuratus*. All of the forementioned organisms were obtained from the Marine Biological Laboratory, Woods Hole, Massachusetts during the months of June and July, with the exception of *Strongylocentrotus purpuratus*, which was procured from the Pacific Bio-Marine Company, Venice, California, during the months of March, April, and May.

For light microscopic analysis, ovarian tissue was fixed in the following fixatives: Ammerman's, Carnoy's, Champy's, and 10% buffered (pH 7.4) formalin. After dehydration, infiltration, and embedding in Paraplast, sections were made of the Ammerman's and Champy's fixed material and stained with Heidenhain's iron hematoxylin, or Mallory's triple stain. Paraplast sections made of tissue fixed in Carnoy's were stained by the periodic acid–Schiff technique, toluidine blue, bromphenol blue, and Alcian blue (41). Tissue fixed in buffered formalin was stained with Perl's Prussian blue for the demonstration of iron (15).

For electron microscopy, tissue was fixed for 2 hr in a 2% solution of glutaraldehyde (pH 7.4) in seawater, or the paraformaldehyde-glutaraldehyde (pH 7.4) mixture recommended by Karnovsky (46, 74). After fixation the tissue was washed in seawater and postfixed in a 1% solution of osmium tetroxide dissolved in seawater. Rapid dehydration of the tissue through graded concentrations of ethanol was followed by infiltration and embedding in Epon (52). 1 μ sections of the Epon-embedded material were stained according to the recommendation of Ito and Winchester (43). Thin sections were stained with uranyl acetate, followed by the lead citrate stain of Venable and Coggeshall (83), and examined with a Philips 200 electron microscope.

For a study on the participation of the cortical granules in the cortical reaction, gametes from *Arbacia* were obtained either by the electrical stimulation technique of Harvey (38) or by injecting 0.5 cc of isosmotic 0.5 M KCl into the lantern coelomic cavity (80). Sperm and egg suspensions were fixed at intervals (3 sec–60 min) after insemination at 18°–20°C. Some of the fertilized eggs were permitted to develop to the two cell stage for study of the spatial relation and configuration of the so-called hyaline layer. The fertilized eggs and the two-cell stages of the embryo were fixed and processed for electron microscopy as outlined above.

OBSERVATIONS

Light Microscopy

The ovaries of sea urchins are aciniform structures. Histological preparations reveal that the outer limits of each ovary consist of a single layer of flat epithelial cells that rest on a basement lamina (*a*, Fig. 1). There are two layers of collagenous connective tissue (84, 85). One of these subtends the basement lamina of the epithelial cells; the other is adjacent to the developing oocytes (*OC*, Fig. 1). These layers of connective tissue are separated from each other by a stratum of smooth muscle fibers (*b*, Fig. 1).

The inner layer of the ovary has been called the germinal epithelium, a terminology about which there exists some controversy. The polemics center around the following questions: (*a*) do oogonia develop in the outer or inner epithelial layer, or (*b*) do they develop in some other organ (19, 20, 75, 79)? For lack of unequivocal information in the

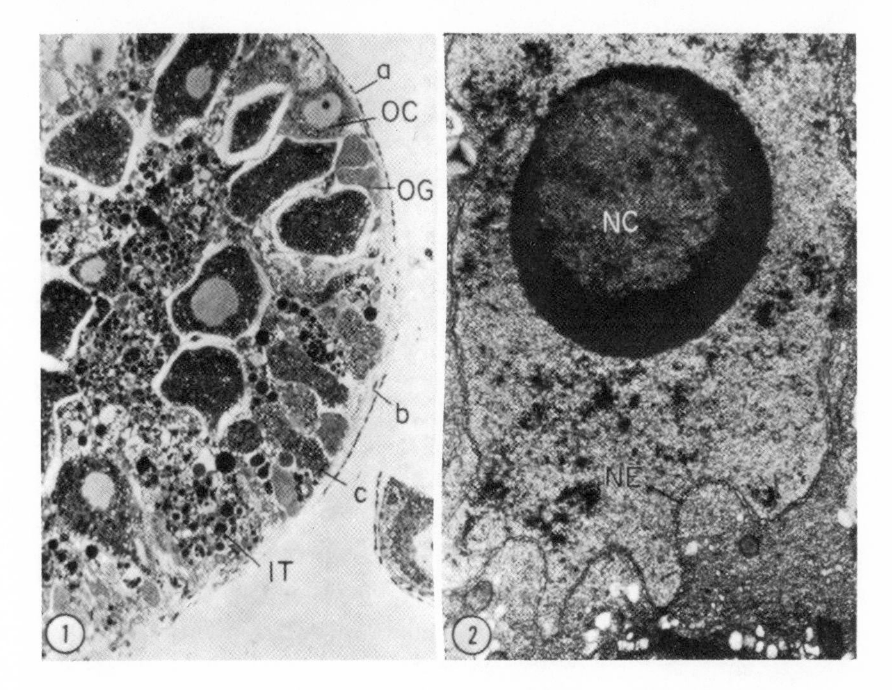

FIGURE 1 A photomicrograph of a portion of an acinus: *a*, outer epithelial layer; *b*, smooth muscle fibers and connective tissue; *c*, germinal layer comprised of oogonia (*OG*) and oocytes (*OC*). Note the interstitial tissue (*IT*). Epon-embedded, toluidine blue–stained. × 250.

FIGURE 2 An electron micrograph of a young oocyte showing its nucleus with a large nucleolus (*NC*) and nuclear envelope (*NE*). × 10,000.

forms investigated, the inner layer of the ovary will be referred to in this paper as the germinal epithelium.

Fig. 1 is a photomicrograph that depicts the germinal epithelium with oogonia (*OG*) and oocytes (*OC*) in different stages of development. At the time of spawning, the eggs are tightly packed together in the central part of the acinus. In this condition, they assume various shapes: elliptical, polygonal, or triangular. Admixed with the developing oocytes is some interstitial tissue (*IT*, Fig. 1) whose cellular processes are closely applied to the surfaces of oocytes (*P*, Fig. 4). Also found among the oocytes and interstitial tissue are various kinds of amebocytes (leukocytes). A large intranuclear crystalloid is frequently found in the amebocytes of *Arbacia*. The present study confirms the observations by Karasaki (45) with respect to the structure and staining qualities of the intranuclear crystalloid.

During the maturation of oogonia, an abundant supply of yolk and echinochrome pigment is formed within the ooplasm. A detailed description of how these deutoplasmic elements develop will not be presented here. Of special interest in the context of this paper, is the appearance of a population of certain spherical bodies within the ooplasm of young oocytes, which occurred before the advent of yolk or pigment bodies. These structures are initially randomly scattered within the ooplasm; however, in the mature egg they are peripherally situated. They are PAS-positive, are metachromatic after staining with toluidine blue, and give a positive reaction for Alcian blue. The

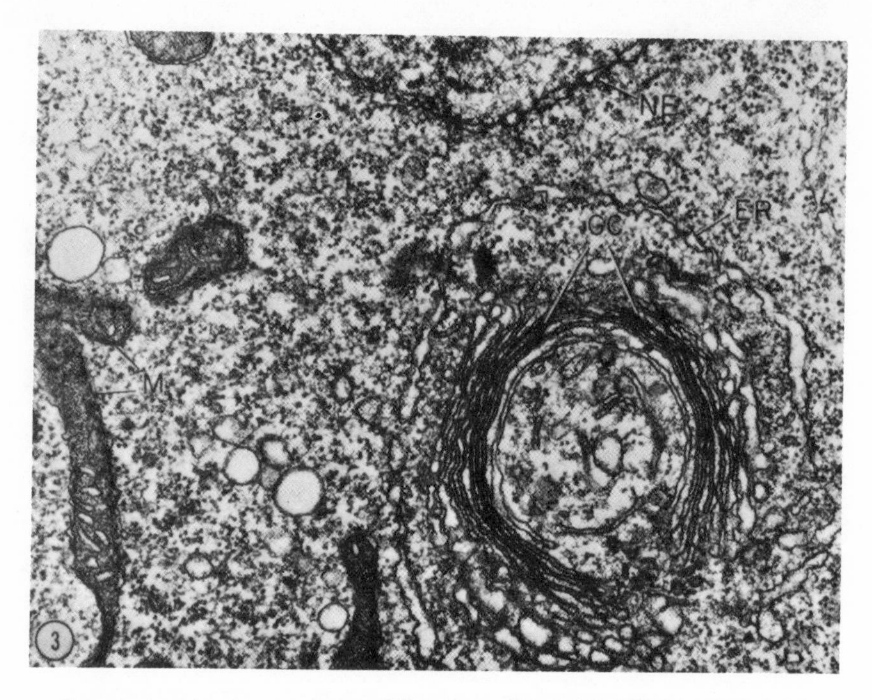

FIGURE 3 A portion of a young oocyte and illustrating nuclear envelope (*NE*), large Golgi complex (*GC*), mitochondria (*M*), and endoplasmic reticulum (*ER*). × 28,000.

aforementioned tinctorial properties indicate the presence of acid mucopolysaccharide. These bodies also give a positive reaction when stained with bromphenol blue, denoting the presence of basic protein, and they are interpreted as cortical granules. A similar reaction was obtained for the primary (vitelline) envelope. Superficial to the primary coat is a jelly-like substance which dissolves when it is in contact with seawater. This material was preserved in neither ovarian nor shed eggs prepared for electron microscopy.

Ultrastructure

OOGONIUM, OOCYTE, AND MATURE EGG: The nucleus of an oogonium is organized like that of early and late oocytes, i.e. it consists of a large bipartite nucleolus (*NC*, Fig. 2) surrounded by a granular nucleoplasm that is limited by a perforated nuclear envelope (*NE*, Fig. 2). In *Arbacia* and presumably in other sea urchins, meïosis of the egg nucleus occurs within the ovary (38, 82). The haploid pronucleus of the mature egg is rather small (Fig. 9) when compared with the diploid nucleus of young and late oocytes.

Located in an adnuclear position within oogonia and very young oocytes is a large Golgi complex (*GC*, Fig. 3). The ooplasm of oogonia contains a host of ribosomal particles and few mitochondria. The rough endoplasmic reticulum makes its initial appearance in very young oocytes (*ER*, Figs. 3–5). As differentiation proceeds, there is an increase in quantity of the previously mentioned organelles, the appearance of the so-called heavy bodies (*HB*, Fig. 9) (see reference 3), and annulate lamellae (*AL*, Fig. 15). In addition to these organelles and inclusions, numerous vesicular bodies also appear within the ooplasm. The interiors of these bodies contain some rodlike structures (*VR*, Figs. 13, 14, 16–18).

CORTICAL GRANULES: Each Golgi com-

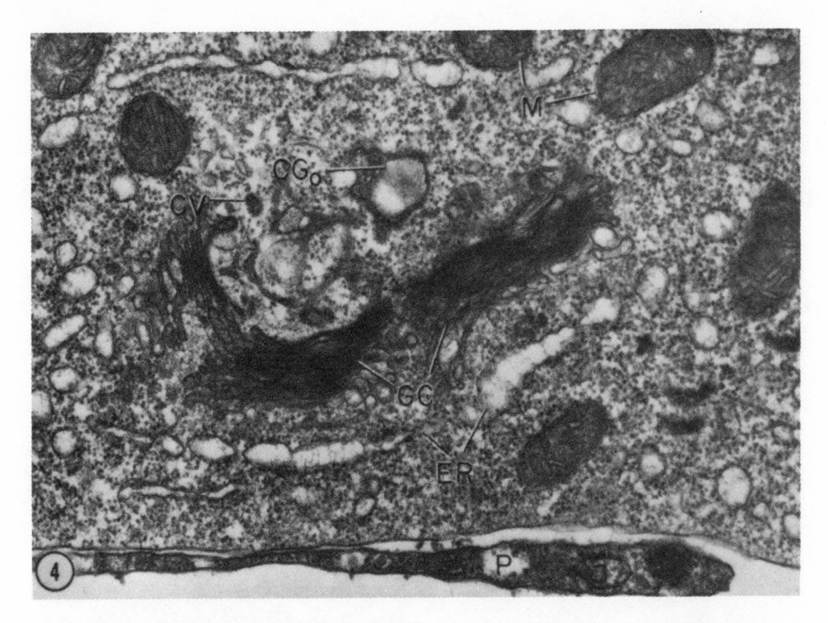

FIGURE 4 An electron micrograph of a small portion of a young oocyte. *GC*, Golgi complexes; *CV*, a coated vesicle; CG_o, a presumptive cortical granule; *ER*, endoplasmic reticulum; *M*, mitochondria. Note the cytoplasmic projection (*P*) of an interstitial cell closely applied to the nonmorphologically specialized oolemma of a young oocyte. × 28,000.

plex consists of a number of saccules arranged in parallel array. Many of the saccules contain a relatively dense substance (*GC*, Figs. 4, 5). The surfaces of the dilated-tip saccules possess a fuzzy coat. In the vicinity of the saccules are mitochondria, cisternae of the rough endoplasmic reticulum, and a host of vesicles of varied sizes and shapes. The vesicles contain a substance whose density is similar to that of the material within the saccules. Some of these vesicles also possess a fuzzy coat; they are thought to be produced by being pinched from the tips of the saccules of the complex. In Figs. 5 and 6 the structures labeled CG_o, and *CG1* are closely associated with the Golgi complex (*GC*). Structures designated as CG_o are here identified as presumptive cortical granules, while those marked CG_l are defined as miniature cortical granules.

The miniature cortical granule is encompassed by a unit membrane and contains two components of different density and consistency (*CG1*, Figs. 5, 6). The dense component, which is triangular in shape, sometimes appears granular; however, it is often reticular. The less dense, usually ovoid portion commonly consists of a filamentous material. As the oocyte continues to differentiate, the cortical granules become randonly distributed within the ooplasm and increase in number and size (in *Arbacia* this size ranges from 0.5 to 1 μ in diameter). The dense component of the mature cortical granule is stellate and marginated by a variable number of less dense ovoid units. When the oocyte develops into a mature egg, the cortical granules are found primarily in the peripheral ooplasm (*CG*, Fig. 9). The tangential section through the periphery of a mature egg, illustrated as Fig. 7 (*CG*), reveals the spatial distribution of the cortical granules and other ooplasmic components. The unit membrane encompassing the contents of the mature cortical granule (*MC*, Fig. 10) is separated from the unit membrane (*OL*, Fig. 10) of the oolemma (see below) by a space of about 200 A.

The cortical granules of the species listed in the

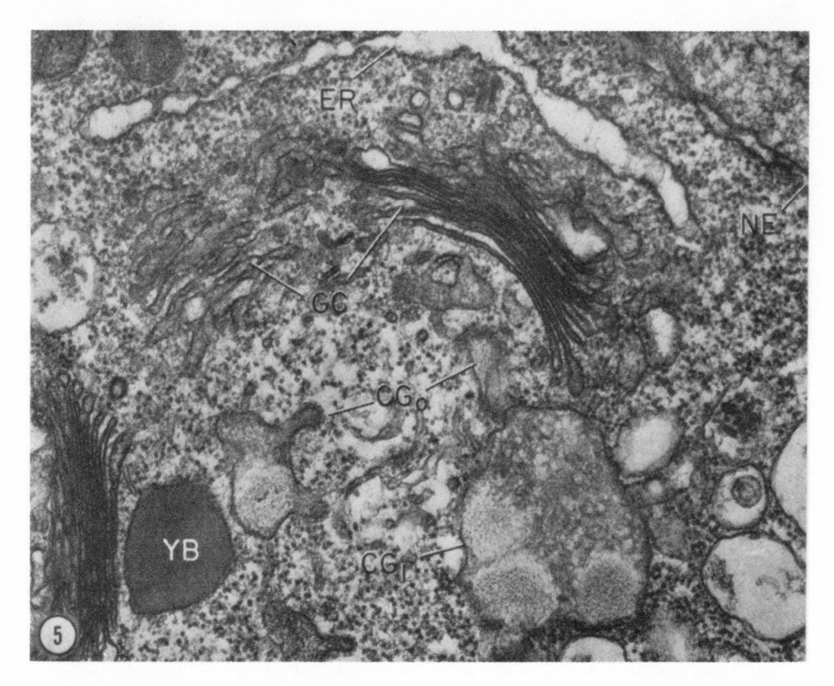

FIGURE 5 Section through a young oocyte which shows nuclear envelope (*NE*), cisternae of endoplasmic reticulum (*ER*), Golgi complex (*GC*), presumptive cortical granules (CG_o), a miniature cortical granule (CG_I), and a yolk body (*YB*). × 40,000.

Materials and Methods section of this paper are formed in a manner similar to that indicated above for *Arbacia*. Although the cortical granules of each investigated species are bounded by unit membranes and contain two structural components, the organization of the internal components varies according to species (see reference 78). For example, Fig. 8 depicts the cortical granule of *Strongylocentrotus purpuratus* which contains a compact filamentous unit that is often eccentrically placed. Emanating from the compact unit (*CS*) are a variable number of lamellar structures (*LS*), each of which is composed of a granular material. Each lamella is associated with its neighbor by fine filaments.

Oolemma

During oocyte differentiation, the oolemma becomes specialized by the formation of microvilli (*MV*, Figs. 9, 12). In *Arbacia*, the microvilli are short and few in number when compared with the large number of rather slender ones produced by the oolemma of the oocytes of the brittle star, *Ophioderma*. It should be pointed out that microvilli are rarely found on those areas of the oolemma overlying a cortical granule in a mature *Arbacia* egg.

Associated with the morphologically specialized surface of the oolemma is the primary (vitelline) envelope composed of a substance which has a matted appearance. In *Arbacia* and *Strongylocentrotus* the primary envelope is produced in scant amounts (*PC*, Fig. 12 inset *c*), whereas in *Asterias* it is produced in rather large quantities.

FERTILIZATION: Inset *a* Fig. 12 is a phase-contrast photomicrograph of a mature living egg. Fig. 9 (inset) shows the initial contact of the sperm with the egg immediately after insemination. After

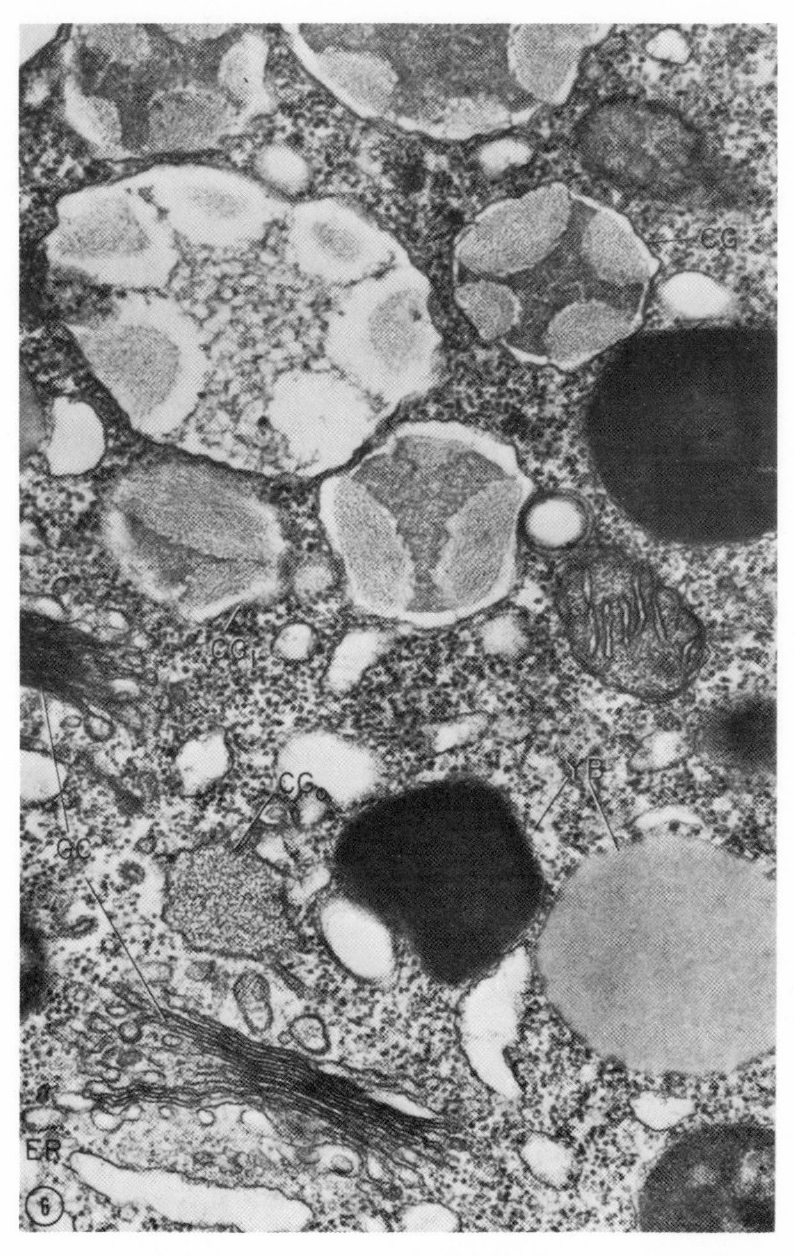

FIGURE 6 A section through the ooplasm of a young oocyte; *GC*, Golgi complex; CG_o, presumptive cortical granules; CG_1, miniature cortical granule, *CG*, cortical granule; *ER*, cisternae of the endoplasmic reticulum; *YB*, yolk bodies. × 42,000.

FIGURE 7 A tangential section through the peripheral ooplasm of a mature egg illustrates cortical granules (*CG*) and mitochondria (*M*). × 42,000.

FIGURE 8 A section of a late oocyte of *Strongylocentrotus purpuratus* illustrates portions of two cortical granules, one of which shows a compact structure (*CS*) associated with lamellar units (*LS*). The lamellar units are associated with each other by fine filaments. Note the unit membrane (*MC*) of the cortical granule. × 90,000.

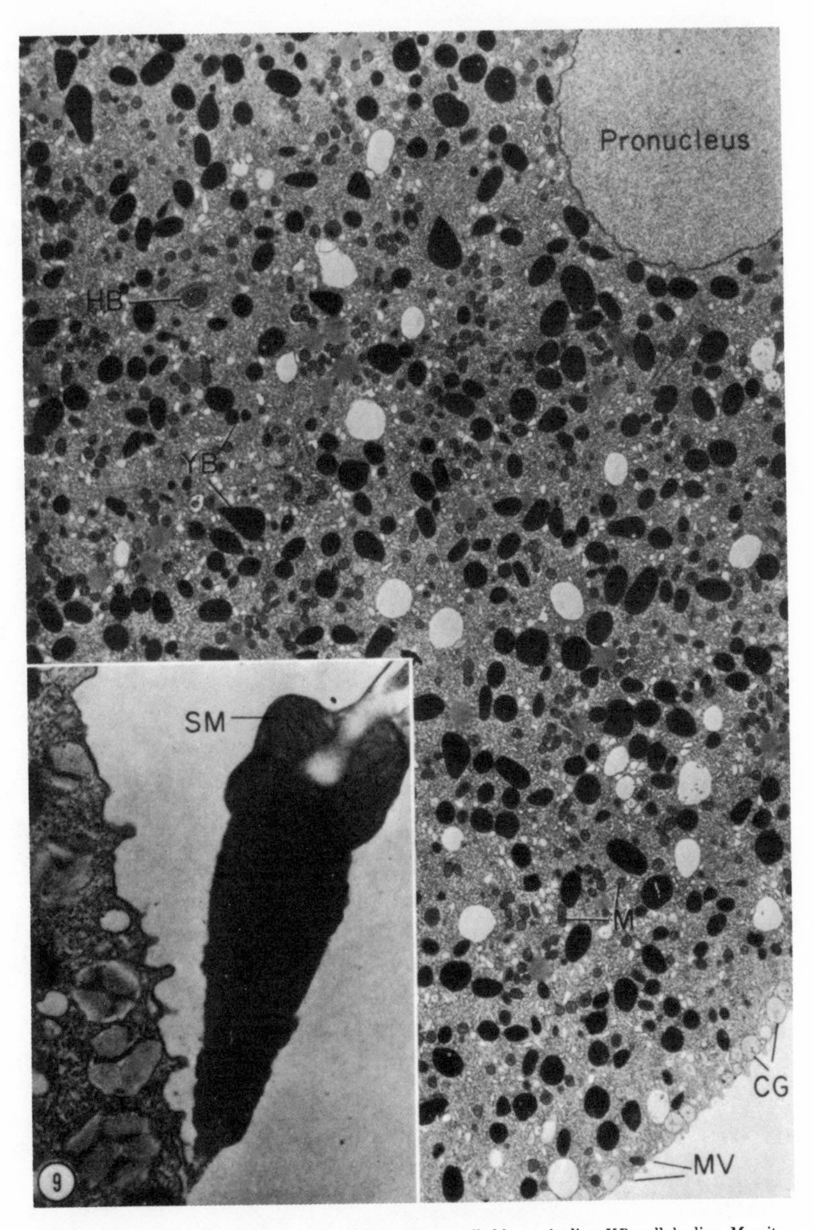

FIGURE 9 A section of a mature egg. Pronucleus; *HB*, so-called heavy bodies; *YB*, yolk bodies; *M*, mitochondria; *CG*, cortical granules; *MV*, microvilli. Inset shows the initial contact of a sperm with the egg's surface. *SM*, sperm mitochondrion. Fig. 9, × 5,000; inset, × 16,000.

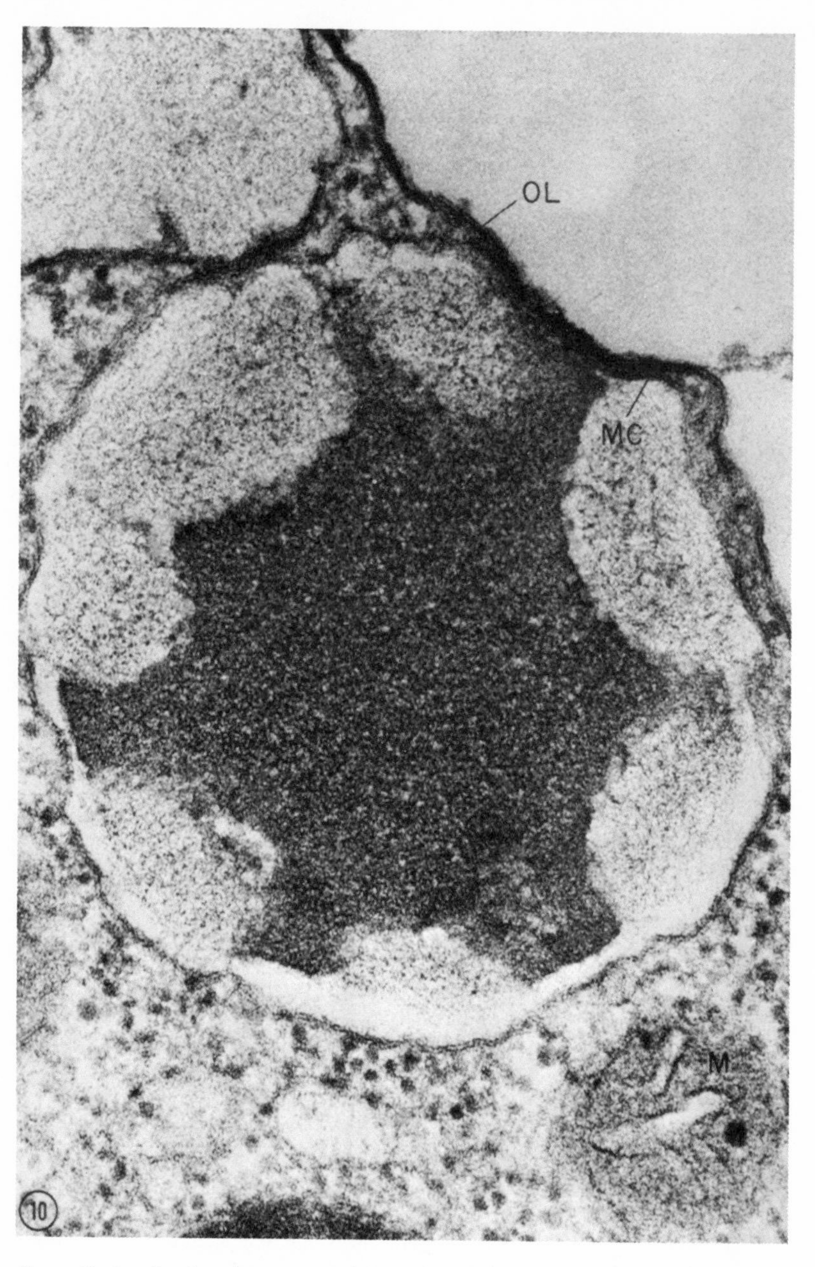

FIGURE 10 A section through a mature cortical granule. *MC*, membrane of cortical granule; *OL*, oolemma; *M*, mitochondrion. × 140,000.

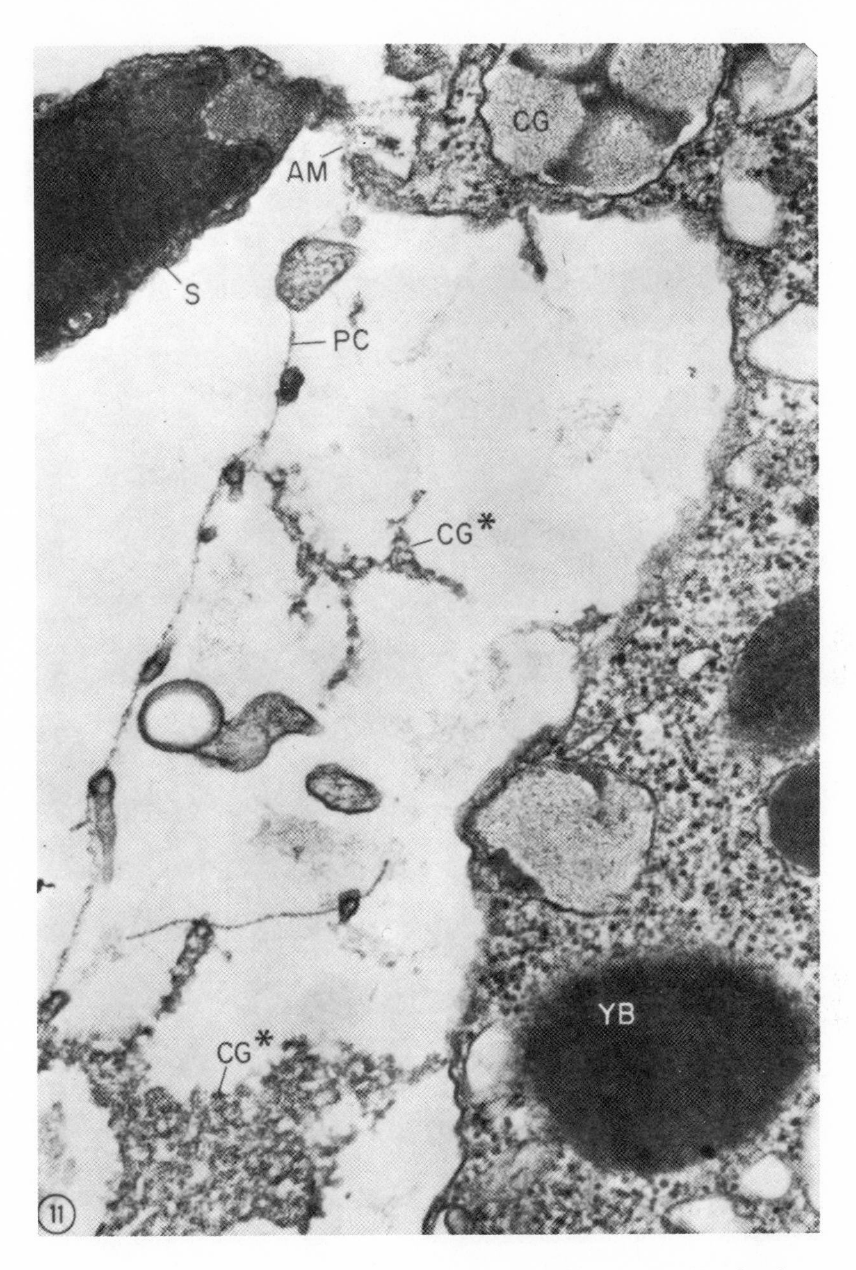

FIGURE 11 A section illustrating the release of acrosomal material (*AM*) by the sperm (*S*) and portions of discharged cortical granules (*CG**). *PC*, primary coat; *CG*, intact cortical granule; *YB*, yolk body. × 42,000.

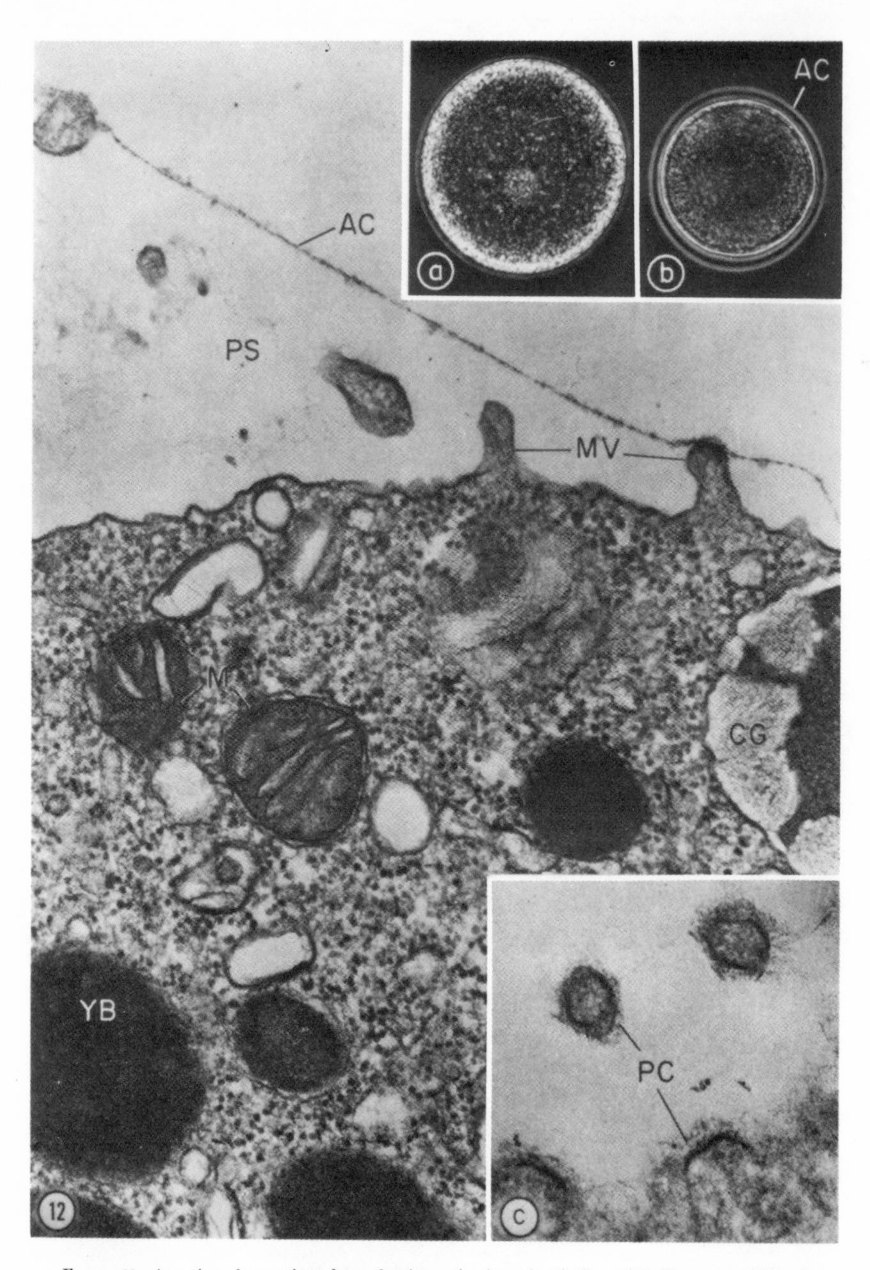

FIGURE 12 A section of an activated egg showing activation calyx (*AC*), perivitelline space (*PS*), microvilli (*MV*), yolk body (*YB*), and a portion of a nonactivated cortical granule (*CG*). Inset *a* is a phase-contrast photomicrograph of a living mature egg. Inset *b* is a phase-contrast photomicrograph of a living fertilized egg showing the complete activation calyx (*AC*). Inset *c* is an electron micrograph of a tangential section of a mature egg illustrating the primary coat (*PC*). *M*, mitochondrion. Fig. 15, × 70,000; inset *a*, × 800; inset *b*, × 700; inset *c*, × 80,000.

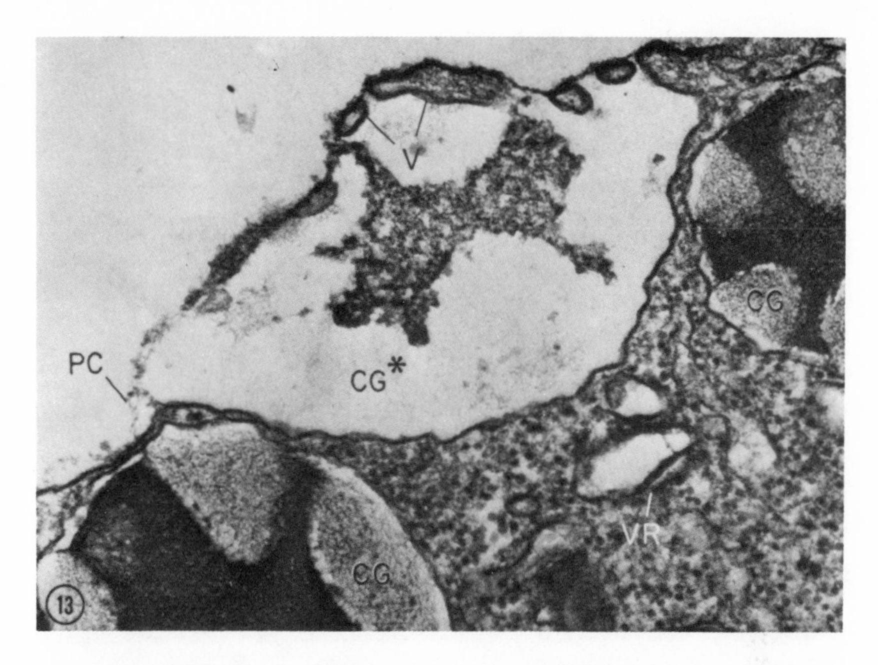

FIGURE 13 A section through a portion of the surface of an activated egg showing the primary coat (*PC*), two intact cortical granules (*CG*) and a partially released one (*CG**). Note the membrane-bounded vesicles (*V*) at the top of the partially released cortical granules and vesicles with rodlike structures (*VR*). × 61,000.

the initial contact of the sperm with the egg, the acrosome undergoes what has been described as the acrosomal reaction (see references 28, 31, 35). Such a reaction was observed in material fixed 3–4 sec after insemination (*AM*, Fig. 11). Shortly after, or at the time of the liberation of the acrosomal substance, there occur (*a*) the formation of the so-called fertilization membrane (see Discussion), (*b*) the release of the contents of cortical granules, and (*c*) the fusion of the sperm with the egg.

FERTILIZATION MEMBRANE: Fig. 12 is an electron micrograph of an egg fixed about 30 sec after insemination. Here the primary envelope becomes disjoined from the surface of the oolemma, thereby forming the fertilization membrane (*AC*). When the membrane is elevated, an area, known as the perivitelline space, is produced between it and the egg (*PS*, Figs. 12, 15–18). The fertilization membrane is not formed over the entire surface of the egg simultaneously. Often, however, it is initiated at the point of sperm-egg fusion and, with time, progresses around the circumference of the egg (see reference 54). About 3 min after union of the gametes the fertilization membrane (*AC*) appears over the entire egg as shown in the phase-contrast photomicrograph featured as inset *b* in Fig. 12. Subsequent to the entrance of the sperm contents into the ooplasm the fertilization membrane becomes thicker. Figs. 17 and 18 (*AC*) illustrate this membrane at 6 min after insemination. The membrane retains this appearance up to about 12 min following insemination (see below).

RELEASE OF THE CONTENTS OF CORTICAL GRANULES: Figs. 13 and 14 show cortical granules in various stages of their dehiscence. The outer portion of the cortical granules show membrane-bounded vesicles (*V*) of varied sizes. Some of these vesicles become closely associated

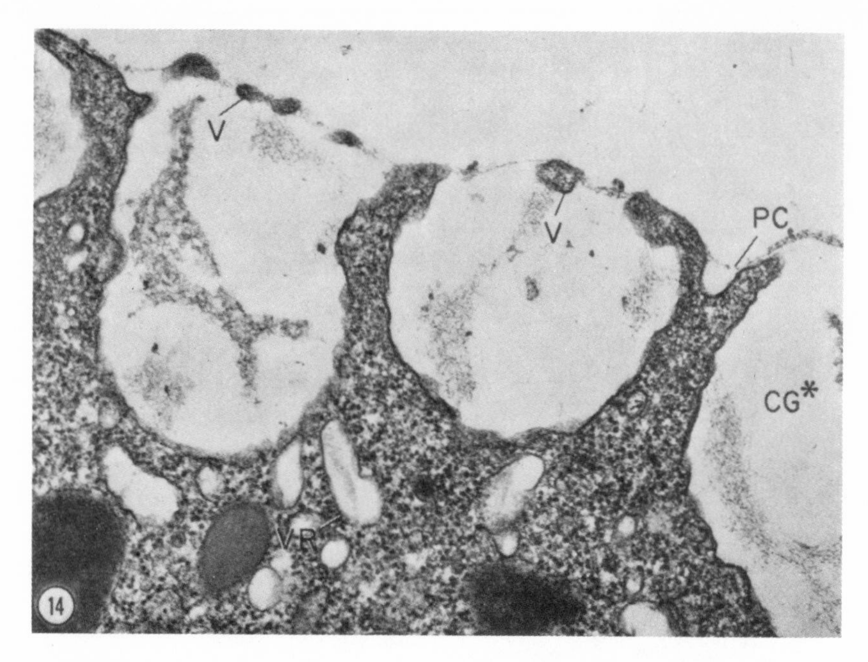

FIGURE 14 A section through an activated egg showing three partially released cortical granules (*CG**). *PC*, primary coat; *V*, membrane-bounded vesicle; *VR*, vesicle containing rodlike structures. × 40,000.

with the inner aspects of the fertilization membrane (*V*, Fig. 16). The primary envelope labeled *PC* in Figs. 13 and 14 was presumably fixed while it was in the process of being detached from the oolemma. Eventually, the contents of the cortical granules (*CG**, Figs. 15, 16) come to lie within the perivitelline space. When the contents of a cortical granule appear within the perivitelline space the major portion of the membrane limiting the cortical granule is now confluent with a portion of the original oolemma. This "new" membrane (see Discussion) possesses some invaginations whose cytoplasmic side is coated (*PV*, Fig. 15 inset). These invaginations are interpreted as initial stages of micropinocytosis. Just prior to the completion of the cortical reaction, i.e. release of the components of the cortical granules into the perivitelline space, the sperm fuses with the egg (*PMS* ↔ *PME*, Fig. 16) (see references 24–26). The egg responds to this fusion by the production of the fertilization cone (*FC*, Fig. 16).

About 7 or 8 min after the initial release of the contents of the cortical granules into the perivitelline space, a stratum is formed known as the hyaline layer (37). Initially, it is composed of what appears to be fine filaments organized in a reticular pattern (*HL*, Figs. 19, 21). During later stages of the maturation of the fertilized egg, the hyaline layer becomes thicker; however, it displays the same structural organization as previously described. Prior to cleavage, the hyaline layer is within the perivitelline space around the fertilized egg. At the two cell stage the hyaline layer follows the contour of the blastomeres; it is not found between blastomeres.

RELEASE OF THE RODLIKE STRUCTURES FROM VESICULAR UNITS: Attention has already been called to the vesicular bodies that contain rodlike structures. During differentiation of the oocyte these structures do not acquire a specific position within the ooplasm; they are randomly distributed. *After* the completion of the cortical reaction many of the vesicular bodies are found closely associated with the new plasma

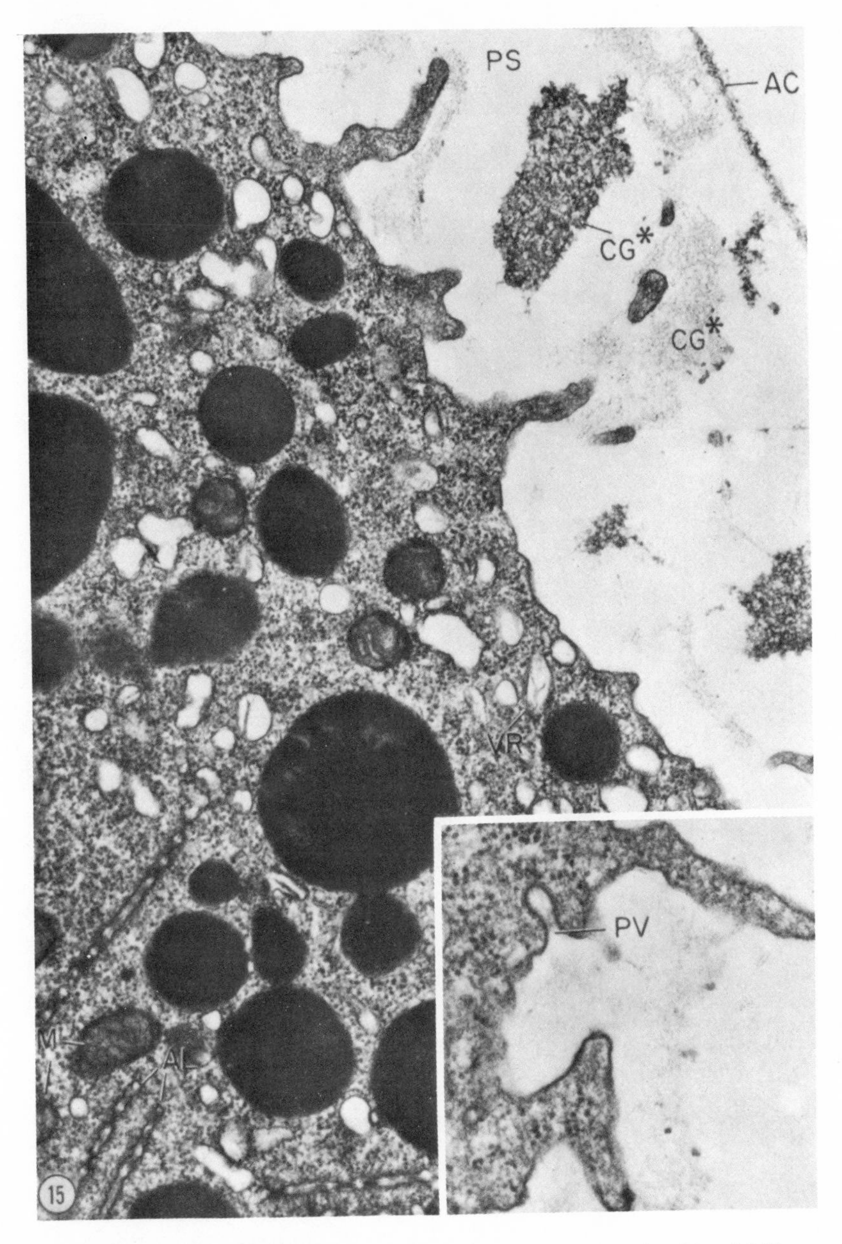

FIGURE 15 A section through the surface of a fertilized egg. *AC*, activation calyx; *PS*, perivitelline space; *CG**, dense and less dense portions of discharged cortical granules; *VR*, vesicles with rodlike structures; *PV*, pinocytotic invagination (see inset); *M*, mitochondria; *AL*, annulate lamellae. Fig. 15, × 27,000; inset, × 42,000.

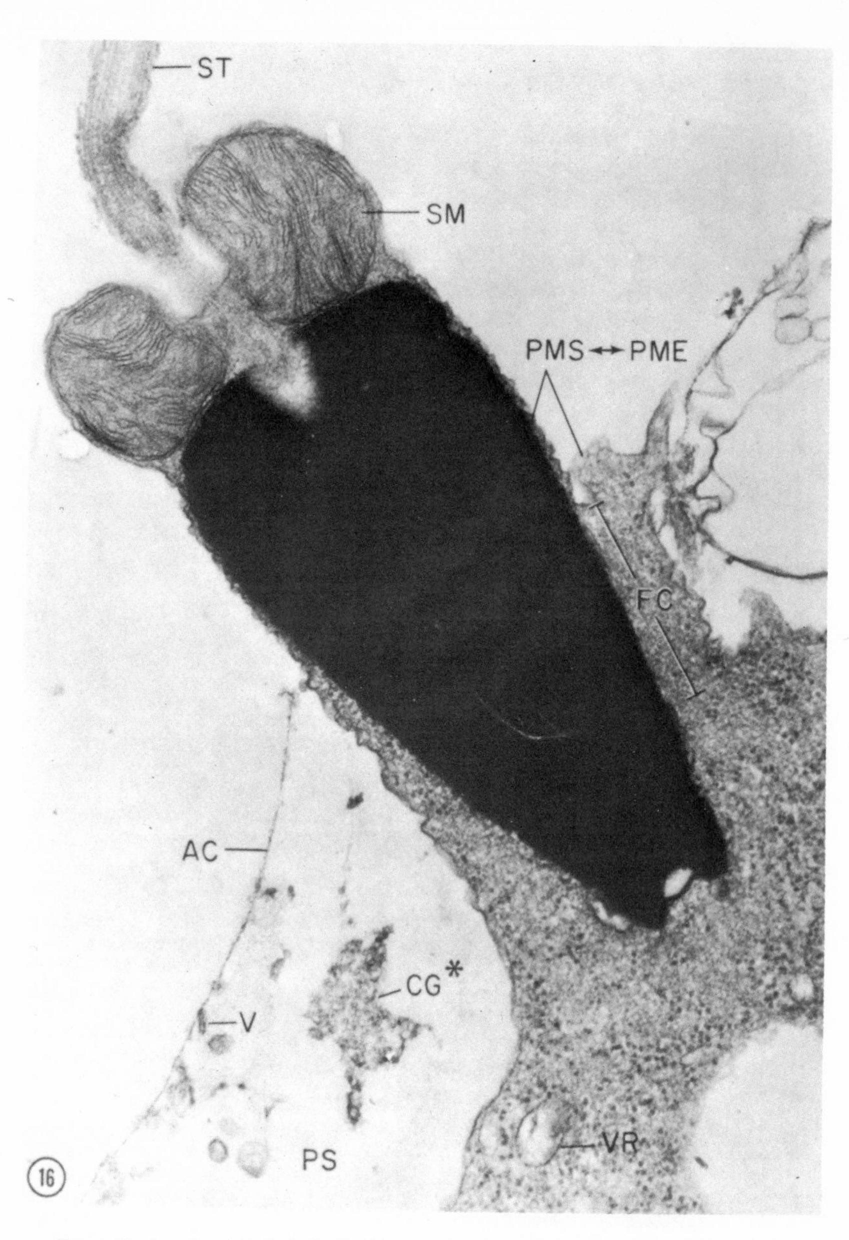

Figure 16 A section through the fertilization cone (*FC*) where the plasma membrane of the activated egg is continuous with that of the sperm (PMS ↔ PME). *ST*, sperm tail; *SM*, sperm mitochondrion; *AC*, activation calyx; *CG**, dense portion of a discharged cortical granule; *V*, membrane-bounded vesicle; *VR*, a vesicle containing rodlike structures; *PS*, perivitelline space. × 42,000.

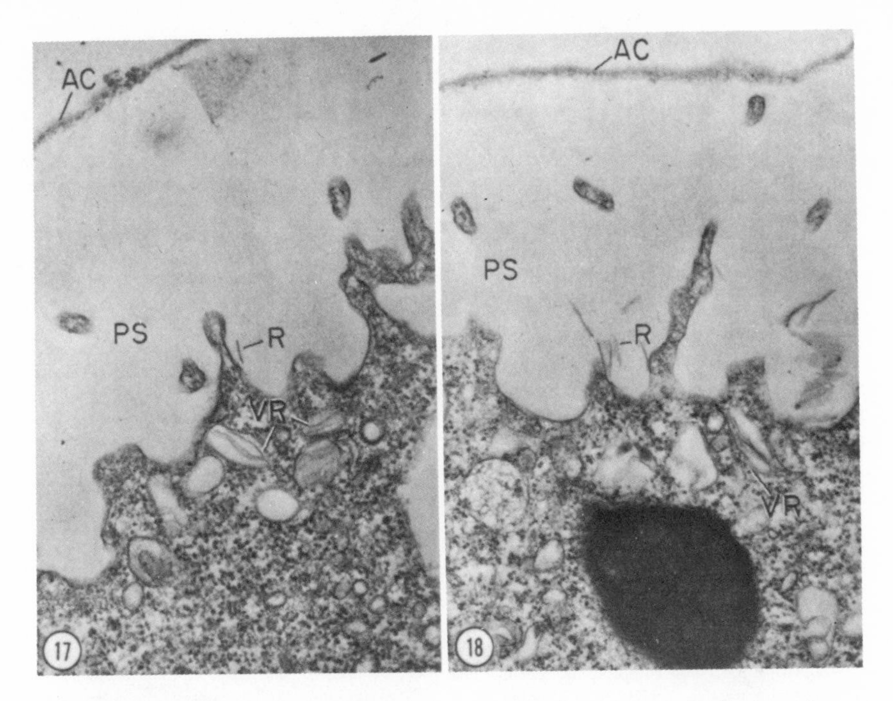

FIGURES 17 and 18 Sections through the periphery of two activated eggs 6 min after fertilization. *AC*, activation calyx; *PS*, perivitelline space; *R*, rods within the perivitelline space; *VR*, vesicles containing rodlike structures. × 42,000.

membrane (*VR*, Figs. 15, 17, 18) of the fertilized egg. In *Arbacia* the rodlike structures display an axial periodicity (*R*, Fig. 20) and were first noted within the perivitelline space 6 min after insemination (*R*, Figs. 17, 18; also see Fig. 19). The rods of *Arbacia* may be equivalent to those of Japanese sea urchins described by Endo (33) and Runnström (72). Presumably the membrane of the vesicles fuses with the membrane of the fertilized egg, and contents of the vesicles are released into the perivitelline space. Extrusion of the rods of these vesicles appears to be akin to the secretory process like that observed in merocrine glands (65). Some of these rods become enmeshed within the constituent (s) of the hyaline layer. The vesicles containing the rods are found not only in oocytes and mature eggs but also within the cytoplasm of blastomeres (two cell stage), and they are apparently released from these cells during this early stage of embryonic differentiation.

CHORION: Fig. 21 is an electron micrograph of the fertilization membrane (*AC*) 12 min (at about 22°C) after insemination. This report is not concerned with the fusion of male and female pronuclei; however, it is important to point out that the illustration presented as Fig. 21 was made from a fertilized egg whose pronuclei were just beginning the process of fusion (Longo, F., and E. Anderson. Unpublished observations). After 14 min the pronuclei have fused and the fertilization membrane appears as two dense lines separated by a homogeneous area (*CH*, Fig. 22). This layer may now be referred to as the chorion (see Discussion). As the zygote (single cell, 2N nucleus) continues to differentiate there is an increase in the thickness of this stratum (*CH*, Fig. 19). This increase is not a simultaneous one, for the stratum appears to thicken in several places. 25–30 min after insemination and during the two cell stage, the chorion (*CH*) appears to be like that shown in Fig. 23. The chorion is a tough layer and is closely associated with the blastomeres (*CH*, Fig. 25); however, it may be easily cut away from the embryo (*CH*, Fig. 24). It is composed of two trilaminar structures (outer *ol*, Fig. 23, and inner *il*) each of which is composed of two 25–30 A dense lines separated by a gap of approximately 50 A. The inner and outer trilaminar structures are separated from each other by a space of about 500–600 A.

DISCUSSION

Origin of Cortical Granules

This study has shown that as oogonia differentiate into oocytes, the first recognizable constituent of the ooplasm is the cortical granule. Staining procedures reveal that these cortical structures contain acid mucopolysaccharide and protein; this finding confirms previous histochemical studies (4, 55, 56). When very young oocytes were examined with the electron microscope the organelle conspicuously associated with cortical granules is the Golgi complex. Each component comprising the Golgi complex, i.e. saccules and companion vesicles, is filled with a rather dense homogeneous substance. A constant feature of the aforementioned organelle is the presence of miniature forms of "mature" cortical granules. Mitochondria are found in the vicinity of the Golgi complex, and a noticeable characteristic is the presence of cisternae of rough endoplasmic reticulum. These morphological observations encourage the suggestion that the Golgi complex plays a major role in the production of cortical granules (8). Here one might think of the genesis of cortical granules as commencing within the vesicular component of the Golgi complex that is presumably derived from the saccules by being pinched from their tips. These vesicles, which are thought to contain the precursor(s), may be viewed as presumptive cortical granules. The presumptive cortical granules subsequently increase in diameter by coalescing with others and thereby produce larger ones. During further differentiation they enlarge and assume a position in the peripheral ooplasm immediately beneath the oolemma.

If the hypothesis set forth is true, namely that the vesicular component of the Golgi complex contains the precursor(s) of the cortical granule, one must assume that either all or a portion of the chemical components are fabricated within the saccules of the Golgi complex. In recent years investigators have concerned themselves with the function of the Golgi complex. In connection with this, Caro (21) and Caro and Palade (22), for example, coupled the techniques of radioautography and electron microscopy to elicit information concerning the function of this organelle in the acinar cells of the pancreas. Their studies revealed that the Golgi complex is a site for protein concentration (also see reference 44). Caro and Palade found that the protein concentrated by the Golgi complex was transferred to this organelle after it had been fabricated within the cisternae of rough endoplasmic reticulum. Utilizing the protocol outlined by Caro and Palade, other investigators have presented evidence that the Golgi complex is not only capable of sequestering protein but is involved in the production of polysaccharides (68, 69). While similar experiments have not been carried out in this study, it does not seem unreasonable to suggest that the protein component of the cortical granule is synthesized by the rough endoplasmic reticulum and transferred to the Golgi complex where it becomes complexed with the polysaccharide fabricated by the Golgi complex. The concept presented above is similar to that reported for the origin of cortical granules in other organisms (1, 12, 15, 47, 77). For different views concerning the origin of cortical granules the reader is referred to references 53 and 73.

Electron micrographs accompanying this paper disclose that vesicles near the saccules of the Golgi complex possess a surface coat. In this connection, it has been demonstrated many times and in a wide variety of cell types that, during the process of micropinocytosis, the pits and invaginations of the plasmalemma which are destined to form vesicles also possess a coat on their cytoplasmic side. The suggestion has been made that the coated pits, routed to become cytoplasmic vesicles, are regions on the plasmalemma that are specialized for the uptake of certain substances, for example protein (5, 71). Moreover, if one examines electron micrographs published by investigators long before attention was called to coated vesicles, one finds coated vesicles commonly associated with the saccules of the Golgi complex. There is relatively little micropinocytotic activity on the oolemma of the

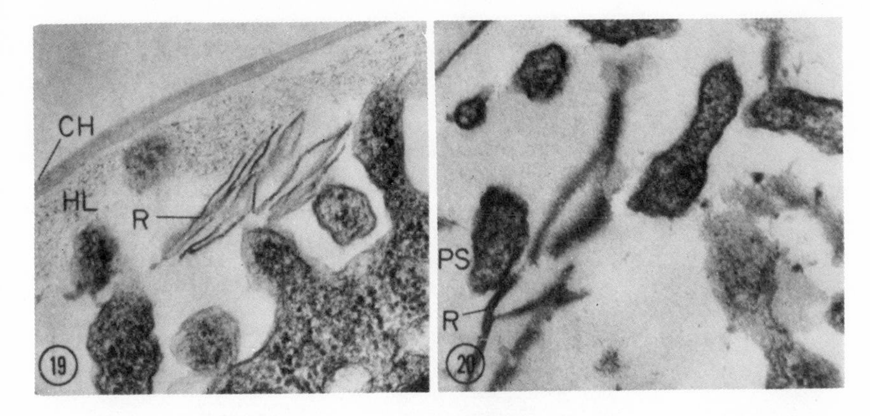

FIGURES 19–20 Small portions of the surface of an egg 30 min after insemination. *CH*, chorion; *HL*, hyaline layer; *R*, rods within the perivitelline space (*PS*). Fig. 19, × 43,000; Fig. 20, × 80,000.

oocytes investigated in this study. Therefore, it is thought that the coated vesicles in the vicinity of the saccules of the Golgi complex do not originate from the oolemma but rather from the Golgi complex. In reference to the latter, Anderson (6) has stated that "it does not seen unreasonable to assume that the flattened sacs of the Golgi complex, after sequestering certain classes of substances, may be induced to undergo evaginations with the production of aveolate and/or nonaveolate vesicles. In other words, *it would appear that the Golgi membranes are selective* and the stimulus(i) and mechanism(s) for the formation of Golgi vesicles may be equivalent to that which operates in the plasma membrane."

Fertilization Membrane

During the differentiation of oocytes of *Arbacia* and other organisms the oolemma acquires a homogeneous, sometimes filamentous coat. This coat in *Arbacia* is composed of acid mucopolysaccharide and is the primary envelope (10). Shortly after the sperm comes into contact with the egg the so-called fertilization membrane is produced (see reference 27). Initially this membrane is none other than the primary envelope. Whether the envelope lifts up from the oolemma or the egg shrinks away from it is obscure. Both interpretations have been offered (57). Of interest is the structure of this layer to which the term membrane has been applied. In reference to the nomenclature "membrane", Bennett (18) has pointed out that the term membrane has been used to indicate many things; for example (*a*) tissue membranes, (*b*) cellular membranes, (*c*) basement membranes (basement laminae), and (*d*) cytoplasmic membranes. Morphologically, the structure that is disassociated from the surface of the egg following sperm activation would be similar to what Bennett has categorized as a basement membrane (basement lamina). Moreover, as a result of a comparative cytological study of material on the surfaces of plasmalemmas, Bennett (17) found that these surface coats are all rich in polysaccharides. He assigned the inclusive term "glycocalyx" to these surface coats, a term that indeed applies to the primary envelope (vitelline envelope) of sea urchin oocytes. It is obvious that the structure of the layer that initially defines the outer limits of the perivitelline space does not conform to the configuration of a unit membrane (70). While the terminology "fertilization membrane" is venerable, the appellation "membrane" does not, morphologically, identify it properly. Since activation is one of the initial phases in the multistep phenomenon of fertilization and since the resulting disjoined structure was the original glycocalyx, it is therefore proposed that the descriptive name "activation calyx" be given to what was formerly called the fertilization membrane.

It is well known that the activation calyx may appear not only by sperm activation but also by

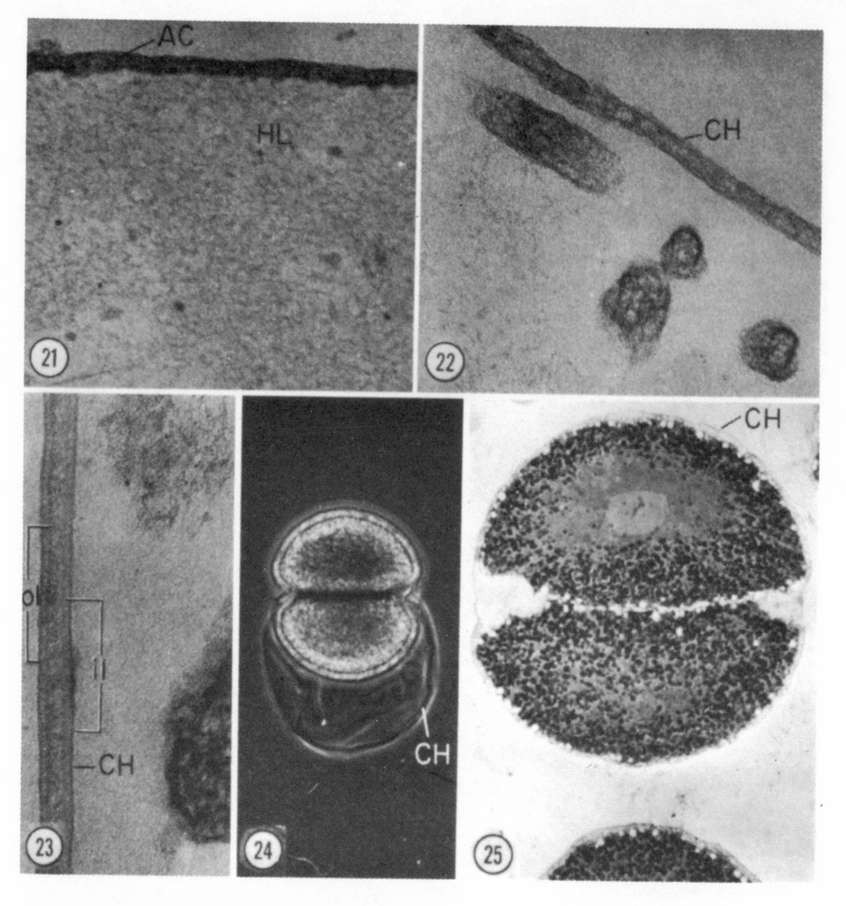

Figures 21–23 Fig. 21 shows the activation calyx (*AC*) 10–12 min after insemination and the hyaline layer (*HL*). Fig. 22 illustrates the chorion (*CH*) 14 min after insemination. 30 min after insemination up to the two-cell stage of the embryo the chorion (*CH*) appears to be like that indicated in Fig. 23. It consists of an outer (*ol*) and an inner (*il*) trilaminar structure separated by a relatively wide space. Figs. 21 and 22, × 105,000; Fig. 23, × 135,000.

Figures 24–25 Fig. 24 is a phase-contrast photomicrograph of a two cell stage embryo showing the chorion (*CH*) partially torn away from the blastomeres. Fig. 25 is a section of an Epon-embedded, toluidine blue–stained embryo which illustrates the intact chorion (*CH*). Fig. 24, × 500; Fig. 25, × 640.

the action of strychnine (59), butyric acid, KCl, NaCl, and distilled water (51, 58, 61). In regard to the formation of the activation calyx by different chemical agents, a significant similarity exists between this phenomenon and the induced lifting, by certain chemical substances, of the pellicle (also an extraneous coat or glycocalyx) of the pigmented protistan ciliate, *Blepharisma undulans.* Nadler (62) noted that when *Blepharisma* is exposed to strychnine, morphine sulphate, and a variety of other chemicals the pellicle is detached from the surface of the organism and discarded. The shedding of the pellicle is accomplished without interfering with the kinetosomes of cilia that are located within the cortical cytoplasm. Shortly after the lifting of the pellicle there is a release of the pigment granules (48). The release of these granules is reminiscent of the release of cortical granules during the cortical reaction by eggs investigated in this study (see below). Thus one sees that a significant parallelism exists between the induced lifting of the primary envelope by chemical agents and the induced lifting of the pellicle of *Blepharisma*. If one removes the activation calyx of the egg of *Arbacia* and subsequently refertilizes it, a second glycocalyx, which is necessary to form the activation calyx, is not produced; however, sperm do enter the egg (81; see also reference 63). On the other hand, once *Blepharisma* sheds its glycocalyx (pellicle) another one may be synthesized. It is understandable why a second glycocalyx is not synthesized by the fertilized egg, since this ability "is lost as an immediate consequence of the fertilization reaction" (49). When one thinks of fertilization, one is reminded of the statement by Moore (58) that "fertilization is a complex series of reactions which, if once completed, exclude all possibilities of repetition." In regard to the induced lifting of the glycocalyx in both the egg and *Blepharisma* one might ask the following question. Is there a physiological parallelism underlying the detachment of the glycocalyx in these two different cell types?

Chorion

Subsequent to its elevation, the activation calyx becomes augmented, i.e., it thickens and hardens. Some investigators have referred to the amplified activation calyx as the fertilization membrane. For example, in their book, Costello et al. (29) state that "this membrane (activation calyx) hardens and thickens during five minutes (after egg insemination) and, after alteration, is called the fertilization membrane." (Words in parentheses are added by the author.) In the electron micrographs presented in this paper, the activation calyx, after the fusion of male and female pronuclei and prior to and during the two cell stage, is composed of an inner and an outer trilaminar structure separated by a 600 A wide interspace. We have found that at about 10–12 min following insemination the male and female pronuclei begin to fuse and that by 14 min a 2N nucleus has been formed. This is the earliest stage of a new generation: hence a zygote. We have already stated elsewhere that if the term chorion is to be retained it should only be applied to the "protective covering surrounding an embryo" (10). Therefore, the augmented activation calyx following the formation of the zygote nucleus should be referred to as the chorion. Whether the thickening and hardening of the activation calyx are due to materials released from the cortical granules, from those vesicles containing the rodlike structures, or from both is unknown. Experiments designed to ascertain the facts are now in progress.

Cortical Reaction

It has already been shown that the cortical granules of the mature egg are found in the peripheral ooplasm. When the sperm activates the mature egg the cortical reaction ensues and progresses in successive stages around the egg (54). The long bibliographic lists appended to the works of Allen (4), Runnström (73), and Tyler and Tyler (82) indicate the interest that biologists have taken in the cortical granules and the cortical reaction. This literature will not be reviewed here. Suffice it to say that during the cortical reaction the contents of the cortical granules are expelled into the perivitelline space (27, 30, 32–34, 67). How might this be accomplished? The data gathered in the present study support the following interpretation. As indicated earlier, the portion of the oolemma associated with the cortical granule is usually devoid of microvilli and thereby leaves a nonmorphologically specialized area of the oolemma associated with the cortical granule. It was also noted that the unit membrane encompassing the cortical granule is separated from the oolemma by a space of about 200 A. Shortly after insemination, the apposing membranes of the cortical granule and oolemma fuse and undergo vesiculation much like that described by Barros et al. (16) between membranes of two different cell types, i.e., the mammalian acrosome reaction. According to Barros et al., "the term membrane vesiculation is used . . . to denote the occurrence of multiple unions between two cellular membranes lying in close apposition, with the formation first of a double-walled fenestrated layer and ultimately of an array of separate membrane-bounded vesicles." As a result of the release of the contents of the cortical granules, a portion of the membrane limiting the cortical granule becomes a part of the plasmalemma of the fertilized egg. Moreover, the over-all surface area of the fertilized egg is increased since it is obvious that more membrane is added than is lost as a result of the union of membranes. There is apparently no increase in cell volume (82). The conclusion, that a portion of the encompassing membrane of the cortical granule fuses with portions of the oolemma in several places, is drawn in spite of the fact that sections depicting the initial fusion have not been obtained. Such an achievement would be extremely difficult since the process is obviously a rather rapid one. The diagrammatic representation illustrated as Fig. 26 recapitulates the events associated with the *cortical reaction.* Other investigators, contributing to the analysis of how the cortical granules release their contents, have presented varied interpretations (12, 13, 33, 47, 60, 86).

All of the cortical granules that are produced by the oocytes are not released during the cortical reaction (see reference 4); some are retained by blastomeres of the gastrula (2). The significance of this is unknown. One might speculate that when the oocyte obtains a full complement of these granules the over-all quantity may exclude the possibility of all of them becoming located adjacent to nonmicrovillous studded areas of the oolemma. It is possible that the nonmicrovillous portion of the oolemma is physiologically different

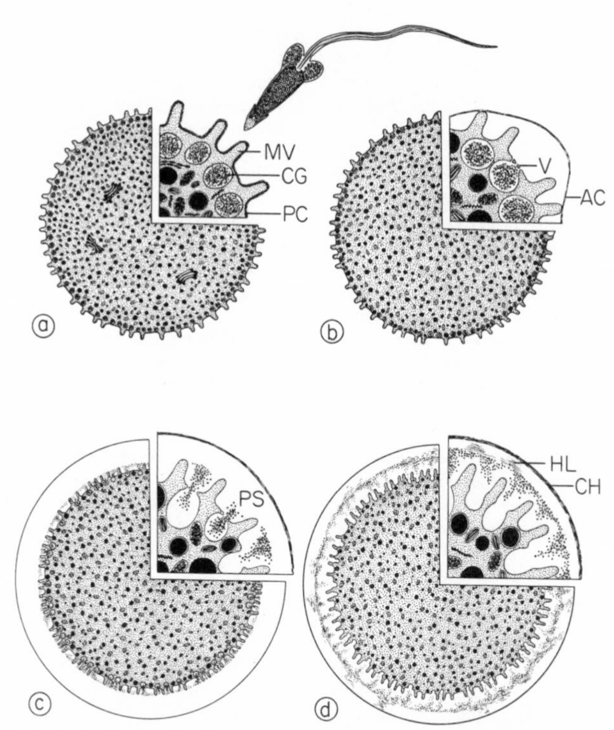

FIGURE 26 A schematic representation of the events associated with the cortical reaction. *a*, Activation of the egg by the sperm: *MV*, a microvillus; *CG*, cortical granule; *PC*, primary coat. *b*, Lifting of primary (vitelline) coat to form the activation calyx (*AC*) and the union of the membrane of the cortical granules with that of the oolemma thereby forming vesicular structures (*V*). *c*, Release of the contents of the cortical granules within the perivitelline space (*PS*). *d*, The thick chorion (*CH*) and hyaline layer (*HL*).

from microvillous areas. If this be the case, one can further conceive that such a difference would permit fusion between that portion of the oolemma and the membrane encompassing the cortical granule. In any event, once the initial steps of fertilization commence one sees that the original oolemma becomes a mosaic (see references 28, 57). In the fertilized egg one can envision that its mosaic plasmalemma might be composed of the following: (*a*) portions of the original oolemma; (*b*) portions of the membrane that encompassed the cortical granules; originally derived from the Golgi complex; (*c*) portions of the plasmalemma of the sperm; and (*d*) portions of the membrane limiting the vesicles that contain the rodlike structures. It is not known how long the mosaic nature of the membrane is retained, nor what significance the mosaic membrane may have on the future of the embryo. It is conceivable that such a mosaic plasmalemma may be important in the initial cleavage of the zygote (9).

Cortical Granules of Other Organisms

It has been well established that the cortical granules in the eggs of a number of different organisms do indeed participate in the cortical reaction (4, 11, 57, 77, 82, 87). In the polychaetous annelid, *Sabellaria*, the contents of a majority of the cortical granules are released when the egg comes into contact with seawater (64, 66). Pasteels (66) suggests that the release of cortical granules is independent of the fertilization phenomenon. In some forms, for example the amphineuran mollusc *Mopalia muscosa* (Anderson, E. Unpublished data.), the pelecypod *Mytilus edulis* (42), and the brachiopod *Terebratalia transversa*,[1] specific bodies come to lie in the peripheral ooplasm when oocytes become mature eggs. In these and other forms there is no visible cortical reaction when the egg is inseminated. Humphreys (42) found that a major portion of the population of cortical granules in *Mytilus* is retained at least to gastrulation. The cortical granules in *Terebratalia* persist, near the surface of the ectodermal cells, until late larval life.[1] In many eggs, the cortical granules are composed of an acid mucopolysaccharide and protein. Long[1] has found that the cortical granules in *Terebratalia* possess some interesting cytochemical properties. They display a positive reaction with bromphenol blue, tyrosine (see reference 50), and alkaline fast green. The tyrosine reaction is blocked by prolonged iodination at high pH, and the coloration with alkaline fast green is blocked by prior diamination with nitrous acid. The granules do not show reaction with the Feulgen reagent, pyronine, periodic acid–Schiff, Alcian blue, oil red 0, methanol fast blue 2S for phospholipids, and Sakaguchi's reaction for tryptophane. On the basis of these results, Long tentatively suggests that the granules of *Terebratalia* contain a histone-like protein.

It is apparent, from what has been presented here, that a number of functions presumably will eventually be found for those structures that come to lie in the peripheral ooplasm of a mature egg and are referred to as cortical granules. In some eggs these structures clearly participate in the cortical reaction and could be classified as the cortical granules of fertilization. When the function of those cortical granules that do not participate in the cortical reaction during fertilization becomes known, perhaps the nomenclature selected could reflect their function.

[1] Long, J. Personal communication.

This investigation was supported by grant GM-08776 from the National Institutes of Health, United States Public Health Service.

The author wishes to thank Mr. and Mrs. L. Musante and Mr. J. Pallozola for their technical assistance.

REFERENCES

1. Adams, E. C., and A. T. Hertig. 1964. Studies on guinea pig oocytes. I. Electron microscope observations on the development of cytoplasmic organelles of primordial and primary follicles. *J. Cell Biol.* **21**:397.
2. Afzelius, B. A. 1956. The ultrastructure of the cortical granules and their products in the sea urchin egg as studied with the electron microscope. *Exptl. Cell Res.* **10**:257.
3. Afzelius, B. A. 1957. Electron microscopy o then basophilic structures of the sea urchin egg. *Z. Zellforsch. Mikroskop. Anat.* **45**:660.
4. Allen, R. D. 1958. The initiation of development. *In* The Chemical Basis of Development. M. D. McElroy and B. Glass, editors. The Johns Hopkins Press, Baltimore. 17.
5. Anderson, E. 1964. Oocyte differentiation and

vitellogenesis in the roach, *Periplaneta americana*. *J. Cell Biol.* **20**:131.

6. ANDERSON, E. 1965. The anatomy of bovine and ovine pineals. Light and electron microscopic studies. *J. Ultrastruct. Res.* **8**(Suppl.):1.
7. ANDERSON, E. 1965. Events associated with differentiating oocytes in two species of Amphineurans (Mollusca), *Mopalia muscosa*, and *Chaetopleura apiculata*. *J. Cell Biol.* **27**:5A.
8. ANDERSON, E. 1966. The origin of cortical granules and their participation in the fertilization phenomenon in Echinoderms (*Arbacia punctulata*, *Strongylocentrotus purpuratus* and *Asterias forbesi*). *J. Cell Biol.* **31**:5A.
9. ANDERSON, E. 1966. The tripartite nature of the oolemma of fertilized Echinoderm eggs (*Arbacia punctulata*, *Strongylocentrotus purpuratus* and *Asterias forbesi*). *J. Cell Biol.* **31**:136A.
10. ANDERSON, E. 1967. The formation of the primary envelope during oocyte differentiation in teleosts. *J. Cell Biol.* **35**:193.
11. AUSTIN, C. R. 1956. Cortical granules in hamster egg. *Exptl. Cell Res.* **10**:533.
12. BALINSKY, B. I. 1960. The role of cortical granules in the formation of the fertilization membrane and the surface membrane of fertilized sea urchin eggs. Symposium on Germ Cells and Development. A. Baselli; Pavia, Italy. 205.
13. BALINSKY, B. I. 1966. Changes in the ultrastructure of amphibian eggs following fertilization. *Acta Embryol. Morphol. Exptl.* **9**:132.
14. BALINSKY, B. I., and R. J. DEVIS. 1963. Origin and differentiation of cytoplasmic structures in the oocytes of *Xenopus laevis*. *Acta Embryol. Morphol. Exptl.* **6**:55.
15. BARKA, T., and P. J. ANDERSON. 1963. Histochemistry. Hoeber Medical Division of Harper and Row, New York.
16. BARROS, C., J. M. BEDFORD, L. E. FRANKLIN, and C. R. AUSTIN. 1967. Membrane vesiculation as a feature of the mammalian acrosome reaction. *J. Cell Biol.* **34**:C1
17. BENNETT, H. S. 1963. Morphological aspects of extracellular polysaccharides. *J. Histochem. Cytochem.* **11**:1.
18. BENNETT, H. S. 1965. Introductory remarks. *Symp. Soc. Cellular Chem.* **14**(Suppl.):7.
19. BLACKLER, A. W. 1958. Contribution to the study of germ-cells in Anura. *J. Embryol. Exptl. Morphol.* **6**:491.
20. BLACKLER, A. W. 1962. Transfer of primordial germ-cells between two subspecies of *Xenopus laevis*. *J. Embryol. Exptl. Morphol.* **10**:641.
21. CARO, L. G. 1961. Electron microscopic radioautography of thin sections: The Golgi zone as a site of protein concentration in pancreatic acinar cells. *J. Biophys. Biochem. Cytol.* **10**:37.
22. CARO, L. G., and G. E. PALADE. 1964. Protein synthesis, storage and discharge in the pancreatic exocrine cell. An autoradiographic study. *J. Cell Biol.* **20**:473.
23. CHAMBERS, R. 1921. Microdissection studies: III-Some problems in the matruation and fertilization of the echinoderm egg. *Biol. Bull.* **41**:318.
24. CHAMBERS, R. 1923. The mechanism of the entrance of sperm into the starfish egg. *J. Gen. Physiol.* **5**:821.
25. CHAMBERS, R. 1930. The manner of sperm entry in the starfish egg. *Biol. Bull.* **58**:344.
26. CHAMBERS, R. 1933. The manner of sperm entry in various marine ova. *J. Exptl. Biol.* **10**:130.
27. CHASE, H. Y. 1935. The origin and nature of the fertilization membrane in various marine ova. *Biol. Bull.* **69**:159.
28. COLWIN, L. H., and A. L. COLWIN. 1967. Membrane fusion in relation to sperm-egg association. *In* Fertilization. C. B. Metz and A. Monroy, editors. Academic Press Inc., New York. 295.
29. COSTELLO, D. P., M. E. DAVIDSON, A. EGGERS, M. H. FOX, and C. HENLEY. 1957. Methods for obtaining and handling marine eggs and embryos. Marine Biological Laboratory, Woods Hole, Mass.
30. DAN, K. 1954. Further study on the formation of the "new membrane" in the eggs of the sea urchin, *Hemicentrotus* (*Strongylocentrotus*) *pulcherrimus*. *Embryologia.* **2**:99.
31. DAN, J. C. 1967. Acrosome reaction and lysins. *In* Fertilization. C. B. Metz and A. Monroy, editors. Academic Press Inc., New York. 237.
32. ENDO, Y. 1952. The role of the cortical granules in the formation of the fertilizatin membrane in eggs from Japanese sea urchins. *Exptl. Cell Res.* **3**:406.
33. ENDO, Y. 1961. Changes in the cortical layer of sea urchin eggs at fertilization as studied with the electron microscope. I. *Clypeaster japonicus*. *Exptl. Cell Res.* **25**:383.
34. ENDO, Y. 1961. The role of the cortical granules in the formation of the fertilization membrane in the eggs of sea urchins II. *Exptl. Cell Res.* **25**:518.
35. FRANKLIN, L. E. 1965. Morphology of gamete membrane fusion and of sperm entry into oocytes of the sea urchin. *J. Cell Biol.* **25**:2.
36. GABE, M., and M. PRENANT. 1949. Contribution à l'histologie de l'ovogenèse chez les polyplacophores. *Le Cellule.* **53**:99.
37. GUSTAFSON, T., and L. WOLPERT. 1967. Cellular movement and contact in sea urchin morphogenesis. *Biol. Rev.* **42**:442.

38. HARVEY, E. B. 1956. The American Arbacia and Other Sea Urchins. Princeton University Press, Princeton.

39. HARVEY, E. N. 1911. Studies on the permeability of cells. *J. Exptl. Zool.* **10**:507.

40. HENDEE, E. C. 1931. Formed components and fertilization in the egg of the sea urchin *Lytechinus variegatus. Tortugas Lab.* (27) and *Carnegie Inst. Wash. Pub. No. 413.* 99.

41. HUMASON, G. L. 1962. Animal Tissue Techniques. W. H. Freeman & Co., San Francisco.

42. HUMPHREYS, W. J. 1967. The fine structure of cortical granules in eggs and gastrulae of *Mytilus edulis. J. Ultrastruct. Res.* **17**:314.

43. ITO, S., and R. J. WINCHESTER. 1963. The fine structure of the gastric mucosa in the rat. *J. Cell Biol.* **16**:541.

44. JAMIESON, J. D., and G E. PALADE. 1967. Intracellular transport of secretory proteins in the pancreatic exocrine cell. *J. Cell Biol.* **34**:577.

45. KARASAKI, S. 1965. Intranuclear crystal within the phagocytes of the ovary of *Aɪbacia punctulata. J. Cell Biol.* **25**:654.

46. KARNOVSKY, M. J. 1965. A formaldehyde-glutaraldehyde fixative of high osmolarity for use in electron microscopy. *J. Cell Biol.* **27**:137A.

47. KEMP, N. E., and N. L. ISTOCK. 1967. Cortical changes in growing oocytes and in fertilized or pricked eggs of *Rana pipens. J. Cell Biol.* **34**:111.

48. KENNEDY, J. R., JR. 1966. The effect of Strychnine and light on pigmentation in *Blepharisma undulans* Stein. *J. Cell Biol.* **28**:145.

49. LILLIE, F. R. 1916. The history of the fertilization problem. *Sceince.* **43**:39.

50. LILLIE, R. D. 1957. Adaptation of the Morel Sisley protein diazotization procedure to the histochemical demonstration of potein bound tyrosine. *J. Histochem. Cytochem.* **5**:528.

51. LOEB, J. 1913. Artificial Parthenogenesis and Fertilization. University of Chicago Press, Chicago.

52. LUFT, J. 1961. Improvements in Epoxy resin embedding methods. *J. Biophys. Biochem. Cytol.* **9**:409.

53. MCCULLOCH, D. 1952. Note on the origin of the cortical granules in *Arbacia punctulata* eggs. *Exptl. Cell Res.* **3**:605.

54. MITCHISON, J. M., and M. M. SWANN. 1952. Optical changes in the membranes of the sea urchin egg at fertilization, mitosis and cleavage. *J. Exptl. Biol.* **29**:357.

55. MONNÉ, L., and S. HARDE. 1951. On the cortical granules of the sea urchin egg. *Arkiv. Zool.* **1**:487.

56. MONNÉ, L., and D. B. SLAUTTERBACK. 1950. Differential staining of various polysaccharides in sea urchin eggs. *Exptl. Cell Res.* **1**:477.

57. MONROY, A. 1965. Chemistry and Physiology of Fertilization. Holt, Reinhart & Winston, Inc., New York.

58. MOORE, C. R. 1916. On the superposition of fertilization on parthenogenesis. *Biol. Bull.* **31**:137.

59. MORGAN, T. H. 1900. Further studies on the action of salt solution and other agents on the eggs of *Arbacia. Archiv. Entwicklungsmech. Roux.* **10**:490.

60. MOSER, F. 1939. Studies on cortical layer response to stimulating agents in the *Arbacia* eggs. I. Response to insemination. *J. Exptl. Zool.* **80**:423.

61. MOSER, F. 1940. Studies on a cortical layer response to stimulating agents in the *Arbacia* egg. III-Response to non-electrolytes. *Biol. Bull.* **78**:68.

62. NADLER, J. E. 1929. Notes on the loss and regeneration of the pellicle in *Blepharisma undulans. Biol. Bull.* **56**:327.

63. NAKANO, E. 1956. Physiological studies on refertilization of the sea urchin egg. *Embryologia.* **3**:139.

64. NOVIKOFF, A. B. 1939. Surface changes in unfertilized eggs of *Sabellaria vulgaris. J. Exptl. Zool.* **82**:217.

65. PALAY, S. L. 1958. The morphology of secretion. *In* Frontiers of Cytology. S. L. Palay, editor. Yale University Press, New Haven. 305.

66. PASTEELS, J. J. 1965. Étude au microscope électronique de la réaction corticale, I. La réaction corticale de fécondation chez *Paracentrotus* et sa chronologie. II. La réaction corticale de l'oeuf vierge de *Sabellaria alveolata. J. Embryol. Exptl. Morphol.* **13**:327.

67. PASTEELS, J. J. 1966. La réaction corticale de fécondation de l'oeuf de *Neries diverscolor,* étudie au microscope électronique. *Acta Embryol. Morphol. Exptl.* **9**:155.

68. REVEL, J. P., and E. D. HAY. 1963. An autoradiographic and electron microscopic study of collagen synthesis in differentiating cartilage. *Z. Zellforsch. Mikroskop. Anat.* **61**:110.

69. REVEL, J. P., and S. ITO. 1967. The surface components of cells. *In* The Specificity of Cell Surfaces. B. D. David and L. Warren, editors. Prentice-Hall, Inc., Englewood Cliffs, N. J. 211.

70. ROBERTSON, J. D. 1967. The organization of cellular membranes. *In* Molecular Organization and Biological Function. M. M. Allen, editor. Harper and Row, Publishers, New York. 65.

71. ROTH, T. F., and K. R. PORTER. 1964. Yolk

protein uptake in the oocyte of the Mosquito *Aedes aegypti*, L. *J. Cell Biol.* **20**:313.

72. Runnström, J. 1948. Further studies on the formation of the fertilization membrane in the sea urchin. *Arkiv. Zool.* **40**:1.
73. Runnström, J. 1966. The vitelline membrane and cortical particles in sea urchin eggs and their function in maturation and fertilization. *In* Advances in Morphogenesis. M. Abercombie and J. Brachet, editors. Academic Press Inc., New York. 221.
74. Sabatini, D. D., K. Bensch, and R. J. Barrnett. 1963. Cytochemistry and electron microscopy. The preservation of cellular ultrastructure and enzymatic activity by aldehyde fixation. *J. Cell Biol.* **17**:19.
75. Smith, L. D. 1966. The role of a "germinal plasm" in the formation of primordial germ cells in *Rana pipiens*. *Develop. Biol.* **14**:330.
76. Szollosi, D. 1962. Cortical granules: a general feature of mammalian eggs? *J. Reprod. Fertility.* **4**:223.
77. Szollosi, D. 1967. Development of cortical granules and the cortical reaction in rat and hamster eggs. *Anat. Record.* **159**:431.
78. Takashima, Y. 1960. Studies on the ultrastructure of the cortical granules in sea urchin eggs. *Tokushima J. Exptl. Med.* **6**:4.
79. Tennent, D. H., and T. Ito. 1941. A study of the oogenesis of *Mespilia globulus* (Linné). *J. Morphol.* **69**:347.
80. Tyler, A. 1949. A simple, non-injurious method for inducing repeated spawning of sea urchins and sand-dollars. *The Collecting Net.* **19**:19.
81. Tyler, A., A. Monroy, and C. Metz. 1956. Fertilization of fertilized sea urchin eggs. *Biol. Bull.* **110**:184.
82. Tyler, A., and B. S. Tyler. 1966. Physiology of fertilization and early development. *In* Physiology of Echinodermata. R. A. Boolootian, editor. Interscience Publishers Inc., New York.
83. Venable, J. H., and R. Coggeshall. 1965. A simplified lead citrate stain for use in electron microscopy. *J. Cell Biol.* **25**:407.
84. Verhey, C. A., and F. H. Moyer. 1967. Fine structural changes during sea urchin oogenesis. *J. Exptl. Zool.* **164**:195.
85. Wilson, L. P. 1940. Histology of the gonad wall of *Arbacia punctulata*. *J. Morphol.* **66**:463.
86. Wolpert, L., and E. H. Mercer. 1961. An electron microscope study of fertilization of the sea urchin egg, *Psammechinus milioris*. *Exptl. Cell Res.* **22**:45.
87. Yamamoto, T. 1961. Physiology of fertilization in fish eggs. *Intern. Rev. Cytol.* **12**:361.

A CYTOLOGICAL STUDY OF ARTIFICIAL PARTHENOGENESIS IN THE SEA URCHIN *ARBACIA PUNCTULATA*

MARTIN I. SACHS and EVERETT ANDERSON

INTRODUCTION

Natural parthenogenesis was first described by Greef in the Echinoderm *Asterias glacialis* (starfish) (18) and is now known to occur in many organisms. On the other hand, artificial parthenogenesis has attracted the attention of researchers since the Hertwigs (37) first gave an account of the basic features of this phenomenon by utilizing chloroform and strychnine as stimulating agents in the sea urchin, *Paracentrotus lividus*. Morgan (68) used various salt solutions including sodium, potassium, and magnesium chloride to artificially activate eggs of the sea urchin *Arbacia punctulata*. It was Loeb (54), however, who was the first to obtain parthenogenetic plutei of *Arbacia punctulata* by using magnesium chloride. Development has been stimulated by physical means such as application of heat or cold (35, 63, 64) and by the utilization of a variety of chemical means, e.g. sodium chloride (33, 54, 55, 64, 68), acids (33, 45, 55), strychnine (69), sucrose (55, 71), saponin (70), and many others (see 31). Cytological studies have been made of events associated with fertilization (2, 57, 58, 65, 67, 88, 94); however, few studies are available concerning those events associated with artificial parthenogenesis at the ultrastructural levels of observation

(9, 59). The present study deals with artificially activated eggs of the sea urchin *Arbacia punctulata* and calls attention to (*a*) the cortical reaction, (*b*) streak formation, and (*c*) nuclear replication. These events are compared with those occurring in the inseminated egg.

MATERIALS AND METHODS

Arbacia punctulata were obtained from The Marine Biological Laboratory at Woods Hole, Massachusetts, during the months of June, July, and August. They were induced to spawn by applying a 10v alternating current across the oral surface (30, 31). The eggs were collected according to the recommendation of Costello et al. (4). Eggs were artificially activated by placing them in seawater made hypertonic by the addition of 30 g of sodium chloride/liter of seawater (44). The time of activation was considered to be the moment the eggs were placed in the hypertonic solution. The eggs were allowed to remain in the hypertonic seawater (19–22°C) for 20 min and were subsequently transferred to fresh seawater. Egg samples were taken at the following intervals: 30 sec, 1, 3, and 5 min, and successive 5-min intervals until 95 min or the initiation of cleavage. Some of the cleaving eggs were permitted to develop to the pluteus stage. Activated eggs, cleaving stages, and plutei were studied by both phase-contrast optics and electron microscopy. The activated eggs from each of the above-mentioned times and the initial cleavage stage were prefixed for 2 hr in a 2% glutaraldehyde-seawater solution or in the glutaraldehyde-paraformaldehyde mixture of Karnovsky (48). After fixation, the specimens were washed in seawater, postfixed for 1 hr in a 1% solution of osmium tetroxide dissolved in seawater, rapidly dehydrated in a graded series of ethanol, infiltrated, and embedded in Epon (62). 1 μ sections, cut on a Porter-Blum MT-2 ultramicrotome, were stained according to the recommendation of Ito and Winchester (43). Thin sections were also obtained with the MT-2 ultramicrotome and stained with uranyl acetate followed by lead citrate (91), and were examined in an RCA EMU-3H electron microscope.

Eggs collected in the manner indicated above were inseminated with the "dry sperm" diluted with seawater (45). The inseminated eggs were fixed for

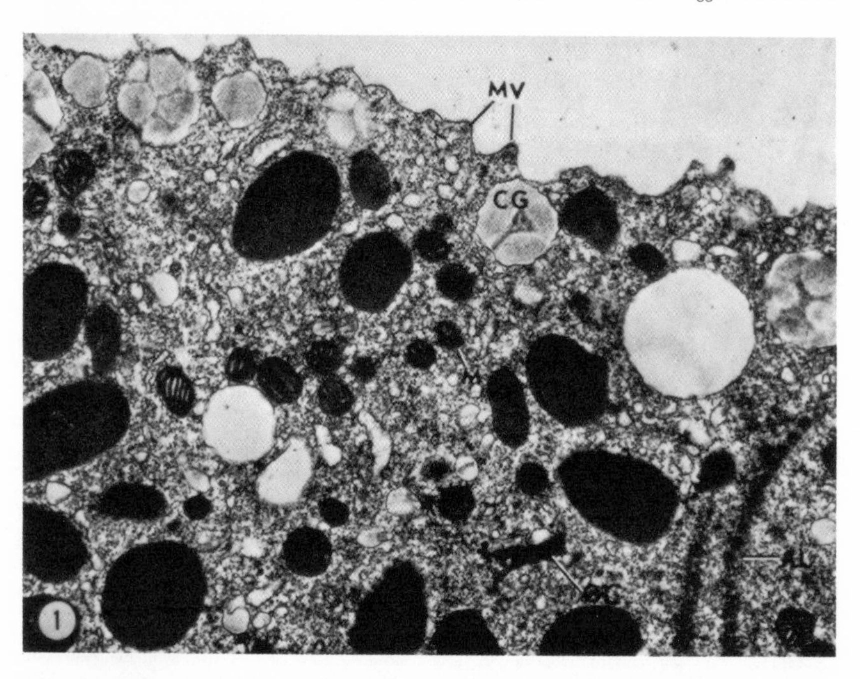

FIGURE 1 An electron micrograph depicting the untreated egg. *MV*, microvilli; *CG*, cortical granule; *AL*, annulate lamellae; *M*, mitochondria; *GC*, Golgi complex. × 8,000.

light and electron microscopy (see above) or were observed with phase-contrast optics at 10-min intervals for 70 min through the initial cleavage.

OBSERVATIONS

Unactivated Egg

The morphology of an unactivated egg is shown in Fig. 1. The oolemma is projected into short microvilli (*MV*). Immediately beneath the oolemma is a population of cortical granules (*CG*) embedded in a matrix of free ribosomes and some vesicles. The ooplasmic components such as yolk droplets, annulate lamellae (*AL*), endoplasmic reticulum, Golgi complexes (*GC*), rod-containing vesicles, and pigment granules are randomly dispersed; the majority of mitochondria (*M*) are randomly distributed, but some are closely associated with lipid droplets.

Activated Egg

CORTICAL CHANGES: When the eggs are treated with hypertonic seawater they undergo a cortical reaction. Figs. 2, 5, 6 show the cortical region of the egg at 1, 5, and 10 min after being exposed to the activating medium. All of the cortical granules that are closely associated with the inner aspect of the oolemma do not fuse with the oolemma simultaneously when activated (Fig. 2, *RCG*); however, they fuse at random loci (Fig. 2 and Fig. 2 *inset*, see arrow). The artificially activated egg produces an activation calyx (Fig. 5, *AC*) and does not produce a protrusion reminiscent of an entrance cone like that of the inseminated egg (2, 57). The fusion of the membrane encompassing the cortical granule with the oolemma produces vesicular structures (Figs. 5, 6, *V*) over the contents of the cortical granules. Upon the completion of membrane fusion, the contents of

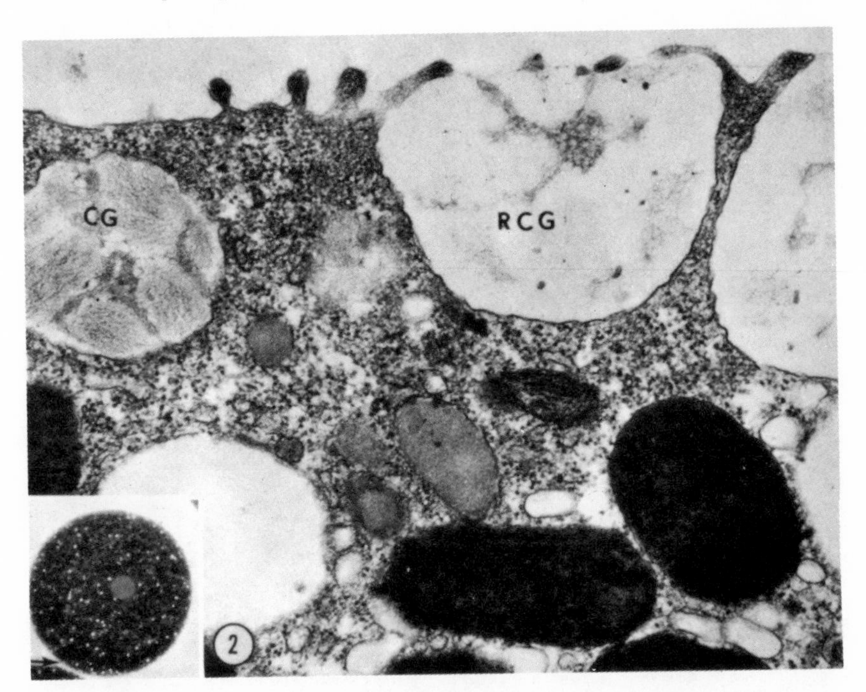

FIGURE 2 The surface of an egg, 1 min following activation with hypertonic seawater. *CG*, cortical granule; *RCG*, released cortical granule. × 25,800. *Inset* is a photomicrograph of an egg 1 min post-activation, demonstrating the release of cortical granules. × 400.

the cortical granules are released, initiating the formation of the perivitelline space (Fig. 6, *PS*). The release of the contents of all the cortical granules does not occur simultaneously, for "pillars" of unreacted cortical ooplasm which contain cortical granules, free ribosomes, and occasional pigment bodies are commonly found (Figs. 5, 6, *P*). The further release of the cortical granules results in the formation of a continuous perivitelline space limited by the "chorion" and the oolemma (Fig. 7, *PS*). In the artificially activated egg, the perivitelline space is smaller than that of the inseminated egg. As demonstrated for the inseminated egg (2), not all cortical granules are released during the initial reaction to the hypertonic seawater.

Within the relatively small perivitelline space, 30 min postactivation, the hyaline layer (Fig. 7, *HL*) may be observed directly beneath the "chorion" (Fig. 7, *C*) (also see 1). It is composed of a mat of fine, electron-opaque, filamentous material. By 65 min, the hyaline layer increases in thickness. Beneath the microvilli the cortical ooplasm is now composed of an accumulation of pigment bodies (Fig. 9, *PB*), few mitochondria, and dense yolk bodies. The cortical ooplasm contains numerous rod-containing vesicles. As in the inseminated eggs, the rods are released subsequent to the release of the contents of the cortical granules. The rodlike structures become associated with the components of the hyaline layer (Fig. 9, *R*).

When cytokinesis is initiated (1½–4½ hr postactivation), the periphery of the embryo is characterized by a well-developed hyaline layer (Fig. 9, *HL*), long microvilli (Fig. 8, *MV*), and an almost continuous stratum of pigment bodies (Fig. 9, *PB*) immediately beneath the plasma membrane.

OOPLASM: The mitochondria show an inter-

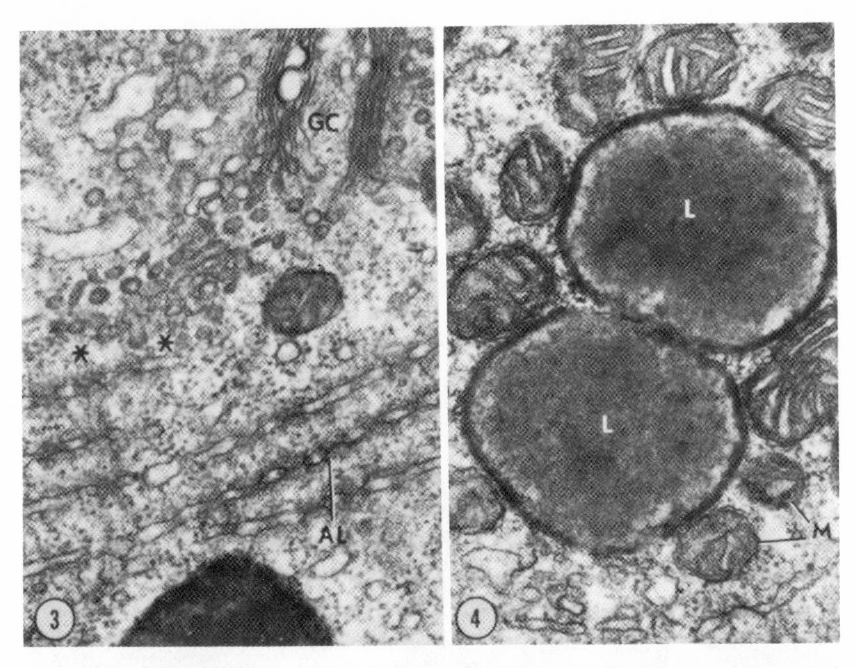

FIGURE 3 An electron micrograph of an egg in hypertonic seawater for 1 min, showing a Golgi complex (*GC*) associated with coated vesicles (**) closely associated with annulate lamellae (*AL*). × 25,000.

FIGURE 4 An electron micrograph of an egg 1 min following activation, depicting mitochondria (*M*) clustered around lipid droplets (*L*). × 38,000.

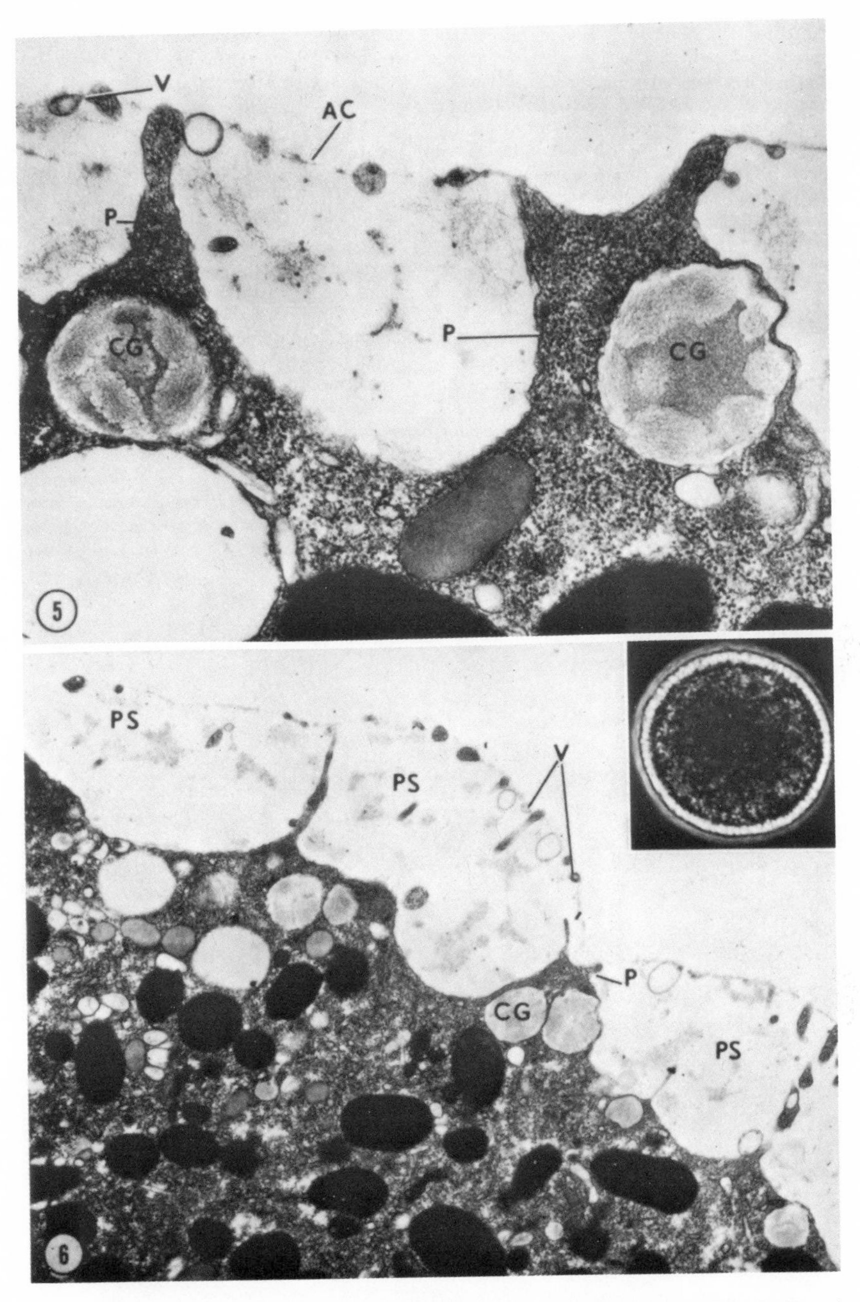

FIGURE 5 An electron micrograph of the surface of an egg, 5 min following activation. *P*, pillars of unreacted ooplasm containing cortical granules (*CG*); *V*, vesicles; *AC*, activation calyx. × 21,000.

FIGURE 6 An electron micrograph of the surface of an egg, 10 min following activation, showing the reduction in number and thickness of the pillars (*P*) which contain cortical granules (*CG*). *V*, vesicles; *PS*, incomplete perivitelline space. × 17,000. The phase-contrast photomicrograph (*inset*) of a living *Arbacia* egg depicts the appearance of the egg at the light microscope level. Note the striated appearance of the perivitelline region. × 450.

nal configuration similar to that of the unactivated egg. At 1 min postactivation, there appears to be a close spatial association between mitochondria (Fig. 4, *M*) and lipid droplets (Fig. 4, *L*). There also appears an intimate association between the Golgi complex (Fig. 3, *GC*) and its associated coated vesicles (Fig. 3, *) and the annulate lamellae (Fig. 3, *AL*) similar to that observed during pronuclear development in the rabbit (58). Centrioles have not been described in the unfertilized egg (2, 92) and have not been observed in the artificially activated eggs until the formation of the aster (9, 80).

FORMATION OF THE STREAK STAGE: According to Harvey (31), in inseminated eggs, a monaster is formed after the fusion of the male and female pronuclei. Subsequently "The rays disappear and the centrosome (probably) divides forming a curved disk over the nucleus . . .". Harvey (31) defines this stage as the *streak stage*. In the case of eggs treated with hypertonic seawater, a streak stage is also formed. Closely associated with the nuclear envelope (Fig. 11, *NE*), prior to the formation of the streak stage (60 min postactivation), are stacks of annulate lamellae (Figs. 10–12, *AL*). Occasionally, one sees intranuclear annulate lamellae (Fig. 11, *IAL*). Concomitant with the organization of the annulate lamellae, centrioles (one, two, or three) may be observed (Figs. 15, 16, 17, *C*) associated with microtubules (Fig. 17, *MT*) and endoplasmic reticulum (Fig. 15, *ER*). Together, the latter organelles form an aster (Figs. 15, 18, *inset a*, *AS*) which is similar to that reported for the sperm (57, also see 26, 27).

The annulate lamellae (Fig. 13, *AL*) become dispersed from their circumnuclear configuration, initiating the elongation of the aster and the formation of the streak. The inset of Figs. 13 and 14 is a phase-contrast photomicrograph of a streak (*ST*) stage 85 min postactivation. The streak is characterized by annulate lamellae (Fig. 14, *AL*) arranged in parallel array encompassed by endoplasmic reticulum and vesicular components. Occasionally, one finds mitochondria amongst the annulate lamellae comprising the streak; however, protein-carbohydrate yolk bodies (Fig. 14, *Y*) are

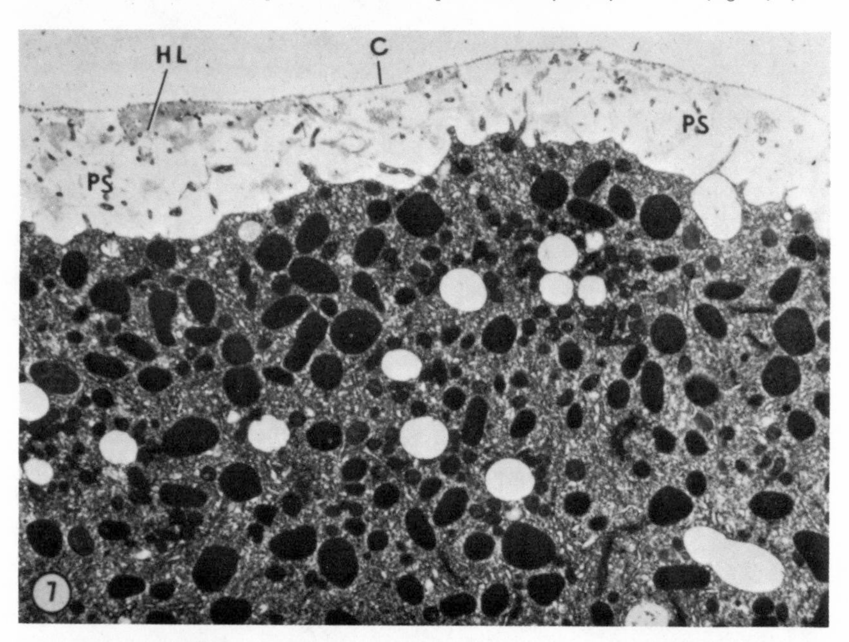

FIGURE 7 An electron micrograph of the surface of an egg 30 min postactivation, *C*, "chorion"; *HL*, hyaline layer; *PS*, perivitelline space. × 6,500.

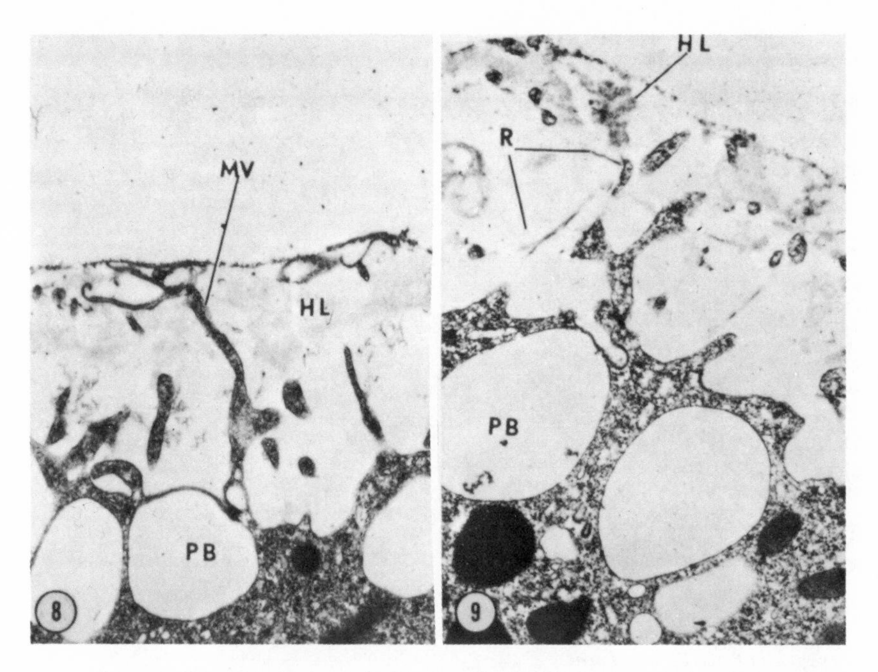

FIGURE 8 The surface of an egg at the time of cytokinesis, depicting the hyaline layer (*HL*), microvilli (*MV*), and pigment bodies (*PB*). × 16,000.

FIGURE 9 A section of an artificially activated egg 65 min postactivation. *HL*, hyaline layer; *R*, rodlike structures; *PB*, pigment bodies. × 25,000.

excluded. The conformation of the annulate lamellae (streak) extends from the tips of the longitudinal axes of the now elliptical nucleus toward the plasmalemma. We have observed that eggs artificially activated with hypertonic seawater may remain in the streak stage for 1–3 hr before dividing. Inseminated eggs remain in the streak stage for only approximately 25 min.

NUCLEUS: The pronucleus contains a granular nucleoplasm in which are suspended nucleolus-like structures (Figs. 10, 15, *NL*). The pronucleus is surrounded by a perforated nuclear envelope. At 85 min postactivation, the pronucleus elongates with a concomitant condensation of its chromatin followed by a breakdown of the pronuclear envelope (Fig. 18, *CH*, and *inset a*).

At metaphase (Fig. 18, *inset b*) and anaphase (Figs. 18, 19, *inset c*) the chromosomes appear as dense masses of granular material embedded within a matrix of ribosomes (Fig. 18, *MR*). The mitotic apparatus (Fig. 18, *inset b* (*SA*), *c*; Fig. 19) is composed of predominantly microtubules (Fig. 19, *MT*) with some endoplasmic reticulum (*ER*) and ribosomes (see 26, 27, 57). At the periphery of the mitotic apparatus may be found mitochondria (Fig. 19, *M*) and yolk bodies.

At telophase, the chromosomes (Fig. 20, *CH*) are elongated in the direction of the centrioles (Fig. 20, *C*) and are often found in intimate association with nucleolus-like bodies (Fig. 20, *NL*). Microtubules (Fig. 20, *MT*), mitochondria, and annulate lamellae are also found among the chromosomes. A perforated envelope forms around the chromosomes, establishing chromosome-containing vesicles (karyomeres) (Fig. 20, *inset*) (see 94). Fusion of the chromosome-containing vesicles and subsequent dispersal of the chromatin results in the formation of two nuclei prior to cytokinesis (Fig. 21). Each of these nuclei contains some dense chromatin material (*CH*).

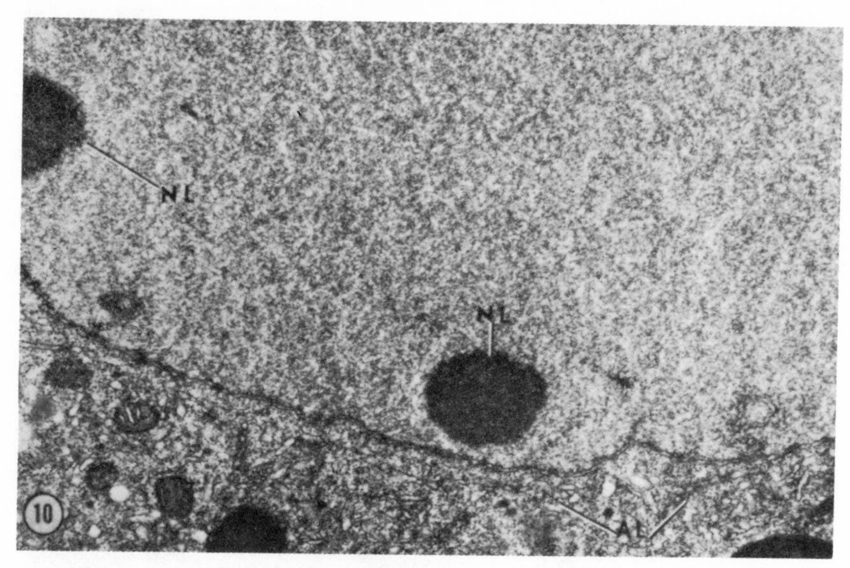

FIGURE 10 A section through the nucleus of an egg 50 min following activation. *NL*, nucleolus-like bodies; *AL*, annulate lamellae. × 22,000.

Development to the Pluteus Larva

First cleavage of artificially activated eggs occurs between 1½ and 4½ hr and results in the two-cell embryo (Fig. 22). Ensuing cleavages result in various multicellular stages (Figs. 23–26). The cells of the multicellular embryo (of which the electron micrographs are not included) are spherical and contain the regularly occurring organelles including centrioles, microtubules, mitochondria, smooth and rough forms of endoplasmic reticulum. Large quantities of yolk are present, but there is a reduction in the amount of lipid. The cells of the morula stage are often found associated by tight junctions.

The blastula contains elongated polarized cells with apically situated nuclei. The cells of the ciliated blastula contain organelles similar to those described for the multicellular stage. The pluteus appears, at the light microscope level, to be identical with that formed from the inseminated egg.

DISCUSSION

Cortical Reaction

Evidence obtained during this study suggests that the complex cortical reaction of the eggs of the sea urchin, *Arbacia*, brought about by treatment with hypertonic seawater, is different from that initiated by insemination (see 2, 13). Artificially activated eggs do not demonstrate the wavelike propagation of cortical granule release seen in inseminated eggs (65). The perivitelline space formed as a result of the cortical reaction is smaller in the artificially activated egg, although the mechanism by which the cortical granule reaction occurs, i.e. fusion and vesiculation as discussed by Anderson (2), appears to be the same for the artificially activated egg and the inseminated egg.

The fact that an increase in tonicity, via the addition of 3% sodium chloride to seawater, induces the cortical reaction suggests that this reaction is due to a change in the water and/or ion content of the egg. Loeb (55) wrote that "It appeared to me that nothing would more clearly demonstrate the sovereign role that electrolytes play in the phenomena of life than by causing, if possible, with their help, unfertilized eggs to develop into larvae." In subjecting *Arbacia* eggs to a hypertonic sodium chloride solution, the actual activating agent in the solution could be a change in water flow, the sodium or chloride ions, or, possibly, a change in the surface of the egg. We choose

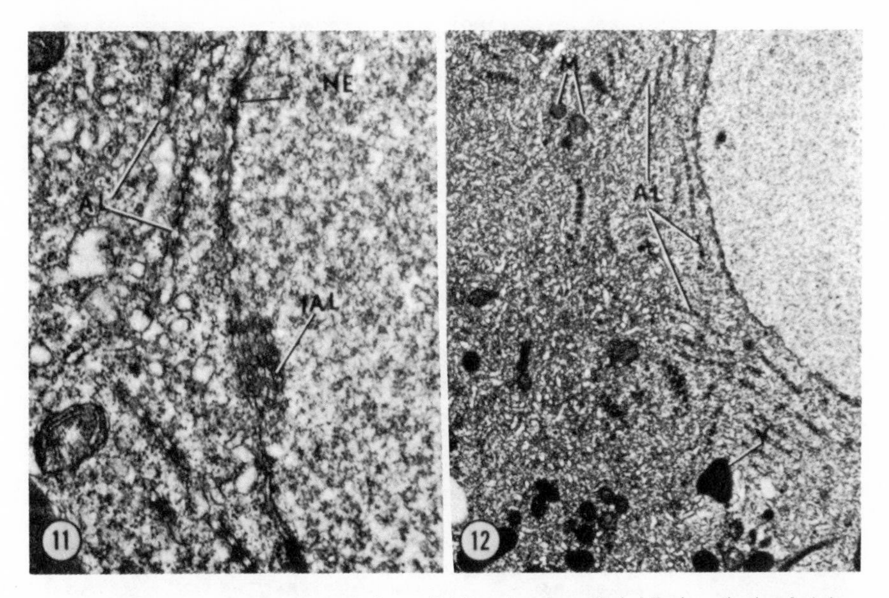

FIGURE 11 An electron micrograph of an artificially activated egg 60 min following activation, depicting annulate lamellae (*AL*) and intranuclear annulate lamellae (*IAL*). Note the nuclear envelope (*NE*). × 29,000.

FIGURE 12 An electron micrograph taken of an egg 75 min postactivation showing annulate lamellae (*AL*) circumferentially located around the nucleus. Note mitochondria (*M*) and yolk (*Y*). × 12,000.

to discuss the possibility of a change in water flow or sodium ion as the activating agent. Discussion favoring the concept that water flow, causing either a decrease or increase in the amount of water in the egg, is responsible for the initiation of the cortical reaction is based largely on the fact that the oolemma acts as a selectively permeable membrane (33, 61, 64). Eggs subjected to hypotonic solution maintain their shape, although there appears to be an increase in volume. When returned to seawater the eggs undergo shrinkage, indicating that the moiety passing through the plasma membrane is water and that the salt content in the egg probably remains constant (61). Many investigators consider that the loss of water from an egg placed in hypertonic seawater is the factor causing what they refer to as the "explosion" of the cortical ooplasm (33, 40, 54, 55, 68). Heilbrunn (33, 34) suggested that treatment with various agents including sodium chloride causes a marked change in the viscosity of the cortical cytoplasm. Recent investigation by Anderson (unpublished data) has demonstrated that there is a change in the position of the cortical granules when treated with sodium chloride or urethane and then centrifuged at high speeds (also see 29). In eggs so treated, the cortical granules abandon their peripheral position and form a stratum. The change in the cortical granule membrane-oolemma relationship does not explain, however, what would dittiate the fusion and subsequent vesiculation process associated with the cortical reaction.

The concentration of the sodium chloride and the exposure time of the eggs to the hypertonic medium are both critical if development is to ensue. Examinations of swelling and shrinkage are only crude indicators of water flow. It would be must fruitful to have the techniques of (*a*) diffusion tracing and (*b*) bulk flow brought to bear on the unactivated and initially activated *Arbacia* egg in order to establish, quantitatively, the actual rate and amount of water flow at activation. These techniques, used on artificial membranes, have demonstrated the ability to measure water flow

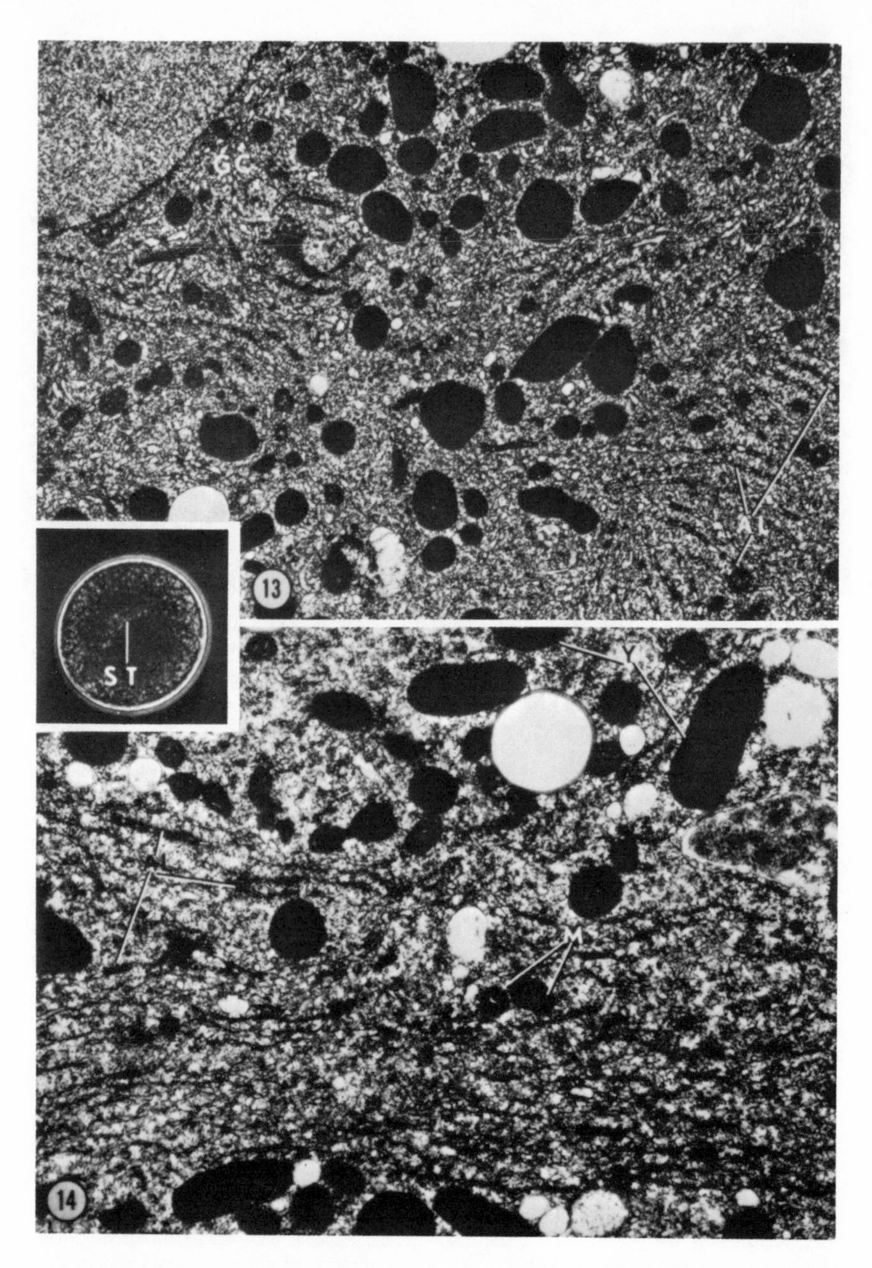

FIGURE 13 A section of an egg 65 min postactivation, showing the initiation of streak formation. *N*, nucleus; *AL*, annulate lamellae; *GC*, Golgi complex. × 11,000.

FIGURE 14 An electron micrograph depicting the components of the streak stage. *AL*, annulate lamellae; *M*, mitochondria; *Y*, yolk. × 11,660. The *inset* is a phase-contrast photomicrograph of the living egg illustrating the streak (*ST*). × 300.

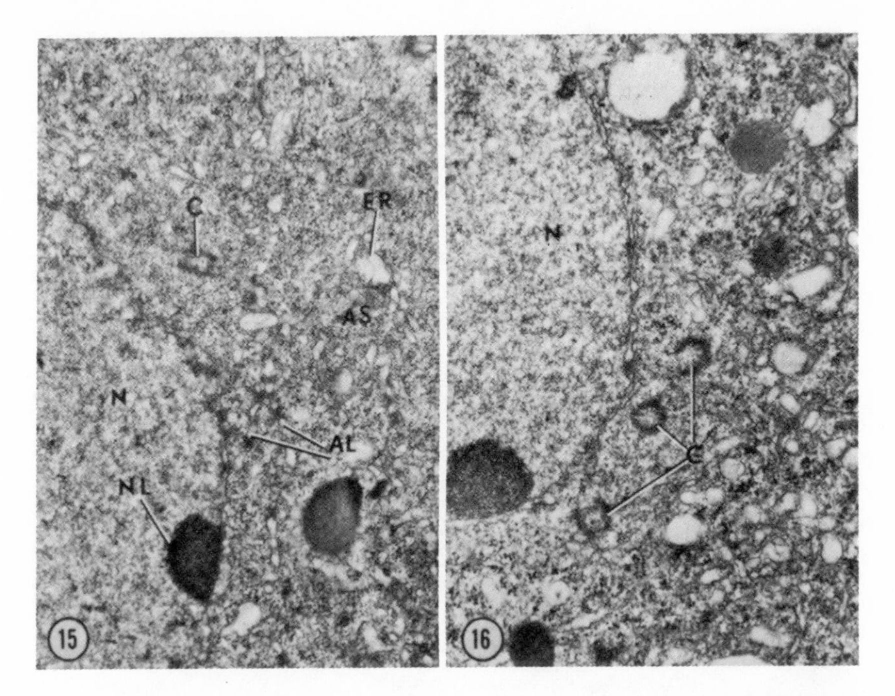

FIGURE 15 A section through the aster (*AS*) (60 min postactivation), which consists of endoplasmic reticulum (*ER*), annulate lamellae (*AL*), and a centriole (*C*). Note the nucleolus-like body (*NL*). × 19,000.

FIGURE 16 An electron micrograph showing three centrioles (*C*) adjacent to the nucleus (*N*). 60 min postactivation. × 18,000.

(16, 22, 38, 72). They have been applied to oocytes of the frog *Rana* and have led investigators to believe that cytoplasmic resistance to water flow is a factor to be considered as well as the physical state of the membrane (7, 55, 60, 76). This is especially significant in many eggs, where the ooplasm is filled with a variety of inclusions, all of which may act as a type of endogenous buffer. This may help to explain the ability of the unactivated egg to maintain its shape when placed in an anisotonic-activating solution.

An alternative hypothesis would be that there is an ion flow into the egg initiated by the increase of Na^{++} and Cl^{-} ions in the hypertonic seawater. In fresh seawater, the sodium ion concentration outside the *Arbacia* egg far exceeds that found in the ooplasm (67, 73, 79, 89). This information, coupled with the findings of Ohnishi (75) that there is an ATP-dependent active transport system in the oolemma, suggests that the oolemma may be not simply a passive barrier to sodium entrance, but instead is equipped to maintain the transmembrane potential via active transport. Ussing (90) and others (6, 36, 51, 86, 93) have stated that active transport, in various cells and tissues responsible for the maintenance of transmembrane potential, can be virtually eliminated by introducing the cells to an osmotic gradient, resulting in a change in the volume of the cell. Thus, when *Arbacia* is subjected to our hypertonic solution the shrinkage may lead to an influx of sodium. The possible relationship of this influx to the initiation of the cortical granule release is unclear.

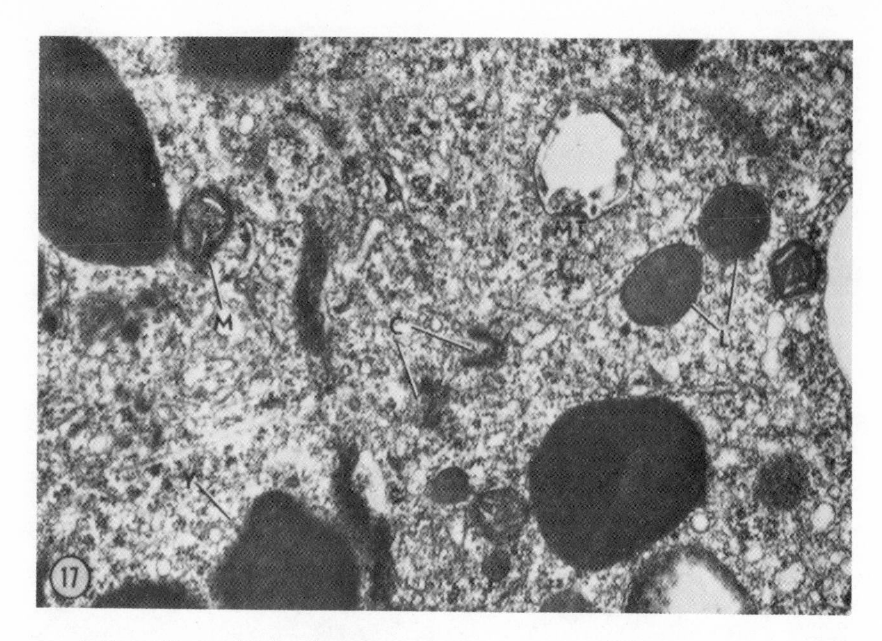

FIGURE 17 A section through an egg 60 min after activation. *C*, centrioles; *MT*, microtubules; *L*, lipid; *Y*, yolk; *M*, mitochondria. × 38,000.

Nucleus

We have seen that the pronucleus (containing the haploid number of chromosomes) can be activated to go through periods of development leading to its division. Division eventually leads to the formation of a pluteus larva. Development, upon activation with hypertonic seawater, proceeds, in a very large percentage of eggs, to the streak stage. Often there is a temporary arrest of the nuclear activity at the streak stage in development, e.g. approximately 1–3 hr. This temporary arrest suggests to us that, in order for development to continue, an "activation" of the nucleus must occur.

In connection with the activation of the nucleus in other systems, some investigators have indicated that an informational transfer between cytoplasm and nucleus results in a change in nuclear activity, for example, DNA replication, RNA synthesis, and protein synthesis (14, 15, 23–25, 39, 47, 49, 74). In our study, the activation of the nucleus could be a direct result of ion or water flow from cytoplasm to nucleus, or possibly an indirect effect of a change in the ion constituency in the cytoplasm, inducing macromolecular synthesis which, in turn, transfers information to the nucleus.

A number of studies have demonstrated a definite relation between the effect of ion and water shifts between the nucleus and the cytoplasm and the activation of nuclear activity. In Dipteran salivary glands, Loewenstein et al. (56) have demonstrated an appreciable resistance of the nuclear envelope. Furthermore, the permeability of the nuclear envelope undergoes changes during development (42; also 41). The change in permeability is accompanied by a change in total DNA content and total protein and is related to the effect of ecdysone. The relationship between the changes mentioned, and the action of the chromosomal puffs, may well be dependent upon ion flow into the nucleus (49, 50, 52, 56). In this connection, Kroeger (49) states that the "genetic loci activated in vitro by ions are also activated by these ions in normal development and that ecdysone exerts it effect on the puffing pattern by stimulating the sodium pump; the consequential change in the internal ion balance of the cell activates the re-

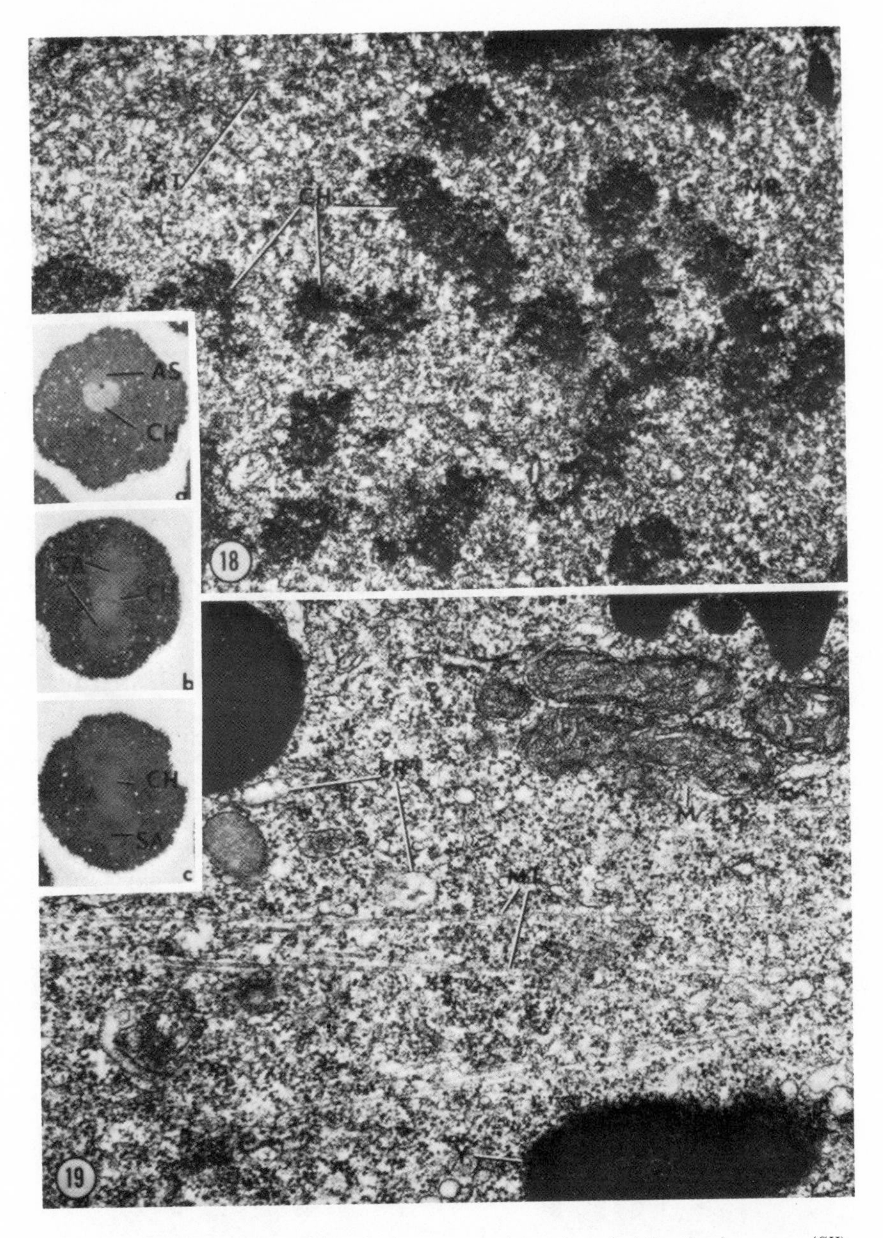

FIGURE 18 An electron micrograph of an egg at early anaphase, depicting the chromosome (*CH*), associated with microtubules (*MT*) and ribosomes (*MR*), 110 min postactivation. × 27,000.

FIGURE 19 A section through a mitotic figure at anaphase, 110 min postactivation, showing mitochrondria (*M*), endoplasmic reticulum (*ER*), yolk (*Y*), microtubules (*MT*). × 34,240.

Insets a, b, c Photomicrographs of artificially activated eggs at 95 minutes (*a*) prophase, 100 min (*b*) metaphase, and 105–115 min postactivation (*c*) anaphase. *CH*, chromosome; *AS*, aster; *SA*, spindle apparatus. *a*, *b*, *c*, × 400.

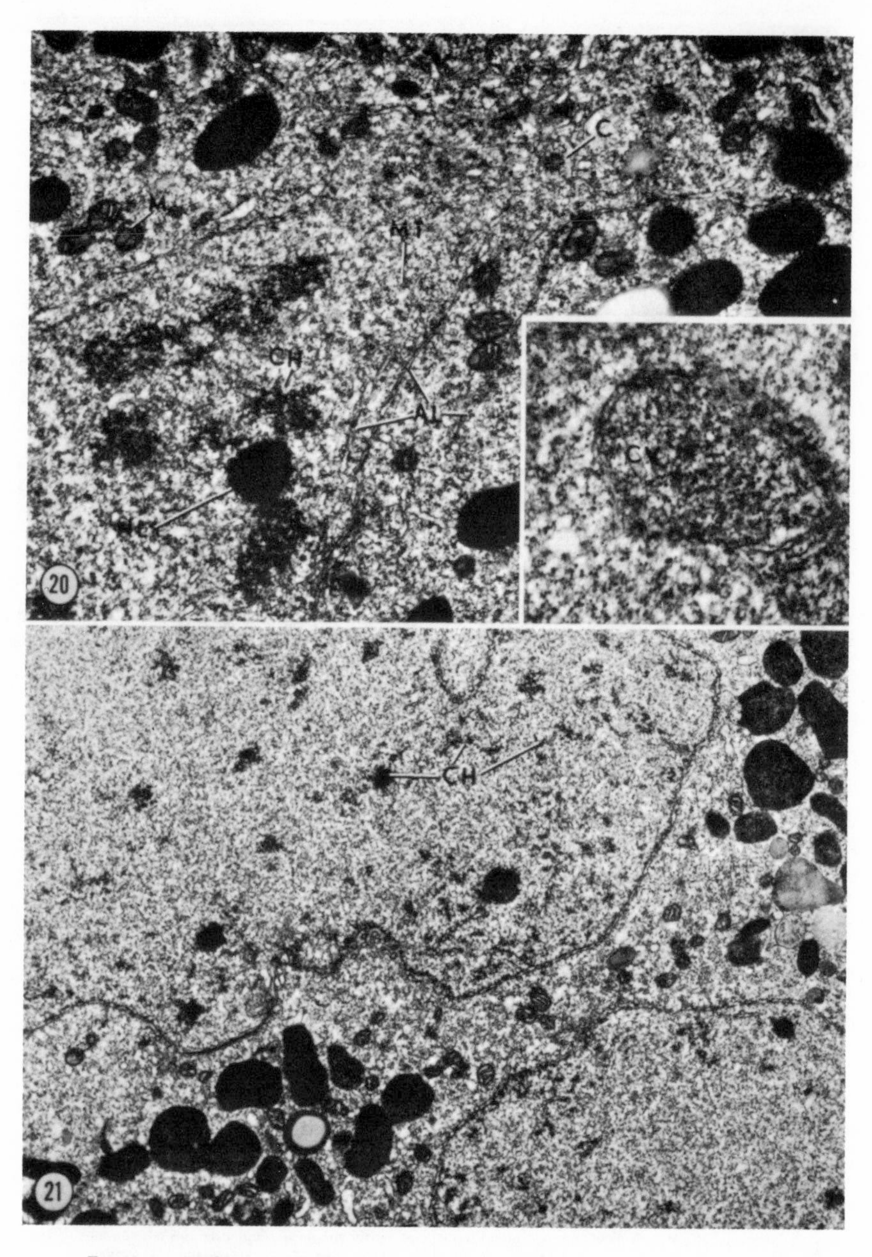

FIGURE 20 An electron micrograph through telophase stage of mitosis of an embryo (110–120 min postactivation). *CH*, chromatin; *C*, centriole; *MT*, microtubules; *AL*, annulate lamellae; *NL*, nucleolus-like body; *M*, mitochondria. *Inset* is an electron micrograph demonstrating the perforated membrane-bounded, chromosome-containing vesicle (*CV*). Fig. 20, × 10,600; *Inset*, × 34,000.

FIGURE 21 A section through an embryo (120 min postactivation) illustrating the membrane-bounded nuclei and chromatin (*CH*). × 17,000.

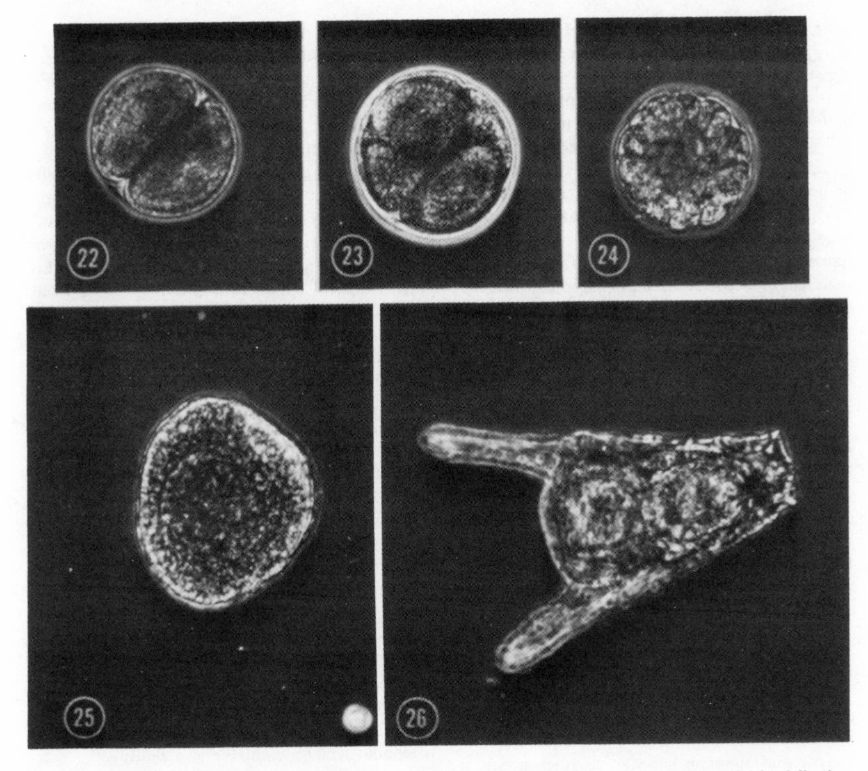

FIGURE 22 A phase-contrast photomicrograph of a two-cell stage, living embryo (3½ hr following activation with hypertonic seawater). × 250.

FIGURE 23 A phase-contrast photomicrograph of a four-cell stage, living embryo (4½ hr following activation with hypertonic seawater). × 400.

FIGURE 24 A phase-contrast photomicrograph of a living multicellular embryo (10–12 hr following activation). × 300.

FIGURE 25 A phase-contrast photomicrograph of a living early gastrula (24–30 hr postactivation). × 500.

FIGURE 26 A phase-contrast photomicrograph of a living, swimming pluteus larva (48 hr following activation). × 500.

spective genes." Further evidence of the effect of ion and water changes in the nucleus on the initiation of nuclear activity have been described by Harris (23–25). In hybrid cells in which erythrocyte nuclei become associated with HeLa cell cytoplasm, Harris (23–25) notes changes in nuclear volume and chromatin dispersion. He suggests that the nuclei are affected by shifts of water and electrolytes across the nuclear envelope. He has found these effects in hybrid cells regardless of species differences. Another example of the possible effects of egg cytoplasm in inducing a change in the nuclear volume and concomitant nuclear activity is in the nuclear transplantation studies. Swelling of the transplanted nuclei is often observed (19, 20, 21, 84) and is found to precede a period of DNA synthesis. The swelling may directly affect the chromosomes or, as Gurdon (20) suggests, may induce the transplanted nuclei to be susceptible to a cytoplasmic factor which thus results in DNA synthesis.

It is not unusual for physical agents, such as ion and water balance, light, heat, etc , to control cellular activity. The ability to artificially synchronize cellular activity in many organisms is a demonstration of the effect of a change in the physical environment of the cell that results in a regulation of cellular activity. Synchronization represents a specialized form of intracellular communication between nucleus and cytoplasm, resulting in a dramatic demonstration of the interdependence of these two cellular compartments.

In many cases, the eggs of some organisms remain diploid after artificial activation. Presumably the diploid condition is achieved by the retention of the second polar body in the final division of maturation (see 12). In the case of *Arbacia*, the egg is shed in the haploid pronuclear stage. We have not been able to ascertain the ploidy state of the nuclei of the artificially activated eggs. It would be of great interest to analyze the replication of DNA prior to first cleavage and in ensuing cleavages to determine if and when the diploid amount of DNA is restored.

Centrioles

The aster in the artificially activated egg is composed of a multitude of vesicular structures embedded in a matrix which also contains microtubules and annulate lamellae oriented around a centriole.

In their investigations, Anderson (2), Longo and Anderson (57), Harris (28), and Verhey and Moyer (92) did not report centrioles in the unactivated egg of *Arbacia*. The possibility exists that centrioles are present in the unactivated egg (see 94), but the alternative possibility, i.e. that one is organized from submicroscopic precursors, is strongly suggested. Evidence for centriole morphogenesis has been demonstrated in the work of Dingle and Fulton (8) and Schuster (81) on the protozoan *Naegleria*.

The artificially activated egg has been demonstrated to contain a centriole prior to the first cleavage (9, 80). The present study illustrates the presence of centrioles before and during the first cleavage, as well as in the four-cell stage embryo, multicellular embryo, and blastula.

In the artificially activated egg, the presence of centrioles and their replication may provide the catalyst for the organization of monomers into those microtubules that constitute the spindle apparatus. Microtubules have been seen emanating from the lateral surface of centrioles, or satellites associated with centrioles (5, 77, 85). Moreover, when the cell is treated with colchicine, low temperature, or hydrostatic pressure, or when the cell is not in mitosis (78), pieces of microtubules may be found associated with the centriole (87). Thus, both the centrioles and microtubules, owing to their consistency in the form they assume and in their temporal appearance within the cell, lend themselves to the assembly theory. The temporal and spatial relationship strongly suggests that the centriole does orient microtubular assembly in the formation of the aster. We have observed, in some artificially activated eggs, the presence of many cytasters (10). We were not able, however, to determine if all these cytasters were centered around centrioles. The fate of these eggs with multiple cytasters was not determined.

Voluminous studies on the origin of basal bodies (possessing centriolar architecture) indicate that procentrioles develop further into basal bodies of cilia (11, 17, 46, 82, 83). The procentriolar structure has also been discussed in the formation of centrioles in the sperm of the water fern *Marsilea* (66). The aforementioned study lends further support to our suggestion that centriole formation in the artificially activated egg is a process of assembly in various steps to produce what may be then morphologically identifiable as a centriole.

Parthenogenesis

The production of an embryo from a female gamete without any genetic contribution from a male gamete, or parthenogenesis (natural or artificial), is found in many phyla, including Echinodermata, Mollusca, Annelida, Arthropoda, Rotifera and Chordata (3, 32). The ubiquity of this phenomenon, together with the fact that in many cases the embryo produced is functionally similar to that produced as a result of fertilization, emphasizes the egg's capability to support development. In the present study, "normal"-appearing plutei developed upon artificial stimulation. The development to the pluteus larva does not assure one that development to the adult sea urchin will follow, but it does suggest that within the machinery of the mature egg the potential for all of the events necessary for larval development is present.

Tyler (88) has stated that "While the discovery

of artificial parthenogenesis did not bring the realization of the early hopes that problems of fertilization would be readily solved, it has greatly enlarged the scope of the attack on the problem of activation by substituting relatively simple chemical and physical agents for the spermatozoon."

This investigation was supported by a grant (HD 04924-09) from the National Institutes of Child Health and Human Development, United States Public Health Service.

BIBLIOGRAPHY

1. Anderson, E. 1967. The formation of the primary envelope during oocyte differentiation in teleosts. *J. Cell Biol.* **35**:193.
2. Anderson, E. 1968. Oocyte differentiation in the sea urchin, *Arbacia punctulata*, with particular reference to the origin of cortical granules and their participation in the cortical reaction. *J. Cell Biol.* **37**:514.
3. Beatty, R. A. 1967. Parthenogenesis in vertebrates. *In* Fertilization. C. B. Metz and A. Monroy, editors. Academic Press Inc., New York. **1**: 413.
4. Costello, D. P., M. E. Davidson, A. Eggers, M. H. Fox, and C. Henley. 1957. Methods for Obtaining and Handling Marine Eggs and Embryos. Lancaster Press, Inc., Lancaster, Pa.
5. deThé, G. 1964. Cytoplasmic microtubules in different animal cells. *J. Cell Biol.* **23**:265.
6. Deyrup, I. 1953. A study of the fluid uptake of rat kidney slices in vitro. *J. Gen. Physiol.* **36**:739.
7. Dick, D. A. T. 1966. Cell Water. Butterworth, Washington, D. C.
8. Dingle, A. D. and C. Fulton. 1966. Development of the flagellate apparatus of *Naeglaria*. *J. Cell Biol.* **31**:43.
9. Dirksen, E. R. 1961. The presence of centrioles in artificially activate sea urchin eggs. *J. Biophys. Biochem. Cytol.* **11**:244.
10. Dirksen, E. R. 1964. The isolation and characterization of asters from artificially activated sea urchin eggs. *Exp. Cell Res.* **36**:256.
11. Dirksen, E. R., and T. T. Crocker. 1966. Centriole replication in differentiating ciliated cells of mammalian respiratory epithelium. An electron microscopic study. *J. Microsc.* **5**: 629.
12. Ebert, J. D. 1965. Interacting Systems in Development. Holt, Rinehart and Winston, Inc., New York.
13. Endo, Y. 1961. Changes in the cortical layer of the sea urchin eggs at fertilization as studied with the electron microscope. I. *Clypeaster japonicus*. *Exp. Cell Res.* **25**:383.
14. Feldherr, C. M. 1962 *a*. The intracellular distribution of ferritin following microinjection. *J. Cell Biol.* **12**:159.
15. Feldherr, C. M. 1962 *b*. The use of colloidal gold for studies of intracellular exchanges in the amoeba *Chaos chaos*. *J. Cell Biol.* **12**:640.
16. Finklestein, A., and A. J. Cass. 1968. Permeability and electrical properties of thin lipid membranes. *J. Gen. Physiol.* **52**:1.
17. Frisch, D. 1967. Fine structure of the early differentiation of ciliary basal bodies. *Anat. Rec.* **157**:245.
18. Greeff, R. 1876. Uber den bau und die entwicklung der Echinodermen. 5. Mittheilung, Parthenogenesis bei den Seesternen. *Sitzungsber. Gesells. Beförd Gesammt. Naturwiss. Marburg.* 83.
19. Gurdon, J. B. 1967. Nuclear transplantation and cell differentiation. *In* Ciba Symposium on Cell Differentiation. A. V. S. De Reuck and J. Knight, editors. Churchill Ltd., London. 65.
20. Gurdon, J. B. 1968. Changes in somatic cell nuclei inserted into growing and maturing amphibian oocytes. *J. Embryol. Exp. Morphol.* **20**:401.
21. Gurdon, J. B., and H. R. Woodland. 1968. The cytoplasmic control of nuclear activity in animal development. *Biol. Rev.* **43**:233.
22. Hanai, T., D. A. Haydon, and W. R. Redwood. 1966. The water permeability of artificial bimolecular leaflets. A comparison of radiotracer and osmotic methods. *Ann. N. Y. Acad. Sci.* **137**:731.
23. Harris, H. 1966. Hybrid cells from mouse and man: A study in genetic regulation. *Proc. Roy. Soc. Series B.* **166**:538.
24. Harris, H. 1968. Nucleus and Cytoplasm. Oxford Press, Oxford, England.
25. Harris, H., E. Sidebottom, D. M. Grace, and M. E. Bromwell. 1969. The expression of genetic information. A study with hybrid animal cells. *J. Cell Sci.* **4**:499.
26. Harris, P. 1961. Electron microscopic study of mitosis in sea urchin blastomeres. *J. Cell Biol.* **11**:419.
27. Harris, P. 1965. Some observations concerning metakinesis in sea urchin eggs. *J. Cell Biol.* **25**:73.
28. Harris, P. 1967. Nucleolus-like bodies in sea urchin eggs. *Amer. Zool.* **7**:753.
29. Harvey, E. B. 1933. Effects of centrifugal force on fertilized eggs of *Arbacia punctulata* as

observed with the centrifuge microscope. *Biol. Bull.* **65**:389.

30. HARVEY, E. B. 1952. Electrical methods of sexing *Arbacia* and obtaining small quantities of eggs. *Biol. Bull.* **103**:284.
31. HARVEY, E. B. 1956. The American Arbacia and Other Sea Urchins. Princeton University Press, Princeton, N. J.
32. HARVEY, E. N. 1910. Methods of artificial parthenogenesis. *Biol. Bull.* **18**:269.
33. HEILBRUNN, L. V. 1915. Studies in artificial parthenogenesis. II. Physical changes in the egg of *Arbacia. Biol. Bull.* **29**:149.
34. HEILBRUNN, L. V. 1924. The colloid chemistry of protoplasm. III. The viscosity of cytoplasm at various temperatures. *Amer. J. Physiol.* **68:** 645.
35. HEILBRUNN, L. V. 1925. Studies in artificial parthenogenesis. IV. Heat parthenogenesis. *J. Exp. Zool.* **41**:243.
36. HEINZ, E. 1967. Transport through biological membranes. *Ann. Rev. Physiol.* **29**:21.
37. HERTWIG, O., and R. HERTWIG. 1887. Über den Befruchtungs-und Teilungsvorgang des tierischen Eies unter dem Einfluss Äusserer Agentien. *Jena Z. Naturforsch.* **13**:120.
38. HUANG, C., and T. E. THOMPSON. 1966. Properties of lipid bilayer membranes separating two aqueous phases: Water permeability. *J. Mol. Biol.* **15**:539.
39. HOROWITZ, S. B. and I. R. FENICHEL. 1968. Analysis of glycerol-^{3}H transport in the frog oocyte by extractive and autoradiographic techniques. *J. Gen. Physiol.* **51**:703.
40. ISHIKAWA, M. 1954. Relation between the breakdown of the cortical granules and permeability to water in the sea urchin egg. *Embryologia.* **2**:57.
41. ITO, S. 1962. Resting and activation potential of the *Oryzias* egg. 2. Change of membrane potential and resistance during fertilization. *Embryologia.* **7**:47.
42. ITO, S., and W. R. LOEWENSTEIN. 1965. Permeability of a nuclear membrane: Changes during normal development and changes induced by growth hormone. *Science* (*Washington*). **150**:909.
43. ITO, S., and R. J. WINCHESTER. 1963. The fine structure of the gastric mucosa in the rat. *J. Cell Biol.* **16**:541.
44. JUST, E. E. 1922. Initiation of development in the egg of *Arbacia.* I. Effect of hypertonic sea water in producing membrane separation, cleavage and top swimming plutei. *Biol. Bull.* **43**:384.
45. JUST, E. E. 1939. Basic Methods for Experiments on Eggs of Marine Animals. P. Blakiston's Son and Co., Inc., Philadelphia, Pa.
46. KALINIS, V. I., and K. R. PORTER. 1969. Centriole replication during ciliogenesis in the chick tracheal epithelium. *Z. Zellforsch.* **100**:7.
47. KANNO, Y., and W. R. LOEWENSTEIN. 1963. A study of the nucleus and cell membranes of oocytes with an intracellular electrode. *Exp. Cell Res.* **31**:149.
48. KARNOVSKY, M. 1965. A formaldehyde-glutaraldehyde fixative of high osmolarity for use in electron microscopy. *J. Cell Biol.* **27**:137a. (Abstr.)
49. KROEGER, H. 1966. Potential differenz und Puff-muster. Electrophysiologische und cytologische Untersuchungen an den Speicheldrüsen von *Chironomus thummi. Exp. Cell Res.* **41**:64.
50. LAUFER, H., and Y. NAKASE. 1965. Salivary gland secretion and its relation to chromosomal puffing in the dipteran *Chironomus Thummi. Proc. Nat. Acad. Sci. U. S. A.* **53**:511.
51. LEAF, A. 1956. On mechanisms of fluid exchanges in vitro. *Biochem. J.* **62**:241.
52. LEZZI, M. 1966. Induktion eines Ecdysonaktivierbaren Puff in isolierten Zellkernen von *Chironomus* durch KCL. *Exp. Cell Res.* **43**:571.
53. LING, G. N., N. M. OCHSENFELD and G. KARREMAN. 1967. Is the cell membrane a universal rate limiting barrier to the movement of water between the living cell and its surrounding medium? *J. Gen. Physiol.* **50**:1807.
54. LOEB, J. 1900. On the artificial production of normal larva from the unfertilized eggs of the sea urchin (*Arbacia*). *Amer. J. Physiol.* **3**:434.
55. LOEB, J. 1913. Artificial Parthenogenesis and Fertilization. University of Chicago Press, Chicago, Ill.
56. LOEWENSTEIN, W. R., Y. KANNO, and S. ITO. 1966. Permeability of nuclear membranes. *Ann. N. Y. Acad. Sci.* **137**:708.
57. LONGO, F. J., and E. ANDERSON. 1968. The fine structure of pronuclear development and fusion in the sea urchin *Arbacia punctulata. J. Cell Biol.* **39**:339.
58. LONGO, F. J., and E. ANDERSON. 1969. Cytological events leading to the formation of the two cell stage in the rabbit: Association of the maternally and paternally derived genomes. *J. Ultrastruct. Res.* **29**:86.
59. LÖNNIG, S. 1967. Studies of the ultrastructure of sea urchin eggs subjected to hypotonic and hypertonic medium. *Arbok. Univ. Bergen. Med. Ser.* **5**:1.
60. LØVTRUP, S. 1963. On the rate of water exchange across the surface of animal cells. *J. Theor. Biol.* **5**:341.
61. LUCKE, B., and M. MCCUTCHEON. 1932. The living cell as an osmotic system and its permeability to water. *Physiol. Rev.* **12**:68.
62. LUFT, J. H. 1961. Improvements in epoxy

resin embedding methods. *J. Biophys. Biochem. Cytol.* **9**:409.

63. MATHEWS, A. P. 1900. Some ways of causing mitotic division in unfertilized Arbacia eggs. *Biol. Bull.* **29**:149.

64. MCCLENDON, J. F. 1909. On artificial parthenogenesis of the sea urchin egg. *Science* (*Washington*). **30**:454.

65. MITCHISON, J. M., and M. M. SWANN. 1952. Optical changes in the membranes of the sea urchin egg at fertilization, mitosis and cleavage. *J. Exp. Biol.* **29**:357.

66. MIZUGAMI, I., and J. GALL. 1966. Centriole replication. II. Sperm formation in the fern, *Marsilea*, and the cycad, *Zamia*. *J. Cell Biol.* **29**:97.

67. MONROY, A. 1965. Chemistry and Physiology of Fertilization. Holt, Rinehart, and Winston, New York.

68. MORGAN, T. H. 1898. The effect of salt solutions on unfertilized eggs of *Arbacia*. *Science* (*Washington*). **7**:222.

69. MORGAN, T. H. 1900. The effect of strychnine on the unfertilized eggs of sea urchins. *Science* (*Washington*). **11**:178.

70. MOSER, F. 1939. Studies on a cortical layer response to stimulating agents in the *Arbacia* egg. II. Response to chemical and physical agents. *J. Exp. Zool.* **80**:447.

71. MOSER, F. 1940. Studies on a cortical layer response to stimulating agents in the *Arbacia* egg. III. Response to chemical and physical agents. *Biol. Bull.* **78**:68.

72. MUELLER, P., and D. O. RHUDIN. 1968. Resting and action potentials in experimental bimolecular lipid membranes. *J. Theor. Biol.* **18**:222.

73. MUIR, C., and H. C. MACGREGOR. 1969. Sodium and potassium in oocytes of *Triturus cristatus*. *J. Cell Sci.* **4**:299.

74. NAORA, H., H. NAORA, M. IZAWA, V. G. ALLFREY, and A. E. MIRSKY. 1962. Some observations on differences in composition between the nucleus and cytoplasm of the frog oocyte. *Proc. Nat. Acad. Sci. U. S. A.* **48**:853.

75. OHNISHI, T. 1963. Adenosine triphosphatase activity relating to active transport in the cortex of sea urchin eggs. *J. Biochem.* **53**:238.

76. PRESCOTT, D. M., and E. ZEUTHEN. 1953. Comparison of water diffusion and water filtration across cell surfaces. *Acta Physiol. Scand.* **28**:77.

77. ROBBINS, E., and GONATAS, N. K. 1964. The ultrastructure of a mammalian cell during the mitotic cycle. *J. Cell Biol.* **21**:429.

78. ROBBINS, E., G. JENTZSCH, and A. MICALI. 1968. The centriole cycle in synchronized Hela cells. *J. Cell Biol.* **36**:329.

79. ROTHSCHILD, L., and H. BARNES. 1953. The inorganic constituents of the sea urchin egg. *J. Exp. Biol.* **30**:534.

80. SACHS, M. I., and E. ANDERSON. 1969. A cytological study of events associated with artificial parthenogenesis in the sea urchin *Arbacia punctulata*. *J. Cell Biol.* **43**:29A. (Abstr.)

81. SCHUSTER, F. 1963. An electron microscope study of the amoeba-flagellate, *Naeglaria gruberi* (Schardinger). *J. Protozool.* **10**:297.

82. SOROKIN, S. P. 1968. Reconstruction of centriole formation and ciliogenesis in mammalian lungs. *J. Cell Sci.* **3**:209.

83. SOROKIN, S. P., and S. J. ADELSTEIN. 1967. Failure of 1100 rads of X radiation to affect ciliogenesis and centriolar formation in cultured rat lungs. *Radiat. Res.* **31**:748.

84. SUBTELNY, S., and C. BRADT. 1963. Cytological observations in the early developmental stages of activated *Rana pipiens* eggs receiving a transplanted nucleus. *J. Morphol.* **112**:45.

85. SZOLLOSI, D. 1964. The structure and function of centrioles and their satellites in the jellyfish, *Phialidium g[illegible]rium*. *J. Cell Biol.* **21**:465.

86. TOSTESON, D. C. 1964. Regulation of cell volume by sodium and potassium transport. *In* The Cellular Functions of Membrane Transport. J. F. Hoffman, editor. Prentice Hall, Englewood Cliffs, N. J. 3.

87. TILNEY, L. G. 1968. Ordering of subcellular units. The assembly of microtubules and their role in the development of cell form. *Develop. Biol. Suppl.* **2**:63.

88. TYLER, A. 1955. Gametogenesis, Fertilization and Parthenogesis. *In* Analysis of Development. B. H. Willier, P. A. Weiss, and V. Hamburger, editors. Saunders, Philadelphia. 170.

89. TYLER, A., C. Y. KAO, and H. GRUNDFEST. 1956. Membrane potential and resistance of the starfish egg before and after fertilization. *Biol. Bull.* **111**:153.

90. USSING, H. H. 1965. Relationship between osmotic reactions and active sodium transport in the frog skin epithelium. *Acta Physiol. Scand.* **63**:141.

91. VENABLE, J. H., and R. COGGESHALL. 1965. A simplified lead citrate stain for use in electron microscopy. *J. Cell Biol.* **25**:407.

92. VERHEY, C. A., and F. H. MOYER. 1967. Fine structural changes during sea urchin oogenesis. *J. Exp. Zool.* **164**:195.

93. WHITTAM, R. 1962. The assymetrical stimulation of a membrane adenosine triphosphatase in relation to active cation transport. *Biochem. J.* **84**:110.

94. WILSON, E. B. 1925. The Cell in Development and Heredity. The Macmillan Company, New York.

INDEX